T0207375

Lecture Notes in Computer Science 11926

More information about this series at http://www.springer.com/series/7407

Evripidis Bampis · Nicole Megow (Eds.)

Approximation and Online Algorithms

17th International Workshop, WAOA 2019
Munich, Germany, September 12–13, 2019
Revised Selected Papers

Editors
Evripidis Bampis
LIP6
Sorbonne University
Paris, France

Nicole Megow
Department for Mathematics
and Computer Science
University of Bremen
Bremen, Germany

ISSN 0302-9743 ISSN 1611-3349 (electronic)
Lecture Notes in Computer Science
ISBN 978-3-030-39478-3 ISBN 978-3-030-39479-0 (eBook)
https://doi.org/10.1007/978-3-030-39479-0

LNCS Sublibrary: SL1 – Theoretical Computer Science and General Issues

This Springer imprint is published by the registered company Springer Nature Switzerland AG
The registered company address is: Gewerbestrasse 11, 6330 Cham, Switzerland

Preface

The 17th Workshop on Approximation and Online Algorithms (WAOA 2019) focused on the design and analysis of algorithms for online and computationally hard problems. Both kinds of problems have a large number of applications in a variety of fields. WAOA 2019 took place in Munich, Germany, during September 12–13, 2019, and was a success: it featured many interesting presentations and provided opportunity for stimulating discussions and interactions. WAOA 2019 was part of the ALGO 2019 event that also hosted ALGOCLOUD, ALGOSENSORS, ATMOS, ESA, and IPEC.

Topics of interest for WAOA 2019 were: graph algorithms, inapproximability results, network design, packing and covering, paradigms for the design and analysis of approximation and online algorithms, parameterized complexity, scheduling problems, algorithmic game theory, algorithmic trading, coloring and partitioning, competitive analysis, computational advertising, computational finance, cuts and connectivity, geometric problems, mechanism design, resource augmentation, and real-world applications. In response to the call for papers, we received 38 submissions. Each submission was reviewed by at least three referees. The submissions were mainly judged on originality, technical quality, and relevance to the topics of the conference. Based on the reviews, the Program Committee selected 16 papers. This volume contains final revised versions of these papers as well as an invited contribution by our invited speaker Laura Sanità. The EasyChair conference system was used to manage the electronic submissions, the review process, the electronic Program Committee discussions, and the collection of the final versions of the papers for the proceedings. It made our task much easier.

We would like to thank all the authors who submitted papers to WAOA 2019 and all attendees of WAOA 2019, including the presenters of the accepted papers. A special thank you goes to the plenary invited speaker Laura Sanità for accepting our invitation and giving a very interesting talk. We would also like to thank the Program Committee members and the external reviewers for their diligent work in evaluating the submissions and their contributions to the electronic discussions. Furthermore, we are grateful to all the local organizer of ALGO 2019, Susanne Albers.

October 2019

Evripidis Bampis
Nicole Megow

The original version of the book was revised: The affiliation of the volume editor Nicole Megow has been corrected. The correction to the book is available at https://doi.org/10.1007/978-3-030-39479-0_17

Organization

Steering Committee

Evripidis Bampis	Sorbonne University, France
Thomas Erlebach	University of Leicester, UK
Klaus Jansen	University of Kiel, Germany
Christos Kaklamanis	University of Patras, Greece
Giuseppe Persiano	University of Salerno, Italy
Martin Skutella	Technical University of Berlin, Germany
Roberto Solis-Oba	The University of Western Ontario, Canada

Program Committee

Evripidis Bampis (Co-chair)	Sorbonne University, France
Marcin Bienkowski	University of Wrocław, Poland
Joan Boyar	University of Southern Denmark, Denmark
Ágnes Cseh	Hungarian Academy of Sciences, Hungary
Bruno Escoffier	Sorbonne University, France
Dimitris Fotakis	National Technical University of Athens, Greece
Zachary Friggstad	University of Alberta, Canada
Naveen Garg	Indian Institute of Technology Delhi, India
Jannik Matuschke	KU Leuven, Belgium
Nicole Megow (Co-chair)	University of Bremen, Germany
Julián Mestre	The University of Sydney, Australia
Ben Moseley	Carnegie Mellon University, USA
Viswanath Nagarajan	University of Michigan, USA
Seffi Naor	Technion, Israel
Aris Pagourtzis	National Technical University of Athens, Greece
Heiko Röglin	University of Bonn, Germany
Andreas S. Schulz	Technical University of Munich, Germany
Nguyen Kim Thang	University of Paris-Saclay, France
Marc Uetz	University of Twente, The Netherlands
Andreas Wiese	Universidad de Chile, Chile
Prudence Wong	The University of Liverpool, UK
Guochuan Zhang	Zhejiang University, China

Additional Reviewers

Haris Angelidakis
Evangelos Bampas
Alexander Birx
Vincent Chau
Lin Chen
Christoph Dürr
Lene Monrad Favrholdt
Carsten Fischer
Kshitij Gajjar
Jinxiang Gan
Marinus Gottschau
Laurent Gourves
Felix Happach
Ben Hermans
Imke Joormann
Gabriel Istrate
Arun Jambulapati
Marcus Kaiser
Tamás Király
Gunjan Kumar
Nikhil Kumar
Marilena Leichter
Hsiang-Hsuan Liu
Yuchen Mao
Mathieu Mari
Christos Nomikos
Qingqin Nong
Denis Pankratov
Vasileios-Orestis Papadigenopoulos
Benedikt Plank
Marc Renault
Xiangkun Shen
Kevin Schewior
Miriam Schlöter
Daniel Schmidt
Paweł Schmidt
Krzysztof Sornat
Abhinav Srivastav
Rob van Stee
Martijn Schoot Uiterkamp
Victor Verdugo
José Verschae
Clara Waldmann
Matthias Walter
Yuyan Wang
Karol Wegrzycki
Chenchen Wu
Chao Xu

On the Hardness of Computing the Diameter of a Polytope (Invited Talk)

Laura Sanità

University of Waterloo, 200 University Ave. W, Waterloo, Canada
laura.sanita@uwaterloo.ca

Abstract. The diameter of a polytope P is the maximum length of a shortest path between a pair of vertices on the 1-skeleton of P, which is the graph where the vertices correspond to the 0-dimensional faces of P, and the edges are given by the 1-dimensional faces of P. Despite decades of studies, it is still not known whether the diameter of a d-dimensional polytope with n facets can be bounded by a polynomial function of n and d. This is a fundamental open question in discrete mathematics, motivated by the (still unknown) existence of a polynomial pivot rule for the Simplex method for solving Linear Programs.

The diameter of a polytope has been studied from many different perspectives, including a computational complexity point of view. In particular, Frieze and Teng in 1994 showed that computing the diameter of a polytope is weakly NP-hard. In this talk, I will show a strengthened hardness result, obtained by exploiting the structure of a well-known polytope in the optimization community: the fractional matching polytope. Eventually, I will also show that the structure of the fractional matching polytope can be used to derive some hardness results concerning the performance of the Simplex method.

Contents

Terrain-Like Graphs: PTASs for Guarding Weakly-Visible Polygons and Terrains

Stav Ashur(✉), Omrit Filtser, Matthew J. Katz, and Rachel Saban

Department of Computer Science, Ben-Gurion University of the Negev, 84105 Beer-Sheva, Israel
{stavshe,omritna}@post.bgu.ac.il, matya@cs.bgu.ac.il, rachelfr@post.bgu.ac.il

Abstract. A graph $G = (V, E)$ is *terrain-like* if one can assign a unique integer from the range $[1..|V|]$ to each vertex in V, such that, if both $\{i, k\}$ and $\{j, l\}$ are in E, for any $i < j < k < l$, then so is $\{i, l\}$. We present a local-search-based PTAS for *minimum dominating set* in terrain-like graphs. Then, we observe that, besides the visibility graphs of x-monotone terrains which are terrain-like, so are the visibility graphs of *weakly-visible polygons* and *weakly-visible terrains*, immediately implying a PTAS for guarding the vertices of such a polygon or terrain from its vertices. We also present PTASs for continuously guarding the boundary of a WV-polygon or WV-terrain, either from its vertices, or, for a WV-terrain, from arbitrary points on the terrain. Finally, we compare between terrain-like graphs and *non-jumping* graphs, and also observe that both families admit PTASs for *maximum independent set*.

1 Introduction

We define a new family of graphs which we name *terrain-like* graphs. Let $G = (V, E)$ be a simple graph. We say that G is *terrain-like* if one can assign a unique integer from the range $[1..|V|]$ to each vertex in V, such that the following property holds: If both $\{i, k\}$ and $\{j, l\}$ are in E, for any $i < j < k < l$, then so is $\{i, l\}$. A well-known subfamily of terrain-like graphs (from which we borrowed the name terrain-like) is the family of visibility graphs of x-monotone (1.5D) terrains. Let $T = (t_1, \ldots, t_n)$ be an x-monotone polygonal line. Then, the *visibility graph* of T is the graph $VG(T) = (V = \{t_1, \ldots, t_n\}, E)$, where $\{t_i, t_j\} \in E$ if and only if t_i and t_j see each other, that is, the line segment connecting them does not intersect the open region bounded from above by T (and from the sides by the vertical lines through t_1 and t_n). T has the following property which is known as the *order claim* [3]: For any four vertices t_i, t_j, t_k, t_l, where t_i is the leftmost among the four, t_j is the second from the left, etc., if t_i sees t_k (i.e., $\{t_i, t_k\} \in E$) and t_j sees t_l (i.e., $\{t_j, t_l\} \in E$), then t_i sees t_l (i.e., $\{t_i, t_l\} \in E$). Thus, $VG(T)$ is terrain-like (assign t_i the number i, for $i = 1, \ldots, n$).

M. J. Katz—Supported by grant 1884/16 from the Israel Science Foundation.

E. Bampis and N. Megow (Eds.): WAOA 2019, LNCS 11926, pp. 1–17, 2020.
https://doi.org/10.1007/978-3-030-39479-0_1

We describe a local-search-based PTAS for computing a minimum dominating set in a terrain-like graph. The PTAS and its proof are similar to the PTAS and proof presented by Gibson et al. [14] for computing a minimum dominating set in the visibility graph of an x-monotone terrain, or, in other words, a minimum guarding set for the terrain vertices (where guards are restricted to vertices of the terrain). However, our PTAS immediately implies such a PTAS in other subfamilies of terrain-like graphs. For example, we show that the family of visibility graphs of polygons which are weakly visible from one of their edges (see below) is also a subfamily of terrain-like graphs. Previously only a constant-factor approximation algorithm for vertex guarding the vertices of such a polygon was known [5].

From this point, the paper develops along two different paths. In the first path we study various guarding problems, while in the second path we study the family of terrain-like graphs and compare it with another family of graphs, namely, *non-jumping* graphs [1,6,19], whose definition is similar in flavor to our definition.

One of our major results in the first path is a PTAS for guarding a polygonal line satisfying a condition which is much weaker than x-monotonicity. Let $P = (p_1, \ldots, p_n)$ be a non-self-intersecting polygonal line, such that from every point p on P, one can see the sky, or, in other words, from every point p on P, one can shoot a ray that hits any horizontal line that lies above P. We present a PTAS for continuously guarding such a polygonal line from arbitrary points on it. Our proof is based on the proof of Friedrichs et al. [12], who presented a PTAS for continuously guarding an x-monotone terrain from arbitrary points on it, but some new observations are required.

A simple graph $G = (V, E)$ is *non-jumping* if one can assign a unique integer from the range $[1..|V|]$ to each vertex in V, such that the following property holds: If both $\{i, k\}$ and $\{j, l\}$ are in E, for any $i < j < k < l$, then so is $\{j, k\}$. The name non-jumping was coined by Ahmed et al. [1] who studied the family of non-jumping graphs, without realizing that this family has been studied before. In particular, Soto and Caro [19] studied the family of graphs that admit a *p-Box-realization*, focusing on the case where the boxes are intervals on the line. This (sub)family of graphs was also studied by Catanzaro et al. [6], under the name *max point-tolerance* graphs. Both Soto and Thraves Caro and Catanzaro et al. observed that this family of graphs is equivalent to the family of non-jumping graphs. Moreover, they, as well as Ahmed et al. showed that every non-jumping graph can be realized as a *monotone L-graph*, i.e., the intersection graph of L-frames whose corners lie on a line of slope -1, and vice versa. This nice observation together with the local-search-based PTAS of Bandyapadhyay et al. [2] for computing a minimum dominating set in monotone L-graphs, implies a PTAS for computing a minimum dominating set in non-jumping graphs.

In Sect. 5 we show that despite the similarity in the definition, the two families (i.e., terrain-like and non-jumping) are very different. In particular, we present an infinite set of graphs which are terrain-like but not non-jumping, and an infinite set of graphs which are non-jumping but not terrain-like. On the other hand, the families of, e.g., outerplanar graphs and convex bipartite graphs are subfamilies of both non-jumping graphs and terrain-like graphs. Due to space limitations, most of the details of this section are not included in this version.

Finally, let $\mathcal{F}$ be *any* family of graphs, which admits a local-search-based PTAS for minimum dominating set. We observe that if the proof for the PTAS is based on the proof scheme of Mustafa and Ray [18] and does not use the fact that the optimal set (which is one of the parts in the constructed bipartite graph) is minimum, then this proof can be applied verbatim to show that $\mathcal{F}$ admits a local-search-based PTAS for maximum independent set. In particular, we conclude that terrain-like graphs, as well as non-jumping graphs, admit a PTAS for maximum independent set.

Additional (Closely) Related Work. Bhattacharya et al. [5] presented a 4-approximation algorithm for vertex guarding the vertices of a weakly-visible polygon and a 6-approximation algorithm for vertex guarding its boundary, where a polygon P is weakly-visible if its boundary is weakly-visible from one of its edges e, i.e., every point on P's boundary is seen by a point on e. Recently, by applying these results, Bhattacharya et al. [4] obtained the first constant-factor approximation algorithms for vertex guarding the vertices or the boundary of a simple polygon, thus settling a conjecture of Ghosh (concerning the former version) from 1987. By an inapproximability result of Eidenbenz et al. [10], the latter version does not admit a PTAS.

As for 1.5D terrain guarding, following a series of constant-factor approximation algorithms [3,8,11,16], Gibson et al. [14] presented a local-search-based PTAS for vertex guarding the vertices of a 1.5D terrain. Their proof is based on the proof scheme of Mustafa and Ray [18]; see also [7]. Friedrichs et al. [12] considered the continuous 1.5D terrain guarding problem, where the goal is to guard the entire terrain either from vertices or from arbitrary points on the terrain. They presented PTASs for these problems, by discretizing them and applying the PTAS from [14]. Finally, King and Krohn [17] proved that the decision version of the 1.5D terrain guarding problem is NP-hard, by a reduction from PLANAR 3-SAT.

Contribution. We see our main contribution in (i) defining the family of terrain-like graphs and identifying the subfamilies of WV-polygons and WV-terrains, (ii) adapting the PTASs of Gibson et al. [14] and of Friedrichs et al. [12] to terrain-like graphs and to WV-polygons and WV-terrains, respectively, thus significantly expanding the class of polygons and polygonal lines that are known to admit PTASs for important versions of the art-gallery problem, (iii) observing that the proof for a local-search-based PTAS for minimum dominating set often constitutes a proof for such a PTAS for maximum independent set, as in the case of terrain-like graphs and non-jumping graphs, (iv) presenting an infinite collection of terrain-like graphs, which are not non-jumping, and vice versa.

2 A PTAS for Minimum Dominating Set in Terrain-Like Graphs

Let $G = (V, E)$ be a terrain-like graph with n vertices. A *minimum dominating set* in G is a minimum-cardinality subset Q of V, such that for each vertex $v \in V$, either $v \in Q$ or v is adjacent to a vertex $u \in Q$ (i.e., $(u, v) \in E$). We present a local-search-based PTAS for computing a minimum dominating set in G, i.e., we present a polynomial-time algorithm that computes a dominating set B in G of size $O(1+\varepsilon) \cdot \mathrm{OPT}$, for any constant $\varepsilon > 0$, where OPT is the size of a minimum dominating set. We prove the bound on the size of B by adapting the proof of Gibson et al. [14], who presented a PTAS for vertex guarding the vertices of an x-monotone (1.5D) terrain, to our more general and abstract setting. The proof of Gibson et al. [14], in turn, is based on the proof scheme of Mustafa and Ray [18], which is used to show that a local-search algorithm is a PTAS (see also [7]).

2.1 Algorithm

Given $\varepsilon > 0$, set $k = \frac{\alpha}{\varepsilon^2}$, for an appropriate constant $\alpha > 0$.

1. $Q \leftarrow V$.
2. Determine whether there exist subsets $S \subseteq Q$ of size at most k and $S' \subseteq (V \setminus Q)$ of size at most $|S| - 1$, such that $(Q \setminus S) \cup S'$ is a dominating set in G.
3. If such S and S' exist, set $Q \leftarrow (Q \setminus S) \cup S'$, and go back to Step 2. Otherwise, return Q.

As usual, the running time of the algorithm is $O(n^{O(1/\varepsilon^2)})$.

2.2 Analysis

Assume that G's vertices were assigned unique indices between 1 and n, such that, for any $i < j < k < l$, if (i, k) and (j, l) are in E, then so is (i, l).

Let R (the red set) be a minimum dominating set and let B (the blue set) be the dominating set obtained by the algorithm above. We say that r *dominates* v and write r *dom* v, for $r \in R$ and $v \in V$, if either $r = v$ or $(r, v) \in E$. We need to prove that $|B| \leq (1+\varepsilon) \cdot |R|$. We may assume that $R \cap B = \emptyset$; otherwise, we prove that $|B'| \leq (1+\varepsilon) \cdot |R'|$, where $R' = R \setminus B$ and $B' = B \setminus R$ (and both R' and B' dominate $V \setminus N[R \cap B]$). We construct a bipartite graph $H = (R \cup B, F)$, and prove that (i) H is planar and (ii) H satisfies the *locality condition*, that is, for any vertex $v \in V$, there exist vertices $r \in R$ and $b \in B$, such that r *dom* v, b *dom* v, and $(r, b) \in F$. By the proof scheme of Mustafa and Ray [18], this implies that $|B| \leq (1+\varepsilon) \cdot |R|$.

For two vertices $u, v \in V$, we say that u *precedes* v (or v *succeeds* u) if $u < v$. For a vertex $w \in V$, if there exists a vertex in $R \cup B$ that dominates w and precedes it, then let $\lambda(w)$ be the first such vertex; see Fig. 1. Similarly, if there

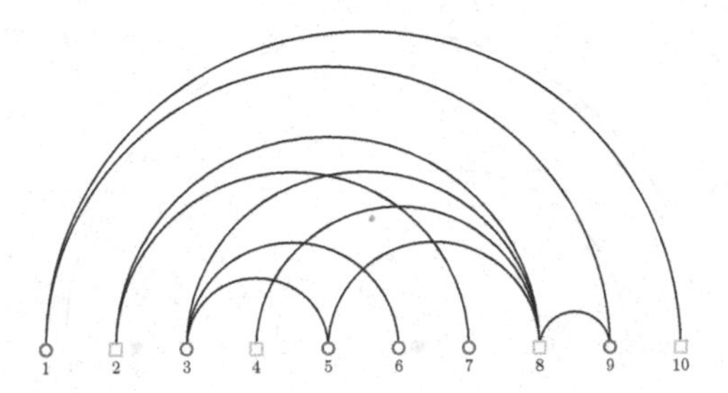

Fig. 1. The vertices in R drawn as red squares and the vertices in B drawn as blue circles; $\lambda(10) = 1$, $\lambda(9) = 1$, $\lambda(8) = 2$, etc. (Color figure online)

exists a vertex in $R \cup B$ that dominates w and succeeds it, then let $\rho(w)$ be the last such vertex. Notice that since $R \cap B = \emptyset$ at least one of the two exists.

Let $A_1 = \{\widehat{\lambda(w)w} \mid w$ a vertex of V for which $\lambda(w)$ is defined$\}$. Let $\widehat{\lambda(x)x}$ and $\widehat{\lambda(y)y}$ be two arcs in A_1. Then, by definition, $x \neq y$. We say that the arcs are *crossing* if either $\lambda(x) < \lambda(y) < x < y$ or $\lambda(y) < \lambda(x) < y < x$. Additionally, let x be a vertex in V. If there exists an arc $\widehat{uv} \in A_1$, such that $u < x < v$, then let $\widehat{\lambda(w)w}$ be the arc among these arcs in which w is the smallest; $\widehat{\lambda(w)w}$ is the arc of A_1 *associated* with x.

Claim 1. *The arcs in A_1 are non-crossing.*

Proof. Let $\widehat{\lambda(x)x}$ and $\widehat{\lambda(y)y}$ be two arcs in A_1, such that $\lambda(x) \neq \lambda(y)$. If $\lambda(x) < \lambda(y)$, then it is impossible that $\lambda(x) < \lambda(y) < x < y$, since this would imply that $(\lambda(x), y) \in E$, which is impossible by the definition of $\lambda(y)$. Similarly, if $\lambda(y) < \lambda(x)$, then it is impossible that $\lambda(y) < \lambda(x) < y < x$.

Constructing H. For each vertex $x \in R \cup B$, do the following. If $\lambda(x)$ is defined and $color(\lambda(x)) \neq color(x)$ (i.e., one is red and the other is blue), add the edge $(\lambda(x), x)$ to F_1. If x has an arc of A_1 associated with it, then let $\widehat{\lambda(w)w}$ be this arc. Now, if $color(\lambda(w)) \neq color(x)$, add the edge $(\lambda(w), x)$ to F_1 (unless, $\widehat{\lambda(w)x} \in A_1$, i.e., $\lambda(w) = \lambda(x)$, and therefore it was already added to F_1). See Fig. 2.

H **is Planar.** We now prove that the bipartite graph $H_1 = (R \cup B, F_1)$ is planar, by describing a planar embedding of H_1 or, more precisely, of the graph $\overline{H}_1 = (V, A_1 \cup F_1)$. We map the vertices in V to equally-spaced points on the x-axis and the edges in $A_1 \cup F_1$ to arcs between pairs of points, see Fig. 2. We claim that the resulting set of arcs is non-crossing, i.e., we have obtained a planar embedding of $\overline{H}_1$ and therefore also of H_1. This follows from Claim 1 and by observing that the edges in $F_1 \setminus A_1$ can be partitioned into a collection of 'fans', where each fan is associated with an arc $\widehat{\lambda(w)w}$ of A_1 and lies below it.

We now define the sets A_2 and F_2 by replacing λ with ρ, that is, $A_2 = \{\widehat{w\rho(w)} \mid w$ a vertex of V for which $\rho(w)$ is defined$\}$ and F_2 is defined w.r.t. A_2.

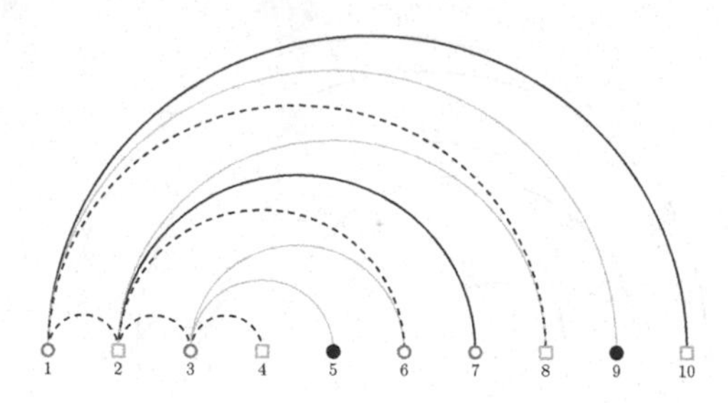

Fig. 2. Referring to Fig. 1. The arcs in A_1 are drawn in non-dashed black and gray, where the former (i.e., $(1, 10)$ and $(2, 7)$) are added as edges to F_1. The other edges added to F_1 are drawn in dashed black. The 'fan' associated with the edge $(2, 7)$ consists of the edges $(2, 3)$ and $(2, 6)$.

We then observe that the bipartite graph $H_2 = (R \cup B, F_2)$ is planar, by describing a planar embedding of $\overline{H}_2 = (V, A_2 \cup F_2)$. Moreover, we claim that the graph $H_1 \cup H_2$ is planar, since we can embed the graph $\overline{H}_1 \cup \overline{H}_2$ by drawing the edges of $\overline{H}_1$ above the x-axis and the edges of $\overline{H}_2$ below the x-axis.

Finally, we define the set F_3 as follows. For each vertex $x \notin R \cup B$, if both $\lambda(x)$ and $\rho(x)$ are defined and $color(\lambda(x)) \neq color(\rho(x))$, then add the edge $(\lambda(x), \rho(x))$ to F_3. The final graph $H = (R \cup B, F)$ where $F = F_1 \cup F_2 \cup F_3$ is planar, since $\overline{H}_1 \cup \overline{H}_2$ is planar and each edge $(\lambda(x), \rho(x)) \in F_3$ can be drawn as the union of the arcs $\widehat{\lambda(x)x} \in A_1$ and $\widehat{x\rho(x)} \in A_2$ (as $x \notin R \cup B$).

H Satisfies the Locality Condition

Lemma 1. *For any vertex $x \in V$, there exist vertices $r \in R$ and $b \in B$, such that r dom x, b dom x, and $(r, b) \in F$.*

Proof. Let x be a vertex of V. We distinguish between two cases:
$x \notin R \cup B$: If both $\lambda(x)$ and $\rho(x)$ are defined and $color(\lambda(x)) \neq color(\rho(x))$, then $(\lambda(x), \rho(x)) \in F_3$ and the condition holds. If both $\lambda(x)$ and $\rho(x)$ are defined but $color(\lambda(x)) = color(\rho(x))$, then there exists a vertex $w \in R \cup B$, such that w *dom* x and $color(w) \neq color(\lambda(x)), color(\rho(x))$. Assume, w.l.o.g., that $\lambda(x) < w < x$ and let z be the first such vertex. Let $\widehat{\lambda(y)y}$ be the arc in A_1 associated with z. Then $\lambda(x) \leq \lambda(y) < z < y \leq x$. Notice that $\lambda(y)$ *dom* x, since, if $y \neq x$, then by the terrain property (applied to $\lambda(y), z, y, x$) $\lambda(y)$ *dom* x. Now, since z is the "first such vertex", $color(z) \neq color(\lambda(y))$, so the edge $(\lambda(y), z) \in F_1$ and the condition holds. If only $\lambda(x)$ is defined, then we proceed as above, that is, we consider the first vertex $z \in R \cup B$, such that z *dom* x and $color(z) \neq color(\lambda(x))$; in this case, $\lambda(x) < z < x$.

$x \in R \cup B$: If $\lambda(x)$ is defined and $color(x) \neq color(\lambda(x))$, then $(\lambda(x), x) \in F_1$ and the condition holds. Similarly, if $\rho(x)$ is defined and $color(x) \neq color(\rho(x))$, then $(x, \rho(x)) \in F_2$ and the condition holds. Otherwise, we conclude w.l.o.g. that there exists a vertex $w \in R \cup B$, such that w *dom* x and $\lambda(x) < w < x$ and $color(w) \neq color(\lambda(x))$. Let z be the first such vertex and proceed exactly as in the previous case.

A More General Version of the Problem. Consider the following slightly more general version of the minimum dominating set problem, which is needed in Sect. 4. Given a terrain-like graph $G = (V, E)$, and two subsets C and W of V, where C is the set of candidate vertices and W is the ground set of vertices, find a minimum-cardinality subset Q of C, such that Q dominates W. It is easy to see that the PTAS above can be easily adapted to this general version. We thus obtain:

Theorem 1. *There exists a PTAS for (general) minimum dominating set in terrain-like graphs. That is, for any $\varepsilon > 0$, there is a polynomial-time algorithm which, given a terrain-like graph $G = (V, E)$ and two sets $C, W \subseteq V$, returns $Q \subseteq C$ such that Q dominates W and $|Q| \leq (1+\varepsilon) \cdot OPT$; here OPT is the size of a minimum subset of C that dominates W.*

2.3 Maximum Independent Set

Let $G = (V, E)$ be a (simple) graph (not necessarily terrain-like) with n vertices. Consider the following local-search algorithm for computing an independent set in G.

Given $\varepsilon > 0$, set $k = \frac{\alpha}{\varepsilon^2}$, for an appropriate constant $\alpha > 0$.

1. $Q \leftarrow \emptyset$.
2. Determine whether there exist subsets $S \subseteq Q$ of size at most $k - 1$ and $S' \subseteq (V \setminus Q)$ of size at least $|S| + 1$ and at most k, such that $(Q \setminus S) \cup S'$ is an independent set in G.
3. If such S and S' exist, set $Q \leftarrow (Q \setminus S) \cup S'$, and go back to Step 2. Otherwise, return Q.

Let R_I be a maximum independent set in G and let $B_I = Q$ be the independent set obtained by the algorithm above. Then B_I is *maximal*, i.e., for any vertex $v \in (V \setminus B_I)$, the set $B_I \cup \{v\}$ is not independent. So, both R_I and B_I are also dominating sets in G, since any maximal independent set is also a dominating set.

Now, assume that G belongs to some family of graphs (e.g., terrain-like graphs), which admits a PTAS for minimum dominating set. More precisely, assume that the algorithm of Sect. 2.1 computes a $(1 + \varepsilon)$-approximation of a minimum dominating set R_D in G, and that the proof for this is based on the proof scheme of Mustafa and Ray [18] and does not use the fact that R_D is minimum. Then, this proof can be applied verbatim to show that $|B_I| \geq (1-\varepsilon) \cdot |R_I|$. In particular, we conclude that terrain-like graphs, as well as non-jumping graphs, admit a PTAS for maximum independent set. We may also conclude, for example, that disk graphs admit a PTAS for maximum independent set, which is of course already known even for pseudo-disks (see [7]), based on the PTAS of Gibson and Pirwani [15] for minimum dominating set in disk graphs.

Remark 1. If G is a non-jumping graph given by a valid realization, then one can compute a maximum independent set in G in polynomial time [6,9]. Similarly, if G is a terrain-like graph given by a valid realization, then one can compute a maximum independent set in G in polynomial time, by adapting the algorithm of Ghosh et al. [13] for computing a maximum hidden vertex set in a polygon weakly visible from a convex edge.

3 Guarding the Vertices

In this section we consider several interesting families of polygons or polygonal chains. For each of these families, we prove that the visibility graph $VG(P)$ of any member P of the family is terrain-like, where $VG(P)$ is the graph whose vertices correspond to the vertices of P and whose edges correspond to the pairs of vertices of P that see each other. But this immediately implies that we have a PTAS for guarding the vertices of P from its vertices, since any vertex guarding set for P's vertices corresponds to a dominating set in $VG(P)$ and vice versa.

3.1 x-monotone (1.5D) Terrains

Let $T = (t_1, \ldots, t_n)$ be an x-monotone polygonal chain. Then, T has the following property which is known as the *order claim* [3]: For any four vertices t_i, t_j, t_k, t_l, where t_i is the leftmost among the four, t_j is the second from the left, etc., if t_i sees t_k and t_j sees t_l, then t_i sees t_l. Thus, $VG(T)$ is terrain-like (assign t_i the number i, for $i = 1, \ldots, n$), and we have a PTAS for vertex guarding T's vertices. Of course, this is not surprising, since such a PTAS was presented several years ago by Gibson et al. [14], and, as mentioned, our proof in Sect. 2 is obtained by adapting their proof for their PTAS.

The next two families of polygons are more interesting, in the sense that a PTAS for vertex guarding the vertices of a member of one of them was not previously known.

3.2 WV-polygons

Let P be a polygon, which is weakly visible from one of its edges $e = uv$. We first assume that the angles at u and v are both convex. In this case, $P \setminus e$ is necessarily contained in one of the open half-planes defined by the line containing e. Without loss of generality, we assume that e is contained in the x-axis, where u is to the left of v, and that $P \setminus e$ is contained in the open half plane above the x-axis. We now prove that (under the convexity assumption) $VG(P)$ is terrain-like.

Let $v_1, \ldots, v_n$ be the vertices of P, in clockwise order, beginning at $v_1 = u$ and ending at $v_n = v$. We assign the number i to vertex v_i, for $i = 1, \ldots, n$. It remains to prove that the required property holds for this numbering. We prove a more general property. For two points a and b on P's boundary, we say that a precedes b (or b succeeds a) and write $a \prec b$ (or $b \succ a$), if when traversing P's boundary clockwise from u, one reaches a before b.

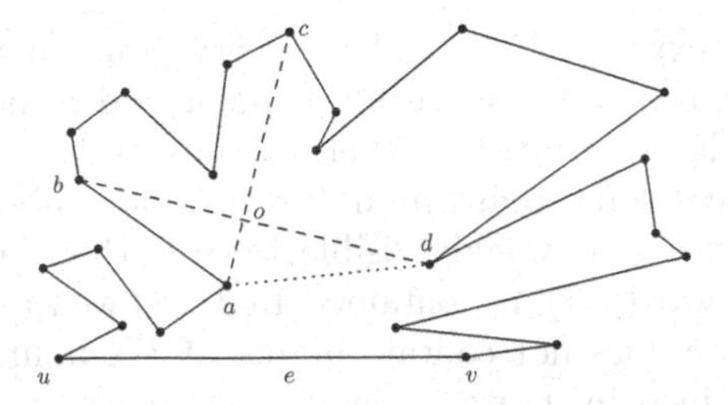

Fig. 3. A polygon weakly visible from $e = (u, v)$. The order claim: $a \prec b \prec c \prec d$, a sees c, b sees $d \Longrightarrow a$ sees d.

Claim 2 (The (clockwise) order claim for 'convex' WV-polygons). *Let a, b, c, d be four vertices (or points on P's boundary) such that $a \prec b \prec c \prec d$, and assume that a sees c and b sees d. Then a must also see d.*

Proof. If a does not see d, then either a or d is not visible from a point on e, see Fig. 3. Indeed, let o denote the intersection point of $\overline{ac}$ and $\overline{bd}$. If the ray from a in the direction of d hits P's boundary before reaching d, then P's boundary enters and leaves the triangle Δaod through the edge ad without intersecting the edges ao and od. If this happens before the boundary 'reaches' a (advancing clockwise from u), then a cannot be seen from e, and if this happens before the boundary 'reaches' d (advancing counterclockwise from v), then d cannot be seen from e.

We have shown that $VG(P)$ is terrain-like, and therefore we have a PTAS for vertex guarding its vertices. We now show that the convexity assumption is not necessary, i.e., we still have such a PTAS, even if the angles at u and v are concave.

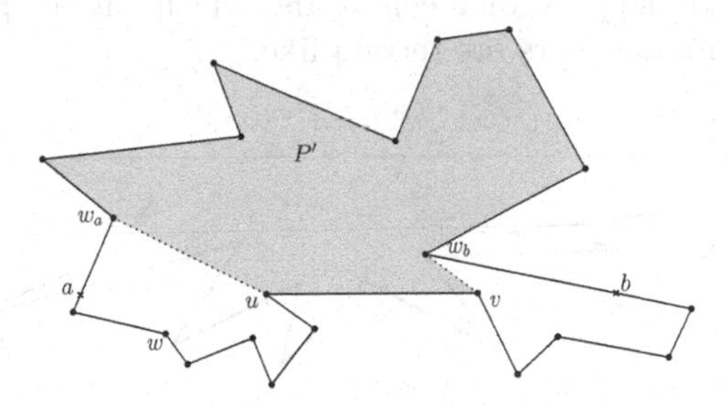

Fig. 4. Removing the convexity assumption.

Removing the Convexity Assumption. We show how to remove the assumption that the angles at u and at v are convex. Assume, e.g., that the angle at u is concave, and let a be the first point on P's boundary (moving clockwise from

u) that lies on the x-axis; see Fig. 4. Then, every point in the open portion of the boundary between u and a is visible from u and is not visible from any other point on the edge $e = uv$. Moreover, for any vertex w in this portion of P's boundary, if w sees some point p on P's boundary above the x-axis, then so does u. Indeed, since P is weakly-visible from e, there exists a point $x \in e$ that sees p. In other words, $\overline{xp}$ is contained in P, as well as $\overline{uw}$ and $\overline{wp}$. Thus, the quadrilateral $uwpx$ does not contain points of P's boundary in its interior, and since $\overline{up}$ is contained in it, we conclude that u sees p. Therefore, we may assume that an optimal guarding set does not include a vertex from this portion. Moreover, we may assume that the size of an optimal guarding set is greater than some appropriate constant, since otherwise we can find such a set in polynomial time. Let w_a be the first vertex following a. We place a guard at u and replace the portion of P's boundary between u and w_a by the edge uw_a. Similarly, if the angle at v is concave, we define the point b and the vertex w_b (by moving counterclockwise from v), place a guard at v, and replace the portion of P's boundary between v and w_b by the edge vw_b. Finally, we apply our local search algorithm to the resulting polygon P', after adjusting k so that together with u and v we still get a $(1+\varepsilon)$-approximation of an optimal guarding set for P.

3.3 WV-terrains: Removing the Monotonicity Assumption

Let $T = (t_1, \ldots, t_n)$ be *any* simple polygonal chain. We say that t_i and t_j see each other, where $i < j$, if either $j = i + 1$, or the directed line segment $t_i t_j$ exits and enters T from the left and does not intersect T (except at t_i and t_j), assuming T is being traversed from t_1 to t_n. Let $VG(T)$ be the visibility graph of T. Then, if $VG(T)$ happens to be terrain-like, then we have a PTAS for vertex guarding T's vertices. This observation is not very useful, since, in general, we do not have an efficient algorithm to determine whether $VG(T)$ is terrain-like. However, we define below a new family of polygonal chains, which is much more general than the family of x-monotone polygonal chains and which has the property that the visibility graphs of its members are terrain-like.

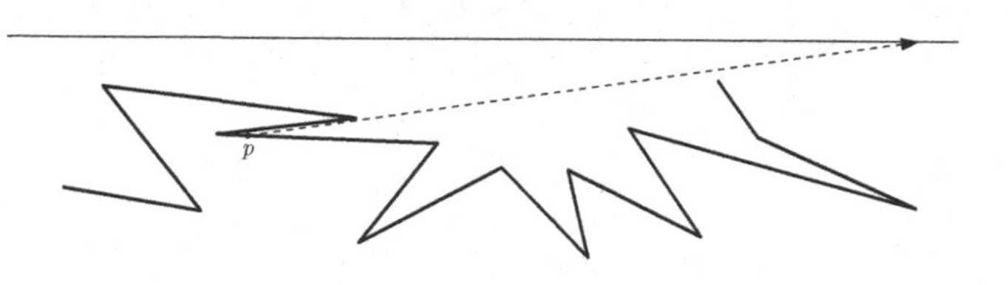

Fig. 5. A WV-terrain.

Definition 1. *A simple polygonal chain T is a WV-terrain, if one can see the sky from every point on T, or, formally, if from every point p on T one can shoot a ray, leaving T to the left, that hits a horizontal line which lies above T without intersecting T (except at p); see Fig. 5.*

Observation. The visibility graph of a WV-terrain is terrain-like.

Proof. Let $T = (t_1, \ldots, t_n)$ be a WV-terrain and let $VG(T)$ be its visibility graph. Let l be any horizontal line lying above T. Since T is a WV-terrain, there exist points a and b on l, where a is to the left of b, such that, one can shoot a ray that hits ab and does not intersect T, from every point on T. Now, consider the polygon $P = (a, t_1, \ldots, t_n, b)$. P is a 'convex' WV-polygon and therefore, by Claim 2, P satisfies the order claim. Finally, since T is contained in P's boundary, we conclude that T also satisfies the order claim.

We will need the following slightly more general statement, whose proof at this point is straight forward.

Theorem 2. *Let T be a WV-terrain, and let C and W be two finite sets of points on T. Then, the visibility graph of $C \cup W$ with respect to T is terrain-like, and thus there exists a PTAS for guarding the points in W by points in C.*

4 Guarding the Boundary

Friedrichs et al. [12] considered two versions of the 1.5D-terrain guarding problem, where the goal is to guard the entire terrain. In the first, the *semi-continuous* version, guards may lie only at vertices, while in the second, the *continuous* version, guards may lie anywhere on the terrain. For both versions, they showed how one can use a PTAS for vertex guarding the vertices of a 1.5D terrain, to obtain a PTAS for guarding the entire terrain. In this section, we provide similar results, both for WV-polygons and WV-terrains in the semi-continuous case, and for WV-terrains in the continuous case. More precisely, we show that the x-monotonicity property of 1.5D terrains is not necessary for their PTASs.

4.1 The Semi-continuous Version

Let P be a polygon which is weakly visible from one of its edges $e = uv$. As shown in the previous section, u and v together see the portion of P's boundary from w_a to w_b (moving counterclockwise from w_a), see Fig. 4. Notice that the portion of P's boundary from w_a to w_b, moving clockwise from w_a, is a WV-terrain $T(w_a, w_b)$. Therefore, given a PTAS for semi-continuously guarding a WV-terrain, we can obtain a PTAS for semi-continuously guarding P, by placing guards at u and v and removing from the witness set computed for $T(w_a, w_b)$ (see below) the witnesses that are already seen by u or v. We thus focus on semi-continuously guarding a WV-terrain.

Let T be a WV-terrain and let V be its set of vertices. We denote by $OPT(V, T)$ the minimum size of a set $Q \subseteq V$ that guards T, i.e., every point p on T is seen by at least one vertex in Q. The underlying idea is to construct a polynomial-size set W of points on T, such that for any set $Q \subseteq V$ it holds that Q guards W if and only if Q guards the entire terrain T. The set W is called a *witness set* of T. Then, we apply the PTAS of Theorem 2 to T, V and W, to obtain a set $Q \subseteq V$, such that Q guards W and $|Q| \leq (1+\varepsilon) \cdot OPT(V, W)$, where $OPT(V, W)$ is the size of an optimal solution for guarding W with vertices from V. Since W is a witness set of T, the set Q guards the entire terrain T. Moreover, $OPT(V, W) \leq OPT(V, T)$, and therefore $|Q| \leq (1 + \varepsilon) \cdot OPT(V, T)$.

Constructing a Witness Set. Let $v \in V$ be a vertex of T. Consider the set of (maximal) pieces of T that are visible from v. This set consists of at most n *visible pieces* of v. Let $W'(v)$ be the set of endpoints of v's visible pieces (see Fig. 6).

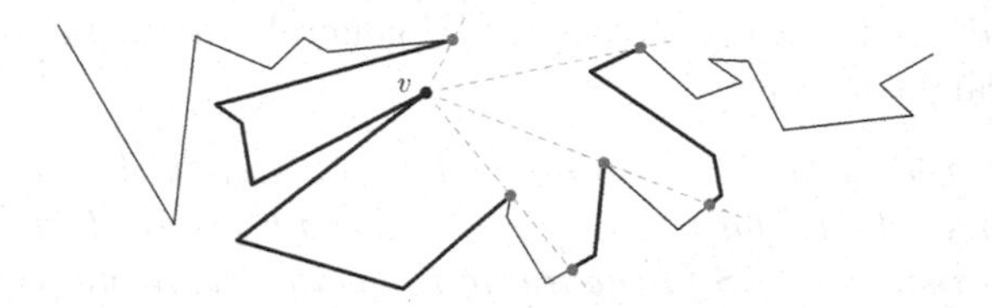

Fig. 6. The vertex v and its three visible pieces and their endpoints.

Set $B = \bigcup_{v \in V} W'(v)$, and let $b_0, b_1, \ldots, b_k$ be the points of B, sorted by their order along T. The points of B divide T into k pieces (since the extreme vertices of T are in B). Denote by T_i the piece between b_{i-1} and b_i, for $i = 1, \ldots, k$, and let p_i be any point in the interior of T_i (i.e. $p_i \in T_i \setminus \{b_{i-1}, b_i\}$). Now, set $W(V) = \{p_1, \ldots, p_k\}$.

Claim 3. *$W(V)$ is a witness set for T.*

Proof. Clearly, $|W(V)| = O(n^2)$, and any set $Q \subseteq V$ that guards T also guards $W(V)$. We need to show that if Q guards $W(V)$, then it also guards T.

Let $g \in Q$ and $p_i \in W(V)$ such that g sees p_i. Notice that p_i belongs to a visible piece of g whose endpoints are in $W'(g)$. Since p_i is not an endpoint, this visible piece must contain T_i (otherwise, one of the endpoints in $W'(g)$ is between b_{i-1} and b_i). Therefore, g sees the entire piece T_i. Since for each point $p_i \in W(V)$, there exists a guard in Q that sees it, we conclude that Q guards the set of pieces $\{T_1, \ldots, T_k\}$, whose union is exactly T.

4.2 The Continuous Version

Recall that in the continuous case, guards may lie anywhere on the polygon's boundary or the terrain. This forces us to restrict our discussion to WV-terrains, since the order claim for WV-polygons is no longer true, once we include the 'seeing' edge uv, see Fig. 7.

We now describe a PTAS based on Theorem 2 for continuously guarding a WV-terrain. Since, after removing the open edge uv from the boundary of a WV-polygon, we remain with a WV-terrain (assuming the angles at u and v are convex), we can also use the PTAS for continuously guarding the boundary of a WV-polygon, where it is forbidden to place guards on the open edge uv.

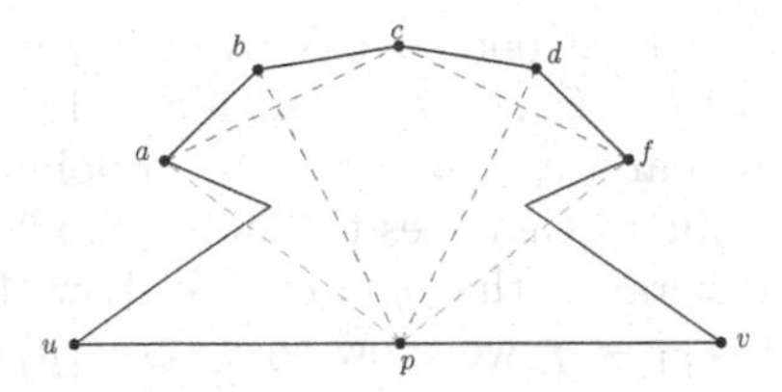

Fig. 7. A point p on the seeing edge uv, and points $a \prec b \prec c \prec d \prec f$. If $c \prec p$, the order claim breaks because a sees c, b sees p, but a does not see p. If $p \prec c$, the claim breaks because p sees d, c sees f, but p does not see f.

Given a WV-terrain T and a (possibly unbounded) set X of points on T, let $OPT(X,T)$ be the minimum size of a set $Q \subseteq X$ that guards T, i.e., every point p on T is seen by at least one guard in Q. Now, the idea is to construct a polynomial-size set C of terrain points (which will contain V), such that $OPT(C,T) = OPT(T,T)$. We call C a *candidate set* of T. Then, we consider a 'new' WV-terrain, T', which is obtained from T by adding the points in $C \setminus V$ to the vertex set of T. Clearly, $OPT(T,T) = OPT(C,T) = OPT(V',T')$, where $V' = C$ is the vertex set of T'. Therefore, by applying the semi-continuous guarding PTAS to T', we obtain a set of guards Q for T, such that $|Q| \leq (1+\varepsilon) \cdot OPT(T,T)$, as required.

As our candidate set, we take the set $C = V \cup W(V)$, where $W(V)$ is as above. Notice that when applying the semi-continuous guarding PTAS to T', we compute a new different witness set $W(V')$, for the vertex set V' of T'.

Lemma 2. $OPT(C,T) = OPT(T,T)$.

Proof. Let $G \subseteq T$ be a guarding set for T, such that $|G| = OPT(T,T)$ and $G \nsubseteq C$. We show that for any guard $g \in G \setminus C$, there exists a point $c \in C$ such that $(G \setminus \{g\}) \cup \{c\}$ is still a guarding set for T. Thus, by replacing each of the guards in $G \setminus C$ by an appropriate point of C, we obtain a guarding set $G' \subseteq C$ for T of size $OPT(T,T)$.

Let g be a guard in $G \setminus C$. Let c_l and c_r be the points in C just before g and just after g, respectively (by the order along T). Since $V \subseteq C$ and $g \notin C$, both c_l and c_r exist. Moreover, g, c_l and c_r lie on the same edge of T. We show below that either $(G \setminus \{g\}) \cup \{c_l\}$ or $(G \setminus \{g\}) \cup \{c_r\}$ is a guarding set for T.

We say that g is *necessary* for an edge e of T, if the set $G \setminus \{g\}$ does not see e completely. Clearly, there exists at least one edge e for which g is necessary.

Assume first that for each edge e for which g is necessary, g sees the entire edge e. Let $e = (v_i, v_{i+1})$ be such an edge. Then, v_i sees g and thus the visible piece of v_i that includes g must also include c_l, which means that c_l sees v_i. Similarly, we conclude that c_l sees v_{i+1}. Now, since c_l sees both v_i and v_{i+1}, it sees the entire edge e. So, in this case, $(G \setminus \{g\}) \cup \{c_l\}$ is a guarding set for T. Symmetrically, c_r sees the entire edge e, so $(G \setminus \{g\}) \cup \{c_r\}$ is also a guarding set for T.

Now, assume there is at least one edge for which g is necessary but g does not see the entire edge. Let $e = (v_i, v_{i+1})$ be such an edge, and let $v_i \prec p \prec v_{i+1}$ be a point that is seen by g and $v_i \prec p' \prec v_{i+1}$ be a point that is not seen by g. Moreover, let $g' \in G$ be a guard that sees the point p'. Clearly, g is not on e, so assume, without loss of generality, that $g \prec v_i$. We show that $(G \setminus \{g\}) \cup \{c_l\}$ is a guarding set for T. (If $v_{i+1} \prec g$, we show that $(G \setminus \{g\}) \cup \{c_r\}$ is a guarding set for T.)

Let q be a point that is seen by g. If q is on the same edge as g (and c_l and c_r), then c_l clearly sees q. If q is not on the same edge as g and succeeds c_r, then, by the order claim (applied to c_l, g, c_r, q), c_l sees q, see Fig. 8. The rest of the proof is devoted to the case where q is not on the same edge as g and precedes c_l. We show that in this case q is also seen by g'.

Observe first that g' necessarily succeeds v_{i+1}, i.e., $v_{i+1} \prec g'$. This is true, since if g' is on $v_i v_{i+1}$, then g' sees the entire edge e, which implies that g is not necessary for e—a contradiction. Assume, therefore, that $g' \prec v_i$. Since both g and g' see a point between v_i and v_{i+1}, and v_i sees v_{i+1}, we conclude, by the order claim, that both g and g' see v_{i+1}. This means that g sees $\overline{pv_{i+1}}$ and g' sees $\overline{p'v_{i+1}}$. Since g does not see p', we have $p' \prec p$, but then $\overline{pv_{i+1}} \subseteq \overline{p'v_{i+1}}$, which again implies that g is not necessary for e—a contradiction.

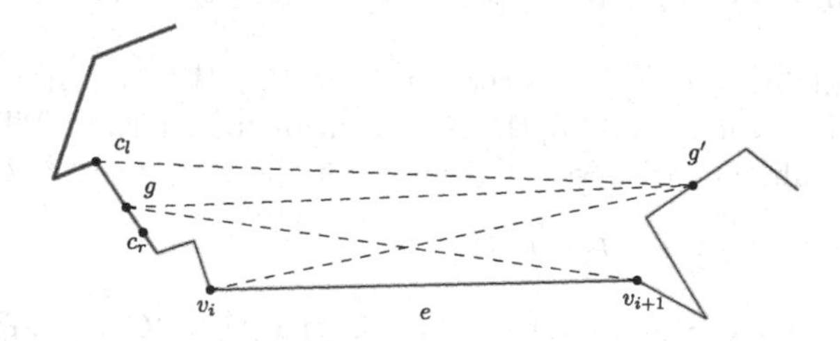

Fig. 8. If q is on g's edge or on any subsequent edge, g sees $q \implies c_l$ sees q. If q is on any edge preceding g's edge, g sees $q \implies g'$ sees q.

Hence, we have $g \prec v_i \prec v_{i+1} \prec g'$. Next, we observe that c_l sees g'. Indeed, g sees v_{i+1} (because g sees p and v_i sees v_{i+1}) and v_i sees g' (because v_i sees v_{i+1} and p' sees g'), so g sees g' (see Fig. 8). Moreover, since c_l sees c_r and g sees g', c_l sees g'.

Finally, since q sees g and c_l sees g', we conclude that q is also seen by g'. Thus, we can replace g with c_l and remain with a guarding set for T.

5 Terrain-Like vs. Non-jumping

We slightly rephrase the definitions given above for non-jumping graphs and terrain-like graphs. Let $G = (V, E)$ be a graph with n vertices. A *labeling* of (the vertices of) G is an injective function $\pi : V \to [n]$. We say that π is a *non-jumping labeling* of G if for any four vertices a, b, c, d such that $\pi[a] < \pi[b] < \pi[c] < \pi[d]$, if both $\{a,c\}$ and $\{b,d\}$ are in E, then so is $\{b,c\}$. The graph G is non-jumping if there exists a non-jumping labeling of G. Similarly, π is a *terrain-like labeling* of G if for any four vertices a,b,c,d such that $\pi[a] < \pi[b] < \pi[c] < \pi[d]$, if both $\{a,c\}$ and $\{b,d\}$ are in E, then so is $\{a,d\}$, and G is terrain-like if there exists a terrain-like labeling of G. Denote by $\mathcal{F}_{\text{NJ}}$ the family of non-jumping graphs and by $\mathcal{F}_{\text{TL}}$ the family of terrain-like graphs.

Several well-known graph families, such as outerplanar graphs, convex bipartite graphs, and complete graphs, are subfamilies of both $\mathcal{F}_{\text{NJ}}$ and $\mathcal{F}_{\text{TL}}$. In this section we investigate the relation between these two graph families, i.e., $\mathcal{F}_{\text{NJ}}$ and $\mathcal{F}_{\text{TL}}$. First, we present a natural infinite family of graphs which are in $\mathcal{F}_{\text{TL}}$ but not in $\mathcal{F}_{\text{NJ}}$, and give a short and simple proof for it. Moreover, the smallest member of this family is a planar graph, which means that there exist planar graphs which cannot be realized as monotone L-graphs (as is also shown in [6]). Then, we present some basic properties of the terrain-like labeling function, and use them to prove that there exists an infinite family of graphs that are in $\mathcal{F}_{\text{NJ}}$ but not in $\mathcal{F}_{\text{TL}}$. Finally, we present a family of graphs which are not in $\mathcal{F}_{\text{TL}} \cup \mathcal{F}_{\text{NJ}}$.

Theorem 3. $\mathcal{F}_{\text{TL}} \nsubseteq \mathcal{F}_{\text{NJ}}$.

Proof. Let $K_n = (V = \{v_1, v_2, \ldots, v_n\}, E)$ be the complete graph on n vertices. For $n \geq 6$, let $K_n^{-3} = (V, E \setminus \{e_1, e_2, e_3\})$, where e_1, e_2, e_3 are any three pairwise-disjoint edges in E. We show that for any $n \geq 6$, $K_n^{-3} \in \mathcal{F}_{\text{TL}} \setminus \mathcal{F}_{\text{NJ}}$. Assume w.l.o.g. that $e_1 = \{v_1, v_2\}$, $e_2 = \{v_3, v_4\}$, and $e_3 = \{v_5, v_6\}$.
$\mathbf{K_n^{-3}} \in \mathcal{F}_{\text{TL}}$: Consider the labeling $\pi[v_i] = i$. For any four vertices $v_{i_1}, v_{i_2}, v_{i_3}, v_{i_4}$ such that $i_1 < i_2 < i_3 < i_4$, we have $\{v_{i_1}, v_{i_4}\} \in E$ since $i_4 - i_1 \geq 3$, and thus π is a terrain-like labeling.
$\mathbf{K_n^{-3}} \notin \mathcal{F}_{\text{NJ}}$: Assume by contradiction that $K_n^{-3} \in \mathcal{F}_{\text{NJ}}$, then there exists a non-jumping labeling π of K_n^{-3}. Assume w.l.o.g that $\pi[v_1] < \pi[v_2]$. We claim that either $\pi[v_1] = 1$ or $\pi[v_2] = n$. Indeed, assume that $\pi[v_i] = 1$ for some $i \neq 1$ and $\pi[v_j] = n$ for some $j \neq 2$. Notice that since e_1, e_2, e_3 are pairwise disjoint, $\{v_i, v_2\}$ and $\{v_1, v_j\}$ are edges of the graph. But $\{v_1, v_2\}$ is not an edge of the graph, so π is not a non-jumping labeling w.r.t. v_i, v_1, v_2, v_j—a contradiction. By symmetry, the above claim holds also for v_3, v_4 and for v_5, v_6, but then π is not an injective function.

As a corollary, we get that not all planar graphs are non-jumping. Indeed, it is easy to verify that K_6^{-3} is planar.

Some Properties of Labeling Functions. In order to prove the opposite direction, i.e., that $\mathcal{F}_{\text{NJ}} \nsubseteq \mathcal{F}_{\text{TL}}$, we need to reveal some of the properties of terrain-like and non-jumping labeling functions. The properties and their proofs can be found in the full version of this paper.

Theorem 4. $\mathcal{F}_{\text{NJ}} \nsubseteq \mathcal{F}_{\text{TL}}$.

Proof. Let C_6 be the cycle graph with vertex set $V = \{v_1, v_2, \ldots, v_6\}$, and P_n the path graph with vertex set $U = \{u_1, u_2, \ldots, u_n\}$, $n \geq 2$. Consider the graph

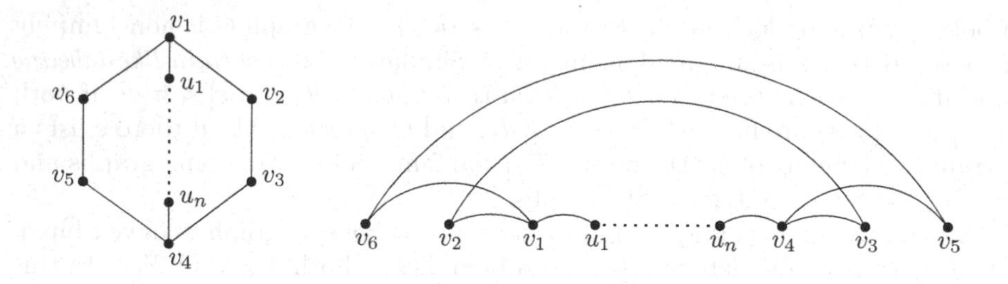

Fig. 9. Left: The graph G. Right: A non-jumping labeling of G, i.e., $\pi[v_6] = 1, \pi[v_2] = 2, \pi[v_1] = 3, \ldots, \pi[v_5] = n + 6$.

$G = (V \cup U, E)$, where $E = E(C_6) \cup E(P_n) \cup \{\{v_1, u_1\}, \{v_4, u_n\}\}$. In other words, G contains an induced cycle on 6 vertices $v_1, v_2, \ldots, v_6$, and an induced path on $n + 2$ vertices $v_1, u_1, u_2, \ldots, u_n, v_4$; see Fig. 9 (left).

Figure 9 (right) shows a non-jumping labeling of G, so G is in $\mathcal{F}_{\mathrm{NJ}}$. The rest of the proof, in which we show that G is not in $\mathcal{F}_{\mathrm{TL}}$, can be found in the full version of this paper.

Finally, does every graph belong either to $\mathcal{F}_{\mathrm{TL}}$ or to $\mathcal{F}_{\mathrm{NJ}}$? The answer is clearly no, but it would be nice to see a concrete and simple example. In the full version of this paper, we present an infinite family of graphs which are neither in $\mathcal{F}_{\mathrm{TL}}$ nor in $\mathcal{F}_{\mathrm{NJ}}$.

References

1. Ahmed, A.R., et al.: L-graphs and monotone L-graphs. arXiv:1703.01544 (2017)
2. Bandyapadhyay, S., Maheshwari, A., Mehrabi, S., Suri, S.: Approximating dominating set on intersection graphs of rectangles and L-frames. In: 43rd International Symposium on Mathematical Foundations of Computer Science, MFCS 2018, 27–31 August 2018, Liverpool, UK, pp. 37:1–37:15 (2018). https://doi.org/10.4230/LIPIcs.MFCS.2018.37
3. Ben-Moshe, B., Katz, M.J., Mitchell, J.S.B.: A constant-factor approximation algorithm for optimal 1.5D terrain guarding. SIAM J. Comput. **36**(6), 1631–1647 (2007)
4. Bhattacharya, P., Ghosh, S.K., Pal, S.: Constant approximation algorithms for guarding simple polygons using vertex guards. arXiv:1712.05492 (2017)
5. Bhattacharya, P., Ghosh, S.K., Roy, B.: Approximability of guarding weak visibility polygons. Discrete Appl. Math. **228**, 109–129 (2017)
6. Catanzaro, D., et al.: Max point-tolerance graphs. Discrete Appl. Math. **216**, 84–97 (2017). https://doi.org/10.1016/j.dam.2015.08.019
7. Chan, T.M., Har-Peled, S.: Approximation algorithms for maximum independent set of pseudo-disks. Discrete Comput. Geom. **48**(2), 373–392 (2012)
8. Clarkson, K.L., Varadarajan, K.R.: Improved approximation algorithms for geometric set cover. Discrete Comput. Geom. **37**(1), 43–58 (2007)
9. Correa, J.R., Feuilloley, L., Pérez-Lantero, P., Soto, J.A.: Independent and hitting sets of rectangles intersecting a diagonal line: algorithms and complexity.

Discrete Comput. Geom. **53**(2), 344–365 (2015). https://doi.org/10.1007/s00454-014-9661-y
10. Eidenbenz, S., Stamm, C., Widmayer, P.: Inapproximability results for guarding polygons and terrains. Algorithmica **31**(1), 79–113 (2001)
11. Elbassioni, K., Krohn, E., Matijević, D., Mestre, J., Ševerdija, D.: Improved approximations for guarding 1.5-dimensional terrains. Algorithmica **60**(2), 451–463 (2011)
12. Friedrichs, S., Hemmer, M., King, J., Schmidt, C.: The continuous 1.5D terrain guarding problem: discretization, optimal solutions, and PTAS. J. Comput. Geom. **7**(1), 256–284 (2016). http://jocg.org/index.php/jocg/article/view/242
13. Ghosh, S.K., Maheshwari, A., Pal, S.P., Saluja, S., Madhavan, C.E.V.: Characterizing and recognizing weak visibility polygons. Comput. Geom. **3**, 213–233 (1993). https://doi.org/10.1016/0925-7721(93)90010-4
14. Gibson, M., Kanade, G., Krohn, E., Varadarajan, K.: Guarding terrains via local search. J. Comput. Geom. **5**(1), 168–178 (2014)
15. Gibson, M., Pirwani, I.A.: Algorithms for dominating set in disk graphs: breaking the $\log n$ barrier. In: de Berg, M., Meyer, U. (eds.) ESA 2010. LNCS, vol. 6346, pp. 243–254. Springer, Heidelberg (2010). https://doi.org/10.1007/978-3-642-15775-2_21
16. King, J.: A 4-approximation algorithm for guarding 1.5-dimensional terrains. In: Correa, J.R., Hevia, A., Kiwi, M. (eds.) LATIN 2006. LNCS, vol. 3887, pp. 629–640. Springer, Heidelberg (2006). https://doi.org/10.1007/11682462_58
17. King, J., Krohn, E.: Terrain guarding is NP-hard. SIAM J. Comput. **40**(5), 1316–1339 (2011). https://doi.org/10.1137/100791506
18. Mustafa, N.H., Ray, S.: PTAS for geometric hitting set problems via local search. In: Proceedings of the 25th Annual Symposium on Computational Geometry, pp. 17–22. ACM (2009)
19. Soto, M., Caro, C.T.: p-box: a new graph model. Discrete Math. Theoret. Comput. Sci. **17**(1), 169–186 (2015). http://dmtcs.episciences.org/2121

A New Lower Bound for Classic Online Bin Packing

János Balogh[1], József Békési[2], György Dósa[3], Leah Epstein[4](✉), and Asaf Levin[5]

[1] Institute of Informatics, University of Szeged, Szeged, Hungary
baloghj@inf.u-szeged.hu

[2] Department of Applied Informatics, Gyula Juhász Faculty of Education, University of Szeged, Szeged, Hungary
bekesi@jgypk.u-szeged.hu

[3] Department of Mathematics, University of Pannonia, Veszprém, Hungary
dosagy@almos.vein.hu

[4] Department of Mathematics, University of Haifa, Haifa, Israel
lea@math.haifa.ac.il

[5] Faculty of Industrial Engineering and Management, The Technion, Haifa, Israel
levinas@ie.technion.ac.il

Abstract. We improve the lower bound on the asymptotic competitive ratio of any online algorithm for bin packing to above 1.54278.

We demonstrate for the first time the advantage of branching and the applicability of full adaptivity in the design of lower bounds for the classic online bin packing problem. We apply a new method for weight based analysis, which is usually applied only in proofs of upper bounds. The values of previous lower bounds were approximately 1.5401 and 1.5403.

1 Introduction

The bin packing problem [13,20] is a well-studied combinatorial optimization problem with origins in data storage and cutting stock. The input consists of items of rational sizes in $(0,1]$, where the goal is to split or pack them into partitions called bins, such that the total size of items for every bin cannot exceed 1. The online bin packing problem [9] is its variant where items are presented one by one, and the algorithm assigns each item to a bin before it can see the

J. Balogh was supported by the European Union, co-financed by the European Social Fund (EFOP-3.6.3-VEKOP-16-2017-00002). J. Békési was supported by the EU-funded Hungarian grant EFOP-3.6.2-16-2017-00015 and by the National Research, Development and Innovation Office of Hungary (NKFIH), grant no. SNN 129178. Gy. Dósa was supported by National Research, Development and Innovation Office – NKFIH under the grant SNN 129364 and by Széchenyi 2020 under the EFOP-3.6.1-16-2016-00015. L. Epstein and A. Levin were partially supported by a grant from GIF - the German-Israeli Foundation for Scientific Research and Development (grant number I-1366-407.6/2016).

E. Bampis and N. Megow (Eds.): WAOA 2019, LNCS 11926, pp. 18–28, 2020.
https://doi.org/10.1007/978-3-030-39479-0_2

next item. This online bin packing is a core problem in online computation and in competitive analysis.

For an algorithm A and an input I, let $A(I)$ be the cost (number of bins) used by A for I. The algorithm A can be an online or offline algorithm, and it can also be an optimal offline algorithm OPT. The absolute competitive ratio of algorithm A for input I is the ratio between $A(I)$ and $OPT(I)$. The absolute competitive ratio of A is the worst-case (or supremum) absolute competitive ratio over all inputs. Given an integer N, we can consider the worst-case absolute competitive ratio over inputs where $OPT(I)$ is not smaller than N. Taking this sequence and letting N grow to infinity, the limit is the asymptotic competitive ratio of A. This measure is the standard one for analysis of the bin packing problem, and it is considered to be more meaningful than the absolute ratio (which is affected by very small inputs).

The current best online algorithm with respect to the asymptotic competitive ratio has an asymptotic competitive ratio no larger than 1.57829 [4], which was found recently by development of new methods of analysis. Previous results were achieved via a sequence of improvements [12,14–16,18,19,22]. In this work, we consider the other standard aspect of the online problem, namely, of establishing lower bounds on the asymptotic competitive ratio that can be achieved by online algorithms.

The first lower bound on the asymptotic competitive ratio was found by Yao [22], and it uses an input with at most three types of items: $\frac{1}{7}+\varepsilon$, $\frac{1}{3}+\varepsilon$, and $\frac{1}{2}+\varepsilon$ (where $\varepsilon > 0$ is sufficiently small). For this input, if the entire input is presented, every bin of an optimal solution has one item of each type (and otherwise there are larger numbers of items in a bin, but all bins are still packed identically). It is possible to start the sequence with smaller items, for example, it can be started with $\frac{1}{1807}+\varepsilon$ and then $\frac{1}{43}+\varepsilon$, which increases the resulting bound. This was discovered by Brown and Liang (independently) [8,17], who showed a lower bound of 1.53635. Van Vliet [21] found an interesting method of analysis and showed that the same approach (the above sequence with additional items) gives in fact a lower bound of 1.5401474. Finally, Balogh, Békési, and Galambos [5] showed that the greedy sequence above is actually not the best one among sequences with batches of identical items, and proved a lower bound of $248/161 \approx 1.5403726$ (see also [6] for an alternative proof). Their sequence starts with decreasing powers of $\frac{1}{7}$ plus epsilon (it can be started with items complementing the other items to 1 but it does not change the bound), and the items following the items of size $\frac{1}{49}+\varepsilon$ are exactly those used by Yao [22]. This result of [5] is the previously best known lower bound.

One drawback of the previous lower bounds is that while the exact input was not determined in advance, the set of sizes used for it was determined prior to the action of the algorithm by the input provider and it was known to the algorithm. Moreover, for classic bin packing, in all previously designed lower bound inputs, sizes of items were slightly larger than a reciprocal of an integer, and optimal solutions consisted of bins with identical packing patterns. The possible item sizes and numbers of items were known to the algorithm, but the stopping point of the input was unknown, and it was based on the action of the algorithm. It seemed unlikely that such examples are indeed the worst-case examples.

Given an input sequence, the online algorithm should perform well on every input prefix while the optimal solution should be optimal only at termination. This sounds weak, but in decades of research this is the unique shortcoming of online algorithms that the community has managed to exploit for proving lower bounds for this core online problem. This holds for bin packing and basically any other packing problem of this type.

We show here that different methods for proving lower bounds and new approaches to sizes of items give an improved lower bound. The contribution of this paper is that this weakness no longer holds. We show for the first time that integrating branching and adaptivity into input constructions improves the resulting lower bound for online bin packing substantially.

Other online problems that are very different from bin packing do not suffer from this weakness. Many problems admit better lower bounds based on adaptivity and branching while the situation with bin packing variants was extremely different as we mentioned above. One such example is makespan minimization (see e.g. [10]).

1.1 New Features of Our Work

Previous lower bound constructions for standard bin packing were defined for inputs without branching. Those are inputs where the possible inputs differ only by their stopping points. Here, we use an input with branching, which makes the analysis harder, as those branches are related (the additional items may use the same existing bins in addition to new bins), but at most one of them will be presented eventually. It is notable that branching was used to design an improved lower bound for the case where the input consists of three batches [3] (where for each one of the batches, all items are presented at once), but it was unknown whether it can be used to design improved lower bounds for standard online bin packing.

It was also not known whether one can exploit methods of constructing fully adaptive inputs, where in some parts of the input every item size is based precisely on previous decisions of the algorithm. Such results were previously proved for online bin packing with cardinality constraints, where (in addition to the constraint on the total size) every bin is limited to containing k items, for a fixed parameter $k \geq 2$ [1,2,7,11]. Thus, in addition to branching we will use the following theorem proved in [2] (see the construction in Sect. 3.1 and Corollary 3 in [2]).

Theorem 1. *Let $N \geq 1$ be a large positive integer and let $k \geq 2$ be an integer. Assume that we are given an arbitrary deterministic online algorithm for a variant of bin packing and a binary condition Con on the possible behavior of an online algorithm for one item (on the way that the item is packed). That is, Con is a logical condition whose value is revealed to the adversary once the item is packed. An adversary is able to construct a sequence of values a_i $(1 \leq i \leq N)$*

such that for any i, $a_i \in \left(k^{-2^{N+3}}, k^{-2^{N+2}}\right)$, and in particular $a_i \in \left(0, \frac{1}{k^4}\right)$ (defining item sizes is done using a given affine linear function of the values a_i and these are not necessarily the item sizes), such that for any item i_1 satisfying Con and any item i_2 not satisfying Con, it holds that $\frac{a_{i_2}}{a_{i_1}} > k$.

Examples for the condition Con can be the following: "the item is packed as a first item of its bin", "the item is packed into a non-empty bin", "the item is packed into a bin already containing an item of size above $\frac{1}{2}$", etc. Here, the condition Con will be that the item is not packed into an empty bin (or a new bin).

Our method of analysis is based on a new type of a weighting function. This kind of analysis is often used for analyzing bin packing algorithms, that is, for upper bounds. It was used for lower bounds by [6] and by van Vliet [21] (where the term weight is not used, and the values given to items are based on the dual linear program, but the specific kind of dual variables and their usage can be adapted to a weighting function). However, those weights were defined for inputs without branching and we extend the use of these weights for inputs with branching for the first time, which adds technical challenges to our work also in the analysis. The advantage of weights is that we do not need to test all packing patterns of an algorithm, whose number can be very large, and thus we obtain a complete and verifiable proof with much smaller number of cases than that of pattern based proofs (see for example [11]).

Our analysis based on this new weighting function makes the proof very short and compact. The very limited case analysis that we carry out in our proof is much less tedious than previous proofs and we provide the full details of it. This compact case analysis is performed using combinatorial insights that are employed to decrease the number of cases by orders of magnitude, and in order to illustrate the power of our analysis we mention that our short proof was constructed by hand without using a computer.

2 The Input

Let $t \geq 3$ be an integer, let $\varepsilon > 0$ be small constant, let M be a large integer and let $N = M \cdot 42^t$ (N is a large integer divisible by $6 \cdot 7^t$). We choose ε such that $\varepsilon < \frac{1}{(2058)^t}$.

Given a specific algorithm ALG, we will analyze it for the set of inputs defined here, where the input depends on the actions of ALG both with respect to stopping the input, but also some of the sizes will be based on the exact action of ALG, and on the previously presented items and their number.

Let $C_t = \frac{1}{6 \cdot 7^{t-1}} - 294\varepsilon$, and for $2 \leq j \leq t-1$, let $C_j = \frac{1+28\varepsilon}{7^j}$. The input starts with batches of N items of the sizes C_j, for every $j = t, t-1, \ldots, 2$, where the input may be stopped after each one of these batches. An item of size C_j is called a C_j–item.

Afterwards, there are N items called A–items. The sizes of A–items will be all strictly larger than $\frac{1+\varepsilon}{7}$ but strictly smaller than $\frac{1+2\varepsilon}{7}$. Any A–item packed as a first item into a bin will be called a large A–item, and any other A–items will be called a small A–item. During the construction, based on the actions of the algorithm, we will ensure that for any large A–item, its difference from $\frac{1+\varepsilon}{7}$ is larger by a factor of more than 4 than the difference from $\frac{1+\varepsilon}{7}$ of any small A–item. The details of attaining this property are given below (see Lemma 1).

Let $\gamma > 0$ be such that the size of every small A–item is at most $\frac{1+\varepsilon+\gamma}{7}$ while the size of every large A–item is above $\frac{1+\varepsilon+4\gamma}{7}$ (where $\gamma < \frac{\varepsilon}{4}$). The input may be stopped after A–items are introduced (the number of A–items is N no matter how many of them are small and how many are large). Let n_L denote the number of large A–items, and therefore there are $N - n_L$ small A–items. Even though the A–items will have different sizes and they cannot be presented at once to the algorithm, we see them as one batch.

If the input is not stopped after the arrival of A–items, there are three options to continue the input (i.e., we use branching at this point). In order to define the three options, we first define the following five items types. A B_{11}–item has size $\frac{1+2\varepsilon}{2}$. A B_{21}–item has size $\frac{1+\varepsilon}{3}$ and a B_{22}–item has size $\frac{1+\varepsilon}{2}$. A B_{31}–item has size $\frac{5-2\varepsilon-3\gamma}{14}$ and a B_{32}–item has size $\frac{7+\gamma}{14} = \frac{1}{2} + \frac{\gamma}{14} < \frac{1}{2} + \frac{\varepsilon}{56}$ (this size is above $\frac{1}{2}$).

The first option to continue is with B_{11}–items, such that a batch of $\frac{N}{3}$ such items arrive. The second option is with a batch of B_{21}–items, possibly followed by a batch of B_{22}–items. In this option, the number of items of each batch is N. The third option is that a batch of B_{31}–items arrive, possibly followed by a batch of B_{32}–items. In the last case, we define the numbers of items based on n_L as follows. The number of B_{31}–items (if they are presented) is $n_{31} = \frac{7N-7n_L}{6}$. The number of B_{32}–items (if they are presented) is $n_{32} = \frac{7N-5n_L}{6}$. This concludes the description of the input (see Fig. 1 for an illustration).

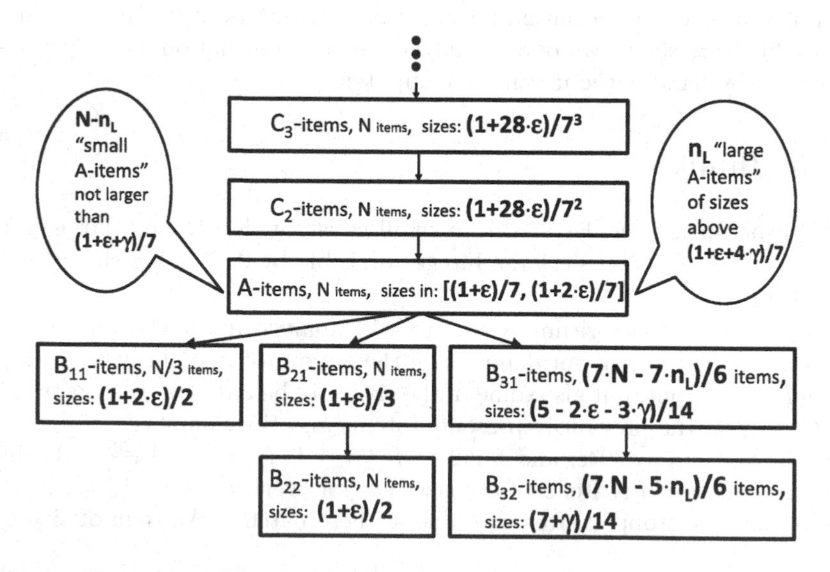

Fig. 1. An illustration of the input. Every box contains a set of items, and the input may be stopped after presenting the items of any box. In cases with branching, at most one path is selected, and any such path may be presented as an input.

We conclude this section by showing that indeed we can construct the batch (or subsequence) of A–items satisfying the required properties.

Lemma 1. *The sizes of A–items can be constructed as described.*

Proof. We use Theorem 1. Condition Con is that the item is packed into a bin that already contains at least one item (this item may be of a previous batch of items). Let $k = \lceil \frac{1}{\varepsilon} \rceil$. The items sizes are $\frac{1+\varepsilon+a_i}{7}$. We find that all item sizes are in $(\frac{1+\varepsilon}{7}, \frac{1+2\varepsilon}{7})$. We also have for two items of sizes $\frac{1+\varepsilon+a_{i_1}}{7}$ and $\frac{1+\varepsilon+a_{i_2}}{7}$, where the second item does not satisfy Con while the first one satisfies Con that $\frac{a_{i_2}}{a_{i_1}} > k$. Let γ be the maximum size of any value a_i of an item satisfying Con. Then, we have $\frac{1+\varepsilon+a_{i_1}}{7} \leq \frac{1+\varepsilon+\gamma}{7}$ and $\frac{1+\varepsilon+a_{i_2}}{7} \geq \frac{1+\varepsilon+k\gamma}{7} > \frac{1+\varepsilon+4\gamma}{7}$, as required. □

In order to give some motivation regarding the sizes of items, note that by $\varepsilon < \frac{1}{(2058)^t}$, we have $\frac{5-2\varepsilon-3\gamma}{14} \geq \frac{5}{14} - \frac{2.75}{14} \cdot \varepsilon > 0.35714$, while $\frac{1+\varepsilon}{3} < 0.33334$.

3 Bounds on the Optimal Costs

In this section we prove upper bounds on the optimal costs of our instances. We denote the optimal cost after the batch of items of sizes C_j is presented by OPT_j (for $j \geq 2$). Similarly, we denote an optimal cost after the batch of A–items by OPT_1.

Lemma 2. *For $t \geq j \geq 2$, we have $OPT_j \leq \frac{N}{6 \cdot 7^{j-1}}$, and $OPT_1 \leq \frac{N}{6}$. Furthermore, let $j \geq 1$, then the total size of one item of each batch up to the batch of C_j–items (if $j \geq 2$) or up to the batch of A–items (if $j = 1$) is at most $\frac{1}{6 \cdot 7^{j-1}} - 293\varepsilon$.*

Proof. First, consider $j \geq 2$. The total size of $t - j + 1$ items, each of a different size out of $C_t, C_{t-1}, \ldots, C_j$ is

$$C_t + \sum_{i=j}^{t-1} C_i = \frac{1}{6 \cdot 7^{t-1}} - 294\varepsilon + \sum_{i=j}^{t-1} \frac{1+28\varepsilon}{7^i}$$

$$= \frac{1}{6 \cdot 7^{t-1}} - 294\varepsilon + \frac{1+28\varepsilon}{7^j} \sum_{i=j}^{t-1} \frac{1}{7^{i-j}}$$

$$= \frac{1}{6 \cdot 7^{t-1}} - 294\varepsilon + \frac{1+28\varepsilon}{7^j} \cdot \frac{7 - 1/7^{t-j-1}}{6}$$

$$= \frac{1}{6 \cdot 7^{t-1}} - 294\varepsilon + (1+28\varepsilon)\frac{1/7^{j-1} - 1/7^{t-1}}{6} < \frac{1}{6 \cdot 7^{j-1}} - 293\varepsilon,$$

as $\sum_{i=j}^{t-1} \frac{1}{7^{i-j}} = \sum_{i=0}^{t-j-1} \frac{1}{7^i} = \frac{1-1/7^{t-j}}{6/7} = \frac{7-1/7^{t-j-1}}{6}$. Thus, it is possible to pack $6 \cdot 7^{j-1}$ items of each size into every bin and get a feasible solution with $\frac{N}{6 \cdot 7^{j-1}}$ bins, so $OPT_j \leq \frac{N}{6 \cdot 7^{j-1}}$.

A similar bound can be used for the input up to the batch of A–items as well. In this case the total size of one item of each size C_j together with one A–item (small or large) is at most $\frac{1}{6\cdot 7^{t-1}} - 294\varepsilon + (1+28\varepsilon)\frac{1/7-1/7^{t-1}}{6} + \frac{1+2\varepsilon}{7} < \frac{1}{6} - 293\varepsilon$. Thus, $OPT_1 \leq \frac{N}{6}$ (by packing six items of each batch into every bin). □

We let $OPT_{11}, OPT_{21}, OPT_{22}, OPT_{31}, OPT_{32}$, denote costs of optimal solutions for the inputs after the batches of B_{11}–items, B_{21}–items, B_{22}–items, B_{31}–items, and B_{32}–items were presented, respectively. In the next lemma we present upper bounds on these optimal costs.

The proof of the next lemma is omitted.

Lemma 3. *We have $OPT_{11} \leq \frac{N}{3}$, $OPT_{21} \leq \frac{N}{2}$, $OPT_{22} \leq N$, $OPT_{31} \leq \frac{7N-5n_L}{12}$, and $OPT_{32} \leq \frac{7N-5n_L}{6}$.*

Next, we prove that the optimal costs are at least M (for all possible inputs). We have

$$C_t = \frac{1}{6\cdot 7^{t-1}} - 294\varepsilon > \frac{1}{6\cdot 7^{t-1}} - 294\cdot\frac{1}{2058^t} > \frac{1}{6\cdot 7^{t-1}+1}$$

as $\frac{1}{6\cdot 7^{t-1}} - \frac{1}{6\cdot 7^{t-1}+1} = \frac{1}{6\cdot 7^{t-1}(6\cdot 7^{t-1}+1)}$ while $294\cdot\frac{1}{2058^t} = \frac{1}{6^{t-1}\cdot 7^{3t-2}}$, and

$$6\cdot 7^{t-1}(6\cdot 7^{t-1}+1) < 6^{t-1}\cdot 7^{t-1}\cdot 7^t < 6^{t-1}\cdot 7^{3t-2}$$

by $t \geq 2$. Thus, as all inputs contain the first batch of C_t–items, and every bin has at most $6\cdot 7^{t-1}$ such items, we get that an optimal solution has at least $\frac{N}{6\cdot 7^{t-1}} > M$ bins.

4 An Analysis Using Weights

In this section we provide a complete analytic proof of the claimed lower bound that we establish using our construction. In fact we verified the tightness of our analysis (for this construction) by solving a mathematical program for some very small values of t but our analytic proof does not use this approach. Our analytic proof is based on assigning weights to items, defining prices to bins using the weights and bin types, and finally using these prices to establish the lower bound.

4.1 The Assignment of Weights to Items

We assign weights to items as follows. For a C_j–item, where $2 \leq j \leq t-1$, we let its weight be $\frac{1}{7^{j-1}}$. The weight of a C_t–item is $\frac{1}{6\cdot 7^{t-2}}$. The weight of a large A–item is denoted by w where $w \in [1, 1.5]$ and we will specify later the exact value of w. The weight of any other item is 1, those are B_{11}–items, B_{21}–items, B_{22}–items, B_{31}–items, B_{32}–items, and small A–items.

4.2 Definition of Bin Types

For a bin packed by the algorithm, we say that it has type j if it has a C_j–item for some $2 \le j \le t$ and no smaller items (i.e., for any k such that $j < k \le t$, it has no C_k–item). We say that it has type 1 if it has an A–item and no smaller items (i.e., it has no C_k–item for all $2 \le k \le t$). We say that it is a double bin if it has a B_{21}–item or a B_{31}–item and no smaller items (i.e., no C_k–item for all $2 \le k \le t$ and no A–item), and we say that it is a single bin if it has only items of sizes above $\frac{1}{2}$, i.e., a B_{11}–item or a B_{22}–item or a B_{32}–item (where every such bin has exactly one item).

4.3 The Price of a Bin Type

We define the price of a bin type as follows. A bin D of a certain type may receive additional items after its first batch of items out of which its first item comes. Moreover, its contents may differ in different continuations of the input (due to branching). Consider the contents of D for all continuations simultaneously (taking into account the situation where these items indeed arrive), and define a set of items $S(D)$ based on this (one can think of $S(D)$ as a virtual bin, which is valid for any possible input). For example, if the bin has one (large) A–item, and in the first continuation it will receive a B_{11}–item, in the second continuation it will receive one B_{21}–item and one B_{22}–item, and in the third continuation it will receive two B_{31}–items, then the set $S(D)$ contains six items (one of size approximately $\frac{1}{7}$, one of size approximately $\frac{1}{3}$, two of sizes approximately $\frac{1}{2}$, and two of sizes approximately $\frac{5}{14}$). The price of D is defined as the total weight of items of $S(D)$ (for the example, this price is $w + 5$). The price of a bin type is the supremum price of any bin of this type.

4.4 Calculating the Prices of the Bin Types

Let W_j denote the price of bin type j, for $1 \le j \le t$, let W_d denote the price of a double bin, and let W_s denote the price of a single bin.

The proofs of the next two lemmas are omitted.

Lemma 4. *For the weights defined above, we have* $W_s = 1$, $W_d = 2$, *and* $W_1 = w + 5$.

Lemma 5. *For the weights defined above, we have* $W_j = 7 - \frac{1}{7^{j-1}}$ *for* $2 \le j \le t - 1$, *and* $W_t = 7$.

4.5 Using the Prices of Bin Types to Establish the Lower Bound on the Asymptotic Competitive Ratio

Let ν_j denote the number of bins opened for C_j–items (bins used for the first time when the batch of C_j–items is presented). Let ν_1 denote the number of bins opened for A–items. Let $\nu_{k\ell}$ denote the number of bins opened for $B_{k\ell}$–items, for $(k,\ell) \in ISB$, where $ISB = \{(1,1),(2,1),(2,2),(3,1),(3,2)\}$. Moreover, as large A–items are exactly those A–items that are packed as first items of their bins, we have $\nu_1 = n_L$.

Let ALG_j denote the cost of the algorithm for the input up to the batch of C_j–items, and let ALG_1 denote the cost of the algorithm up to the batch of A–items. Let $ALG_{k\ell}$ denote the cost of the algorithm up to the batch of $B_{k\ell}$–items for $(k,\ell) \in ISB$.

Let R be the asymptotic competitive ratio of ALG, and let f be a function such that $f(n) = o(n)$ and for any input I it holds that $ALG(I) \leq R \cdot OPT(I) + f(OPT(I))$.

We have $ALG_j \leq R \cdot OPT_j + f(OPT_j)$ for $1 \leq j \leq t$. We also have $ALG_{k\ell} \leq R \cdot OPT_{k\ell} + f(OPT_{k\ell})$ for $(k,\ell) \in ISB$.

Let W denote the total weight of all items (for all branches, such that every possible item is counted exactly once). Since $\frac{1}{6 \cdot 7^{t-2}} + \sum_{j=2}^{t-1} \frac{1}{7^{j-1}} = \frac{1}{6}$, we have

$$\begin{aligned}
W &= N \cdot \left(\frac{1}{6 \cdot 7^{t-2}} + \sum_{j=2}^{t-1} \frac{1}{7^{j-1}}\right) + w \cdot n_L + (N - n_L) + \frac{N}{3} + 2N + n_{31} + n_{32} \\
&= \frac{N}{6} + (w-1)n_L + \frac{10N}{3} + \frac{7N - 7n_L}{6} + \frac{7N - 5n_L}{6} \\
&= w \cdot n_L - 3 \cdot n_L + \frac{35N}{6}.
\end{aligned}$$

Lemma 6. *We have*

$$\begin{aligned}
W &\leq \sum_{j=1}^{t} W_j \nu_j + W_d(\nu_{21} + \nu_{31}) + W_s(\nu_{11} + \nu_{22} + \nu_{32}) \\
&= \sum_{j=1}^{t} W_j \nu_j + \nu_{11} + 2\nu_{21} + \nu_{22} + 2\nu_{31} + \nu_{32}.
\end{aligned}$$

Proof. The weight of every item is included in the price of exactly one bin used by the algorithm. Thus, the total weight is equal to the total price of bins. Given the supremum prices, we get an upper bound on the total price. This proves the inequality, the equality holds by substituting the values of W_d and W_s. □

Let $n'_L = \frac{n_L}{N}$, and $W' = \frac{W}{N} = w \cdot n'_L - 3 \cdot n'_L + \frac{35}{6}$.

The proof of the next lemma is omitted.

Lemma 7. *For any value of n_L ($0 \leq n_L \leq N$) and for any value of w ($1 \leq w \leq 1.5$), we have*

$$R \geq \frac{W'}{2133/588 - 1.25 n'_L + \frac{1}{7 \cdot 48 \cdot 49^{t-2}} + \frac{1}{48 \cdot 49} + w/7},$$

and therefore

$$R \geq \frac{w \cdot n'_L - 3 \cdot n'_L + \frac{35}{6}}{8533/2352 - 1.25 n'_L + w/7}.$$

Theorem 2. *We have* $R \geq \frac{1363-\sqrt{1387369}}{120} \approx 1.5427809064729$. *That is, there is no online algorithm for bin packing with asymptotic competitive ratio strictly smaller than* $\frac{1363-\sqrt{1387369}}{120} \approx 1.5427809064729$.

Proof. Let $r = \frac{1363-\sqrt{1387369}}{120} \approx 1.5427809064729$ and let

$$w = \frac{\sqrt{1387369} - 1075}{96} \approx 1.07152386690879,$$

where $w = 3 - 1.25 \cdot r$.

We have

$$R \geq \frac{w \cdot n'_L - 3 \cdot n'_L + \frac{35}{6}}{8533/2352 - 1.25n'_L + w/7},$$

and we show that this expression is equal to r (for any n'_L, where $0 \leq n'_L \leq 1$). The denominator is positive as

$$8533/2352 - 1.25n'_L + w/7 > 8533/2352 - 1.25 + 1/7 > 2,$$

by $n'_L \leq 1$ and $w \geq 1$. Thus, it is equivalent to showing

$$w \cdot n'_L - 3 \cdot n'_L + \frac{35}{6} = r(8533/2352 - 1.25n'_L + w/7),$$

which is equivalent to

$$n'_L(w - 3 + 1.25 \cdot r) + \frac{35}{6} - r(8533/2352 + w/7) = 0.$$

Indeed $w - 3 + 1.25 \cdot r = 0$, by the choice of w and r. Additionally,

$$\frac{35}{6} - r(8533/2352 + w/7) = \frac{35}{6} - r(8533/2352 + (3 - 1.25 \cdot r)/7) = 0,$$

by the choice of r. □

Remark 1. We note that our choice of w and r are optimal in the sense that the lower bound of Lemma 7 cannot be used to prove a higher lower bound on R using other values of w for the formula which we obtained. This can be observed by solving the corresponding mathematical program of maximizing (over the possible values of w) of minimizing (over the possible values of n'_L) of the ratio function defined using these two parameters that we establish in Lemma 7. This part of the analysis is the only one we can replace with a solution of a (small) mathematical program. However, since this claim is not necessary for establishing the correctness of the lower bound we omit this mathematical program and its (analytic) solution.

References

1. Babel, L., Chen, B., Kellerer, H., Kotov, V.: Algorithms for on-line bin-packing problems with cardinality constraints. Discrete Appl. Math. **143**(1–3), 238–251 (2004)
2. Balogh, J., Békési, J., Dósa, G., Epstein, L., Levin, A.: Online bin packing with cardinality constraints resolved. In: Proceedings of the 25th European Symposium on Algorithms (ESA2017), pp. 10:1–10:14 (2017)
3. Balogh, J., Békési, J., Dósa, G., Galambos, G., Tan, Z.: Lower bound for 3-batched bin packing. Discrete Optim. **21**, 14–24 (2016)
4. Balogh, J., Békési, J., Dósa, G., Epstein, L., Levin, A.: A new and improved algorithm for online bin packing. In: Proceedings of the 26th European Symposium on Algorithms (ESA 2018), pp. 5:1–5:14 (2018)
5. Balogh, J., Békési, J., Galambos, G.: New lower bounds for certain classes of bin packing algorithms. Theoret. Comput. Sci. **440**, 1–13 (2012)
6. Békési, J., Dósa, G., Epstein, L.: Bounds for online bin packing with cardinality constraints. Inf. Comput. **249**, 190–204 (2016)
7. Blitz, D.: Lower bounds on the asymptotic worst-case ratios of on-line bin packing algorithms. Technical report 114682, University of Rotterdam (1996). M.Sc. thesis
8. Brown, D.J.: A lower bound for on-line one-dimensional bin packing algorithms. Coordinated Science Laboratory Report no. R-864 (UILU-ENG 78–2257) (1979)
9. Csirik, J., Woeginger, G.J.: On-line packing and covering problems. In: Fiat, A., Woeginger, G.J. (eds.) Online Algorithms. LNCS, vol. 1442, pp. 147–177. Springer, Heidelberg (1998). https://doi.org/10.1007/BFb0029568
10. Epstein, L.: A survey on makespan minimization in semi-online environments. J. Sched. **21**(3), 269–284 (2018)
11. Fujiwara, H., Kobayashi, K.M.: Improved lower bounds for the online bin packing problem with cardinality constraints. J. Comb. Optim. **29**(1), 67–87 (2015)
12. Heydrich, S., van Stee, R.: Beating the harmonic lower bound for online bin packing. In: Proceedings of 43rd International Colloquium on Automata, Languages, and Programming (ICALP 2016), pp. 41:1–41:14 (2016)
13. Johnson, D.S.: Near-optimal bin packing algorithms. Ph.D. thesis, MIT, Cambridge (1973)
14. Johnson, D.S.: Fast algorithms for bin packing. J. Comput. Syst. Sci. **8**, 272–314 (1974)
15. Johnson, D.S., Demers, A., Ullman, J.D., Garey, M.R., Graham, R.L.: Worst-case performance bounds for simple one-dimensional packing algorithms. SIAM J. Comput. **3**, 256–278 (1974)
16. Lee, C.C., Lee, D.T.: A simple online bin packing algorithm. J. ACM **32**(3), 562–572 (1985)
17. Liang, F.M.: A lower bound for on-line bin packing. Inf. Process. Lett. **10**(2), 76–79 (1980)
18. Ramanan, P., Brown, D.J., Lee, C.C., Lee, D.T.: Online bin packing in linear time. J. Algorithms **10**, 305–326 (1989)
19. Seiden, S.S.: On the online bin packing problem. J. ACM **49**(5), 640–671 (2002)
20. Ullman, J.D.: The performance of a memory allocation algorithm. Technical report 100, Princeton University, Princeton (1971)
21. van Vliet, A.: An improved lower bound for online bin packing algorithms. Inf. Process. Lett. **43**(5), 277–284 (1992)
22. Yao, A.C.C.: New algorithms for bin packing. J. ACM **27**, 207–227 (1980)

Improved Deterministic Strategy for the Canadian Traveller Problem Exploiting Small Max-(s,t)-Cuts

Pierre Bergé[1(✉)] and Lou Salaün[2,3]

[1] LRI, Université Paris-Sud, Université Paris-Saclay, Orsay, France
Pierre.Berge@lri.fr
[2] Bell Labs, Nokia Paris-Saclay, 91620 Nozay, France
lou.salaun@nokia-bell-labs.com
[3] LTCI, Telecom ParisTech, Université Paris-Saclay, 75013 Paris, France

Abstract. The k-Canadian Traveller Problem consists in finding the optimal way from a source s to a target t on an undirected weighted graph G, knowing that at most k edges are blocked. The traveller, guided by a strategy, sees an edge is blocked when he visits one of its endpoints. A major result established by Westphal is that the competitive ratio of any deterministic strategy for this problem is at least $2k+1$. REPOSITION and COMPARISON strategies achieve this bound.

We refine this analysis by focusing on graphs with a maximum (s,t)-cut size $\mu_{\max}$ less than k. A strategy called DETOUR is proposed and its competitive ratio is $2\mu_{\max} + \sqrt{2}(k - \mu_{\max}) + 1$ when $\mu_{\max} < k$ which is strictly less than $2k+1$. Moreover, when $\mu_{\max} \geq k$, the competitive ratio of DETOUR is $2k+1$ and is optimal. Therefore, DETOUR improves the competitiveness of the deterministic strategies known up to now.

Keywords: Canadian traveller problem · Competitive analysis · Online algorithms

1 Introduction

Related Work. The *k-Canadian Traveller Problem* (k-CTP) was defined by Papadimitriou and Yannakakis [8] and is PSPACE-complete [1,8]. Given an undirected weighted graph G and two of its vertices $s, t \in V$, the objective is to make a traveller walk from s to t on graph G in the most efficient way despite the existence of some blocked edges $E_* \subsetneq E$. Parameter k is an upper bound of the number of blocked edges: $|E_*| \leq k$. The traveller does not know which edges are blocked when he begins his walk. He discovers a blocked edge $e = (u, v)$ when he visits one of its endpoints u or v.

The traveller traverses graph $G = (V, E, \omega)$, where $n = |V|$ and $m = |E|$. Edge weights are given by the function $\omega : E \to \mathbb{Q}^+$. Our objective is to make the traveller reach target t with a minimum cost (also called distance), which is

E. Bampis and N. Megow (Eds.): WAOA 2019, LNCS 11926, pp. 29–42, 2020.
https://doi.org/10.1007/978-3-030-39479-0_3

the sum of the weights of edges traversed. A pair (G, E_*) is called a *road map*. All the road maps considered are feasible: there is an (s,t)-path in $G \backslash E_*$, the graph G deprived of the obstructed edges E_*.

A solution to the k-CTP is an online algorithm, called a *strategy*. Its quality can be assessed with competitive analysis [4]. Roughly speaking, the competitive ratio is the quotient between the distance actually traversed by the traveller and the distance he would have traversed, knowing which edges are blocked before beginning his walk. Westphal [10] proved that no deterministic strategy achieves a competitive ratio better than $2k+1$. Said differently, for any deterministic strategy A, there is at least one k-CTP road map for which the competitive ratio of A is at least $2k+1$.

Two strategies proposed in the literature reach this optimal ratio: REPOSITION [10] and COMPARISON [11]. REPOSITION makes the traveller traverse the shortest (s,t)-path. If there is a blocked edge (u,v) on this path, the traveller discovers it when he visits vertex u. Then, he comes back to s passing through the same path. The process starts again on $G \backslash E'_*$, the graph G deprived of the blocked edges E'_* identified until now. COMPARISON is based on a different principle: when the traveller discovers a blockage (u,v) and stands on vertex u, he compares the shortest (u,t)-path $P^{(u,t)}_{\min}$ (cost $\omega^{(u,t)}_{\min}$) of $G \backslash E'_*$ with its shortest (s,t)-path $P^{(s,t)}_{\min}$ (cost $\omega^{(s,t)}_{\min}$). If $\omega^{(s,t)}_{\min} \leq \omega^{(u,t)}_{\min}$, the traveller moves as in REPOSITION. If $\omega^{(s,t)}_{\min} > \omega^{(u,t)}_{\min}$, the traveller traverses the path $P^{(u,t)}_{\min}$, etc.

Randomized strategies, *i.e.* strategies in which choices of direction depend on a random draw, were also studied. Westphal [10] proved that there is no randomized strategy achieving a ratio lower than $k+1$. Bender *et al.* [2] studied graphs composed only of vertex-disjoint (s,t)-paths and proposed a polynomial-time strategy of ratio $k+1$. A slight revision of that strategy is reported in [9]. To the best of our knowledge, there is no polynomial-time randomized strategy achieving a competitive ratio smaller than $2k+1$ on general graphs. Such a strategy would not be memoryless [3].

Contributions. Our work exclusively concerns deterministic strategies. We establish a relationship between the size $\mu_{\max}$ of the largest minimal (s,t)-cut of a graph G and the competitive ratio that can be obtained on G, for any configuration of blocked edges. Concretely, the competitive ratio of deterministic strategies on graphs where $\mu_{\max} < k$ is studied.

According to the proof of Lemma 2.1 in [10], for any value $\mu \in \mathbb{N}^*$, there is at least one graph (made up of vertex-disjoint (s,t)-paths only) such that $\mu_{\max} = \mu$ and no deterministic strategy has a competitive ratio less than $2k+1$ on it if $\mu_{\max} \geq k$. In this study, we focus on graphs fulfilling $\mu_{\max} < k$: we assess the competitive ratio of strategies REPOSITION and COMPARISON under this condition. We devise a more competitive strategy called DETOUR. We list our contributions:

- For any value $\mu_{\max} \geq 4$, we prove that there is at least one graph with $\mu_{\max} < k$ for which both REPOSITION/COMPARISON strategies are $(2k+1)$-competitive.

- We propose a polynomial-time strategy called DETOUR with competitive ratio $2\mu_{\max} + \sqrt{2}(k - \mu_{\max}) + 1$ when $\mu_{\max} < k$. It outperforms the competitive ratio of the existing deterministic strategies. In brief, ratio $2k+1$ is widely defeated by a deterministic strategy on graphs G satisfying $\mu_{\max} < k$.

Strategy DETOUR is also $(2k+1)$-competitive when $\mu_{\max} \geq k$. For this reason, it becomes the best deterministic strategy known for the k-CTP because it performs as well as REPOSITION/COMPARISON when $\mu_{\max} \geq k$ and better than them when $\mu_{\max} < k$.

The organization of this article follows. In Sect. 2 we remind some definitions related to online algorithms, paths and cuts. Section 3 contains the proof that REPOSITION and COMPARISON are $(2k+1)$-competitive, even if $\mu_{\max} < k$. The DETOUR strategy is described in Sect. 4 and its competitive ratio is evaluated. We conclude this study in Sect. 5 and provide some directions for future research.

2 Preliminaries

We present the definition of the competitive ratio and some notions associated with paths and cuts.

Competitive Ratio. For any set of blocked edges $E'_* \subseteq E_*$, $\omega_{\min}(G, E'_*)$ is the cost of the shortest (s,t)-path in graph $G \backslash E'_*$. Value $\omega_{\text{opt}} = \omega_{\min}(G, E_*)$ is the *optimal offline cost* for the road map (G, E_*). Concretely, this corresponds to the distance the traveller would have traversed if he had known the blockages in advance.

The competitive ratio is defined in [4]. We denote by $\omega_A(G, E_*)$ the distance traversed by the traveller guided by strategy A on graph G from source s to target t with blocked edges E_*. The competitive ratio $c_A(G, E_*)$ of A over a road map (G, E_*) is defined as $c_A(G, E_*) = \frac{\omega_A(G, E_*)}{\omega_{\text{opt}}}$. The competitive ratio c_A of A is thus:

$$c_A = \max_{(G,E_*)} c_A(G, E_*) \tag{1}$$

Similarly, we say strategy A is c_A-competitive for a family $\mathcal{F}$ of graphs (for example, $\mathcal{F} = \{G : \mu_{\max} < k\}$) if it is the maximum of value $c_A(G, E_*)$ over road maps (G, E_*) such that $G \in \mathcal{F}$.

Paths. A *simple path* P is a sequence of pairwise different vertices $v_1 \cdot v_2 \cdots v_i \cdot v_{i+1} \cdots v_\ell$, with departure v_1 and arrival v_ℓ, such that two successive vertices (v_i, v_{i+1}) are adjacent in G. All paths mentioned in this article are simple. To improve readability, we abuse notations: $v_1 \in P$ and $(v_1, v_2) \in P$ mean that vertex v_1 and edge (v_1, v_2) are on path P, respectively. If vertices u and v belong to path P, then $P^{(u,v)}$ denotes the section of path P between vertices u and v. Any path is naturally associated with a direction, from the departure to the arrival. We say the *successor* of edge e in P is the edge arriving just after e in P. The *descendants* of e are all edges arriving after e in P, *i.e.* edges further than u from the departure of P. The *predecessor* and the *ancestors* are defined symmetrically.

Graphs may contain several shortest (s,t)-paths. Our algorithm in Sect. 4 requires to compute one of the shortest (s,t)-paths of any graph in a deterministic way. To achieve it, a solution is to associate any vertex with an identifier in $\{1,\ldots,n\}$. If two paths have the same distance, we compare their lexicographic order. Dijkstra's algorithm [6] is adapted to this extra criterion: for any vertex v, it stores the shortest path from the start point to v with the smallest lexicographic order. Whenever we refer to "the shortest (u,v)-path", for any vertices u and v, this process is executed.

Cuts. A set $X \subseteq E$ is an edge (s,t)-cut if source s and target t are separated in graph G deprived of edges X. We say that cut X is *minimal* if none of its proper subsets $X' \subsetneq X$ is an (s,t)-cut. Let $\mu_{\max}$ be the maximum cardinality of a minimal (s,t)-cut:

$$\mu_{\max} = \max_{\substack{X \text{ minimal} \\ (s,t)-\text{cut}}} |X| \,. \tag{2}$$

Any (s,t)-cut X where $|X| > \mu_{\max}$ is not minimal. If X is a minimal (s,t)-cut, graph $G\backslash X$ contains exactly two connected components: one, denoted $R(X,s)$, contains all vertices reachable from s and another one, denoted $R(X,t)$, all vertices reachable from t. Largest minimal (s,t)-cuts $X_{\max}$, $|X_{\max}| = \mu_{\max}$, are called max-(s,t)-cuts throughout our study.

3 Competitive Ratio of Existing Strategies When $\mu_{\max} < k$

We study the family of graphs satisfying $\mu_{\max} < k$. We assess, on such instances, the competitiveness of the two best deterministic strategies known for now in the literature. Indeed, REPOSITION and COMPARISON are $(2k+1)$-competitive for general graphs. We prove that they do not benefit from the inequality $\mu_{\max} < k$. We begin with REPOSITION strategy.

Theorem 1. *For any $k > 4$, there is a road map $(G_k, E_{*,k})$, $\mu_{max} = 4$, such that the competitive ratio of* REPOSITION *on $(G_k, E_{*,k})$ is $2k+1$: $c_{\text{rep}}(G_k, E_{*,k}) = 2k+1$.*

Proof. The road map $(G_k, E_{*,k})$ is drawn in Fig. 1. Graph G_k has a horizontal axis of symmetry Δ. On each side, there are $\lceil \frac{k}{2} \rceil$ diamond graphs, *i.e.* cycles of length 4, put in series. They are surrounded by two edges, one of weight 1 incident to s and one of weight $\varepsilon \ll 1$ incident to t. For any diamond graph above Δ, three of its edges are weighted with ε and the bottom left one is weighted with 3ε. All the top right edges are blocked (red edges in Fig. 1). All diamonds below Δ are identical, except for the one closest to s (weights 2ε, ε, 4ε, and ε, see Fig. 1). If k is even, as in Fig. 1, the top right edges of all diamonds are blocked. If k is odd, there is no blockage on the diamond below Δ which is the closest to t. In this way, there are always k blocked edges in $E_{*,k}$ and the max-(s,t)-cut size of G_k is $\mu_{\max} = 4$. Let $g(k) = 2\lceil \frac{k}{2} \rceil \in \{k, k+1\}$. The cost of the shortest (s,t)-path in G_k is $1 + (g(k)+1)\varepsilon$.

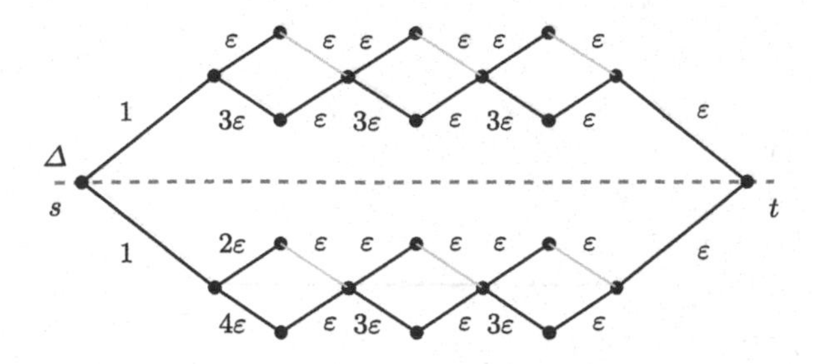

Fig. 1. Graph G_6 and blocked edges $E_{*,6}$ in red (Color figure online)

Guided by REPOSITION, the traveller traverses the shortest (s, t)-path which is above Δ and is blocked in the first diamond (distance $1+\varepsilon$). Set E'_* denotes the blocked edges discovered during the execution: for now, $|E'_*| = 1$. The traveller comes back to s (distance $1 + \varepsilon$). The shortest (s, t)-path in graph $G \backslash E'_*$ is now below axis Δ and its cost is $1 + (g(k) + 2)\varepsilon$ as it contains an edge of weight 2ε. The traveller traverses this path and is blocked in the first diamond below Δ (distance $1 + 2\varepsilon$). Then, the current shortest (s, t)-path in $G \backslash E_*$ is above Δ and its cost is $1 + (g(k) + 3)\varepsilon$, etc. In summary, the traveller is blocked k times traversing paths with cost larger than $1 + \varepsilon$ in two directions. The total distance traversed d_{rep} satisfies $d_{\text{rep}} \geq 2k(1 + \varepsilon) + \omega_{\text{opt}} \geq (2k + 1)(1 + \varepsilon)$. As $\omega_{\text{opt}} = 1 + (2g(k) + 1)\varepsilon$, the competitive ratio of REPOSITION c_{rep} is thus:

$$c_{\text{rep}} \geq (2k+1)\frac{1+\varepsilon}{1+(2g(k)+1)\varepsilon} \overset{\varepsilon \to 0}{\longrightarrow} 2k+1.$$

As ε may tend to zero, there always is a road map on which REPOSITION achieves a ratio $2k + 1 - \delta$ for any arbitrarily small value $\delta > 0$. □

This result remains true for any value $\mu_{\max} > 4$ as we can artificially add (s, t)-paths disjoint from G_k which make $\mu_{\max}$ increase. It suffices to assign a sufficiently large cost to these paths, so that REPOSITION never makes the traveller traverse them. Now we focus on COMPARISON strategy.

Theorem 2. *For any $k > 3$, there is a road map $(G'_k, E'_{*,k})$, $\mu_{max} = 3$, such that the competitive ratio of* COMPARISON *on $(G'_k, E'_{*,k})$ is $2k+1$: $c_{\text{comp}}(G'_k, E'_{*,k}) = 2k + 1$.*

Proof. Road map $(G'_k, E'_{*,k})$ is drawn in Fig. 2. Axis Δ' is represented to facilitate the description of G'_k. Above Δ', $k-1$ diamonds graphs are put in series and are surrounded as in G_k (see Theorem 1). On each diamond, the edge weights are ε, except for the bottom left edges weighted with value 1. The top left edges are blocked. Moreover, the edge incident to t above Δ' is also blocked, so $\left|E'_{*,k}\right| = k$. Below Δ', there is an open (s, t)-path with cost $1 + 2k\varepsilon$. The shortest (s, t)-path in G'_k is above Δ' and its cost is $1 + (2k - 1)\varepsilon$. Graph G'_k is such that $\mu_{\max} = 3$.

Guided by COMPARISON, the traveller traverses the shortest (s, t)-path and is blocked when he arrives on the first diamond (distance 1). Then, the cost

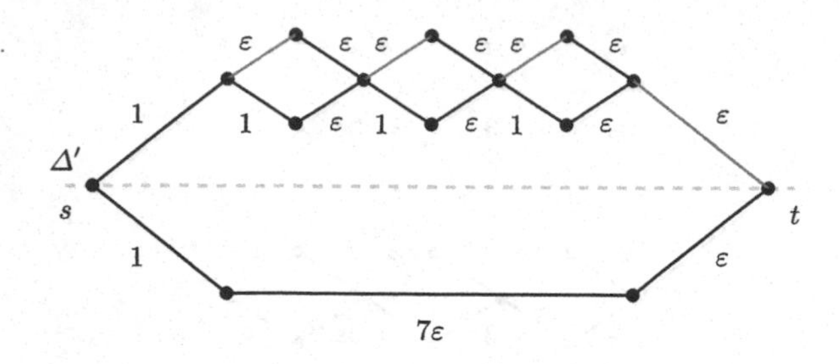

Fig. 2. Graph G'_4 and blocked edges $E'_{*,4}$ in red (Color figure online)

of the shortest (s,t)-path in $G \backslash E'_*$, i.e. $1+2k\varepsilon$, is compared with the shortest distance between the current position of the traveller and t, i.e. $1+(2k-2)\varepsilon$. Since $1+(2k-2)\varepsilon < 1+2k\varepsilon$, the traveller chooses to take the shortest path between its current position and t, which is above Δ'. He meets a second blockage when arriving on the second diamond (distance $1+\varepsilon$). Then, he makes the same decision and traverses the diamonds above Δ'. Eventually, when he meets the last blockage incident to t, he travels back to s and finally passes through the optimal offline path, below Δ'. The total distance traversed is $d_{\text{comp}} = 2+2(k-1)(1+\varepsilon)+1+2k\varepsilon$. The competitive ratio c_{comp} of COMPARISON strategy on the road map $(G'_k, E'_{*,k})$ follows:

$$c_{\text{comp}} = \frac{2+2(k-1)(1+\varepsilon)+1+2k\varepsilon}{1+2k\varepsilon} \overset{\varepsilon \to 0}{\longrightarrow} 2k+1.$$

Making ε tend to zero terminates the proof. □

The existence of a deterministic strategy achieving a ratio less than $2k+1$ on graphs fulfilling $\mu_{\max} < k$ is still an open question after the results established in Theorems 1 and 2. Indeed, we showed that the existing strategies cannot defeat their global competitive ratio on this particular family of graphs. In the remainder, we devise a strategy outperforming REPOSITION and COMPARISON when $\mu_{\max} < k$.

4 DETOUR Strategy

We first introduce in Subsect. 4.1 a parameterized strategy called α-DETOUR. It takes as input graph G, source s, target t, and a parameter $\alpha \in (0,1)$. In Subsect. 4.2, we provide an upper bound of its competitive ratio. This bound is minimized for $\alpha = \frac{\sqrt{2}}{2}$ and is $2\mu_{\max} + \sqrt{2}(k - \mu_{\max}) + 1$ in this case. Strategy DETOUR mentioned earlier corresponds to $\frac{\sqrt{2}}{2}$-DETOUR. Finally, we provide the execution time of DETOUR strategy and discuss some of its properties in Subsect. 4.3.

4.1 Description of α-Detour Strategy

We present the α-DETOUR strategy in Algorithm 1. Variable pos keeps track of the traveller's current position. The idea is to perform successively two phases: an *exploration* followed by a *detour-backtracking*. The exploration starts when the traveller is on source s (line 8). He traverses the shortest (s,t)-path $P_{\min}^{(s,t)}$ called the *exploration path*. Its cost $\omega_{\min}^{(s,t)} = \omega_{\min}(G, E'_*)$ is stored in $\omega_{\exp}$ (line 7). At this point, there are two possibilities:
(1) The traveller reaches t and the execution terminates (line 13).
(2) The traveller arrives at $\mathsf{pos} = u$ and discovers a blocked edge $(u,v) \in P_{\min}^{(s,t)}$. Then, the detour-backtracking phase begins.

Each exploration followed by a detour-backtracking phase can be seen as a depth-first search (DFS). When the traveller is blocked on $P_{\min}^{(s,t)}$, we ask whether an α-*detour*, *i.e.* a (pos,t)-path with cost at most $\alpha\omega_{\exp}$, exists. If an α-detour exists, the traveller traverses the shortest path $P_{\min}^{(\mathsf{pos},t)}$ from the current position pos to target t (line 9). Obviously, its cost satisfies $\omega_{\min}^{(\mathsf{pos},t)} \leq \alpha\omega_{\exp}$. Otherwise, the traveller backtracks to the vertex before $\mathsf{pos} = u$ on the exploration path (lines 14–16).

Algorithm 1. The α-DETOUR strategy

```
1:  Input: graph G, source s, target t, parameter α ∈ (0,1)
2:  E'_* ← ∅; G' ← G\E'_*; pos ← s; u_0 ← s; ω_exp ← ω_min(G, ∅);
3:  stack ← Empty Stack; V_stack ← ∅;
4:  while true do
5:      u_0 ← pos;
6:      if u_0 = s then
7:          ω_exp ← ω_min(G, E'_*);
8:          traverse the shortest (s,t)-path P_min^(u_0,t) in G';
        else
9:          traverse the shortest (u_0,t)-path P_min^(u_0,t) in G'\V_stack;
        endif
10:     update pos;
11:     push the vertices visited in P_min^(u_0,t) except pos on stack;
12:     update E'_*, G', and V_stack;
13:     if pos = t then break;
14:     while pos ≠ s and there is no P_min^(pos,t) in G'\V_stack such that
        ω_min^(pos,t) ≤ αω_exp do
15:         pos ← pop(stack);
16:         V_stack ← V_stack\{pos};
        end
    end
```

As in a DFS, we use a stack to remember the previous vertices for backtracking. We denote by V_{stack} the set of vertices in the stack. We do not allow an

α-detour $P_{\min}^{(\mathsf{pos},t)}$ to pass through any vertex $v \in V_{\mathsf{stack}}$, since the section $P_{\min}^{(v,t)}$ will be considered later on when $\mathsf{pos} = v$. The vertices of an exploration path traversed by the traveller are naturally put in stack. Moreover, when the traveller is blocked on an α-detour $P_{\min}^{(\mathsf{pos},t)}$, the vertices of $P_{\min}^{(\mathsf{pos},t)}$ from pos to the endpoint of the blocked edge are put in stack. Finally, if the traveller backtracks to s, the algorithm goes back to the exploration phase. At this moment, the stack is empty.

Recall that E'_* represents the set of discovered blocked edges. Variable G' contains the graph G deprived of the discovered blockages E'_* at any moment of the execution. At each iteration of the while loop, the variables are updated as follows: if the path $P_{\min}^{(u_0,t)}$ currently traversed (lines 8–9) does not contain any blockage, then the traveller reaches t, i.e. $\mathsf{pos} \leftarrow t$. In this case, the algorithm terminates since the destination is reached. Otherwise, let $P_{\min}^{(u_0,t)} = u_0 \cdots u_i \cdot u_{i+1} \cdots u_r \cdot t$, where (u_i, u_{i+1}) is its first blocked edge. The traveller's position is updated from u_0 to u_i (line 10). Then, we update E'_* with the newly discovered blockages including (u_i, u_{i+1}), and $G' \leftarrow G \backslash E'_*$ (line 12). In addition, we push the traversed vertices $u_0, \ldots, u_{i-1}$ on the stack (except u_i) and update accordingly $V_{\mathsf{stack}} \leftarrow V_{\mathsf{stack}} \cup \{u_0, \ldots, u_{i-1}\}$. In case there is no α-detour $P_{\min}^{(u_i,t)}$ in $G' \backslash V_{\mathsf{stack}}$, the algorithm backtracks by popping u_{i-1} from the stack and setting $\mathsf{pos} \leftarrow u_{i-1}$ (lines 14–16).

If $\alpha = 0$, the algorithm does not take any detour. As a consequence, 0-DETOUR is equivalent to REPOSITION, as both procedures perform an exploration phase followed by backtracking without taking any detour. In the following, we provide an upper bound of α-DETOUR's competitive ratio.

4.2 Competitive Analysis

We denote by $P_1, \ldots, P_\ell$ the exploration paths $P_{\min}^{(s,t)}$ such that the distance from s to the blocked edge discovered on it is greater than α multiplied by their own cost, *i.e.* $\alpha\omega_i$. In other words, the distance d_i traversed by the traveller on the exploration paths P_i, $1 \leq i \leq \ell$, satisfies $d_i \geq \alpha\omega_i$. Paths P_i are sorted in order to fulfil $\omega_1 \leq \cdots \leq \omega_\ell$. The exploration paths $P_1, \ldots, P_{\ell-1}$ are blocked, while path P_ℓ can be open. If P_ℓ does not contain any blockage, then the algorithm terminates after the traveller traverses it.

Let us partition $P_1, \ldots, P_\ell$ into two sequences $S_1 = P_1, \ldots, P_{h-1}$ and $S_2 = P_h, \ldots, P_\ell$ such that $2\alpha\omega_{h-1} < \omega_\ell \leq 2\alpha\omega_h$. In the particular case where $\omega_\ell \leq 2\alpha\omega_1$, then $h = 1$ and the two sets are $S_1 = \emptyset$ and $S_2 = P_1, \ldots, P_\ell$. We denote by $G[P_h, \ldots, P_\ell]$ the subgraph of G induced by paths $P_h, \ldots, P_\ell$, *i.e.* containing only the vertices and edges of paths P_i, $h \leq i \leq \ell$.

Theorem 3. *The max-(s,t)-cut size induced on graph $G[P_h, \ldots, P_\ell]$ is at least $\ell - h + 1$.*

Proof. We denote by b_i the blocked edge discovered on P_i, for $i \in \{h, \ldots, \ell\}$. We construct inductively a set $\{e_h, \ldots, e_\ell\}$ of edges satisfying the following induction hypotheses, for all $i \in \{h, \ldots, \ell\}$:

$H_1(i)$: $\{e_h,\ldots,e_i\}$ is a minimal (s,t)-cut of $G[P_h,\ldots,P_i]$,
$H_2(i)$: Either $e_i = b_i$ or e_i is an ancestor of b_i in P_i,
$H_3(i)$: For $j \in \{i+1,\ldots,\ell\}$, P_j cannot pass through e_i.

Basis: For $i = h$, $G[P_h,\ldots,P_i]$ contains only one path P_h. We choose $e_h = b_h$, which fulfils $H_2(h)$. Since any edge of P_h is a max-(s,t)-cut of $G[P_h]$, it satisfies $H_1(h)$. Statement $H_3(h)$ is also true, as e_h is blocked.

Inductive Step: Assume that $H_1(i)$ to $H_3(i)$ are true for a certain integer i in $\{h,\ldots,\ell-1\}$. We will construct e_{i+1} and prove the induction hypotheses $H_1(i+1)$ to $H_3(i+1)$. For simplicity, we denote sets $R(\{e_h,\ldots,e_i\},s)$ and $R(\{e_h,\ldots,e_i\},t)$ in graph $G[P_h,\ldots,P_i]$ by $R_i(s)$ and $R_i(t)$, respectively.

Let $P_{i+1}^{(v_0,v_p)} = v_0 \cdot v_1 \cdots v_p$ be the longest section in P_{i+1}, starting from $v_0 = s$, such that $v_0,\ldots,v_p \in R_i(s)$ and $p \in \mathbb{N}$. Section $P_{i+1}^{(v_0,v_p)}$ contains at least vertex $v_0 = s$. For $j \in \{h,\ldots,i\}$, all ancestors of e_j in P_j belong to $R_i(s)$, and all descendants belong to $R_i(t)$. Therefore, according to $H_2(i)$, all exploration paths' sections of the form $P_j^{(s,u)}$ are open and equal to the shortest path from s to u, for $u \in R_i(s) \cap P_j$ and $j \in \{h,\ldots,i\}$. In particular, since $P_{i+1}^{(v_0,v_p)}$ is the shortest (v_0,v_p)-path, we deduce that it is open as v_p belongs to some P_j by definition of $R_i(s)$.

According to $H_3(i)$, $P_{i+1}^{(v_p,t)}$ is a new path connecting $R_i(s)$ to $R_i(t)$, which does not traverse any edge of the cut $\{e_h,\ldots,e_i\}$. Furthermore, no vertex in $P_{i+1}^{(v_{p+1},t)}$ belongs to $R_i(s)$. Indeed, suppose for the sake of contradiction that $u \in P_{i+1}^{(v_{p+1},t)}$ and $u \in R_i(s)$. There would exist $j \in \{h,\ldots,i\}$, such that $P_j^{(s,u)}$ is the shortest (s,u)-path, and all its vertices belong to $R_i(s)$. This contradicts with the fact that $P_{i+1}^{(s,u)}$ is also the shortest (s,u)-path and $v_{p+1} \notin R_i(s)$, by definition. Let $v_{p'}$ be the first vertex of P_{i+1} belonging to $R_i(t)$, *i.e.* $v_{p'} \in R_i(t)$ and $p < p'$. Such a vertex exists as t is a candidate. We derive that $P_{i+1}^{(v_p,v_{p'})}$ is the unique path both connecting $R_i(s)$ to $R_i(t)$ and avoiding cut X. Figure 3 represents cut $\{e_h,\ldots,e_i\}$, path P_{i+1} and its vertices v_p and $v_{p'}$.

We fix e_{i+1} differently depending on the position of b_{i+1}. We already proved that $b_{i+1} \notin P_{i+1}^{(v_0,v_p)}$, the remaining cases are:

- If $b_{i+1} \in P_{i+1}^{(v_p,v_{p'})}$, then we set $e_{i+1} = b_{i+1}$. As $e_{i+1} \in E_*$, $H_3(i+1)$ is true.
- Otherwise, if $b_{i+1} \in P_{i+1}^{(v_{p'},t)}$, we choose $e_{i+1} = (v_{p'-1}, v_{p'})$. We prove that the cost of the current shortest $(s,v_{p'})$-path, $P_{i+1}^{(s,v_{p'})}$, is at least $\alpha\omega_h$. Indeed, as vertex $v_{p'}$ belongs to a certain path $P_{j'}$, $j' \in \{h,\ldots,i\}$, the cost of $P_{i+1}^{(s,v_{p'})}$ is at least the cost of $P_{j'}^{(s,v_{p'})}$. If we have $\omega_{i+1}^{(s,v_{p'})} \leq \alpha\omega_h$, the distance traversed by the traveller on $P_{j'}$ is less than $\alpha\omega_h \leq \alpha\omega_{j'}$, as $v_{p'} \in R_i(t)$. This contradicts with the fact that $P_{j'} \in \{P_h,\ldots,P_\ell\}$. Moreover, after the $(i+1)$-th detour-backtracking phase, all remaining open $(v_{p'},t)$-paths are longer than $\alpha\omega_{i+1} \geq \alpha\omega_h$. Therefore, the cost of any exploration (s,t)-path passing through $v_{p'}$ is greater than $\alpha\omega_h + \alpha\omega_{i+1} \geq 2\alpha\omega_h$. This is impossible since the last exploration

path P_ℓ satisfies $\omega_\ell \leq 2\alpha\omega_h$. As a consequence, no exploration path passes through $v_{p'}$ and $H_3(i+1)$ is true.

Both cases fulfil naturally $H_2(i+1)$. It only remains to prove statement $H_1(i+1)$. We showed that $P_{i+1}^{(v_p,v_{p'})}$ is the only path connecting $R_i(s)$ to $R_i(t)$, and $e_{i+1} \in P_{i+1}^{(v_p,v_{p'})}$. Thus, $\{e_h, \ldots, e_{i+1}\}$ is an (s,t)-cut of $G[P_h, \ldots, P_{i+1}]$. If we re-open edge e_{i+1}, path $P_{i+1}^{(v_p,v_{p'})}$ connects $R_i(s)$ to $R_i(t)$. If we re-open e_j, $j < i+1$, there is a path in $G[P_h, \ldots, P_i]$ which connects $R_i(s)$ to $R_i(t)$ independently of $P_{i+1}^{(v_p,v_{p'})}$, according to the minimality of $\{e_h, \ldots, e_i\}$ in $H_1(i)$. As a consequence, no proper subset of $\{e_h, \ldots, e_{i+1}\}$ is an (s,t)-cut. Cut $\{e_h, \ldots, e_{i+1}\}$ is minimal.

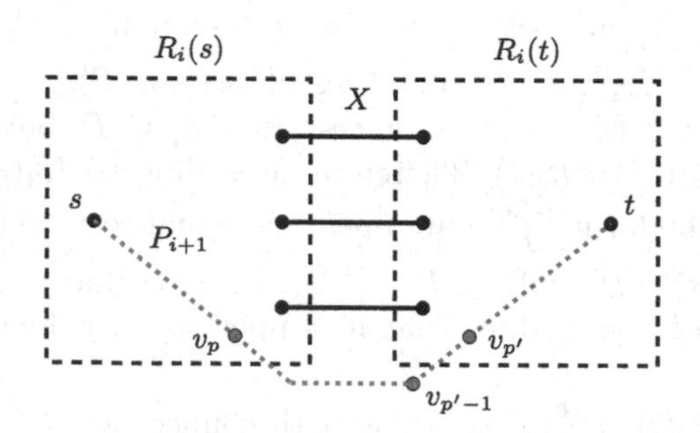

Fig. 3. Cut $X = \{e_h, \ldots, e_i\}$, path P_{i+1}, and vertices v_p, $v_{p'-1}$, $v_{p'}$

In summary, we derive by induction that $\{e_h, \ldots, e_\ell\}$ is a minimal (s,t)-cut of $G[P_h, \ldots, P_\ell]$. The size of the max-(s,t)-cut is at least $\ell - h + 1$. □

The following lemma states that the max-(s,t)-cut size of graph $G[P_h, \ldots, P_\ell]$ cannot exceed the max-(s,t)-cut size of the bigger graph G.

Lemma 1. *The max-(s,t)-cut size on graph $G[P_h, \ldots, P_\ell]$ is less than or equal to the max-(s,t)-cut size μ_{max} of the original graph G.*

Proof. Let X be one of the max-(s,t)-cuts in graph $G[P_h, \ldots, P_\ell]$. Cut X is minimal, so no subset $X' \subsetneq X$ is an (s,t)-cut. If X is an (s,t)-cut in G, then it is also minimal in G as none of its subsets can be an (s,t)-cut. Therefore, $|X| \leq \mu_{\max}$.

Suppose now that X is not an (s,t)-cut in G. We denote by Y the max-(s,t)-cut in graph G deprived of edges X, *i.e.* $G \backslash X$. Set $X \cup Y$ is thus a minimal (s,t)-cut in graph G as Y is minimal in $G \backslash X$. So, $|X| \leq |X \cup Y| \leq \mu_{\max}$. In both cases, the max-(s,t)-cut size in $G[P_h, \ldots, P_\ell]$ is at most $\mu_{\max}$. □

According to both Theorem 3 and Lemma 1, a relationship exists between values ℓ, h, and $\mu_{\max}$, which is $\ell - h + 1 \leq \mu_{\max}$.

After traversing an exploration path P_i, the traveller performs a detour-backtracking phase. The number of blockages discovered during this i-th detour-backtracking phase is denoted by q_i. We analyse the cost of traversing P_i and performing the i-th detour-backtracking phase in Lemma 2.

Lemma 2. *The total cost of both the i-th exploration phase and the i-th detour-backtracking phase is not greater than* $(2 + 2\alpha q_i)\omega_i$.

Proof. The stack in Algorithm 1 ensures that each edge is only traversed twice: first time when moving towards t on an exploration path or a detour, and a second time when backtracking. The exploration path costs ω_i and each detour costs no more than $\alpha\omega_i$. Besides, the number of detours is at most q_i. Hence, the total cost is at most $2\omega_i + q_i 2\alpha\omega_i$, which concludes the proof. □

We denote by k_1 (resp. k_2) the number of blocked edges discovered during the exploration and detour-backtracking phases associated with paths $P_1, \ldots, P_{h-1}$ (resp. $P_h, \ldots, P_\ell$). Let k_3 be the number of blockages discovered during the other phases, so that $k_1 + k_2 + k_3 = k$. We derive in Theorem 4 an upper-bound on the competitive ratio as a function of k_1, k_2, k_3, and α.

Theorem 4. *The competitive ratio of* α-DETOUR *is upper-bounded by:*

$$\frac{k_1}{\alpha} + 2\mu_{max} + 2\alpha(k_2 + k_3 - \mu_{max}) + 1. \tag{3}$$

Proof. Since path P_ℓ is the shortest (s,t)-path of a certain graph $G \backslash E'_*$ where $E'_* \subseteq E_*$, the offline optimal cost satisfies

$$\omega_{\text{opt}} \geq \omega_\ell. \tag{4}$$

According to Lemma 2, the distance traversed during the exploration and detour-backtracking phases of $P_1, \ldots, P_{h-1}$ is not greater than

$$\sum_{j=1}^{h-1} (2 + 2\alpha q_j)\omega_j \leq 2\omega_{h-1} \sum_{j=1}^{h-1} (1 + q_j) = 2k_1\omega_{h-1}. \tag{5}$$

Inequality (5) comes from the fact that $\omega_1 \leq \cdots \leq \omega_{h-1}$ and $\sum_{j=1}^{h-1} (1 + q_j) = k_1$.

We evaluate the cost of the phases associated with $P_h, \ldots, P_\ell$. Path P_ℓ is either open and traversed in one direction only (Case 1) or it is blocked and the traveller reaches t via a detour (Case 2).

Case 1: If P_ℓ does not contain any blockage, then the algorithm terminates after traversing it. This final exploration phase costs ω_ℓ. We have $q_\ell = 0$ and $k_2 = \sum_{j=h}^{\ell-1} (1 + q_j)$. Given Lemma 2, the cost of the h-th to ℓ-th phases is less than:

$$\begin{aligned}
\sum_{j=h}^{\ell-1} (2 + 2\alpha q_j)\omega_j + \omega_\ell &= \sum_{j=h}^{\ell-1} (2\alpha + 2\alpha q_j)\omega_j + \sum_{j=h}^{\ell-1} (2 - 2\alpha)\omega_j + \omega_\ell, \\
&\leq 2\alpha k_2\omega_\ell + (2 - 2\alpha)(\ell - h)\omega_\ell + \omega_\ell, &(6) \\
&< 2\alpha k_2\omega_\ell + (2 - 2\alpha)\mu_{\max}\omega_\ell + \omega_\ell, \\
&= 2\alpha(k_2 - \mu_{\max})\omega_\ell + 2\mu_{\max}\omega_\ell + \omega_\ell. &(7)
\end{aligned}$$

We deduce Inequality (6) from $\omega_h \leq \cdots \leq \omega_\ell$. By applying Theorem 3 and Lemma 1 on $S_2 = P_h, \ldots, P_\ell$, we derive that $\ell - h \leq \mu_{\max} - 1 < \mu_{\max}$ in Inequality (7).

Case 2: Suppose that P_ℓ is blocked. The ℓ-th exploration and detour-backtracking phases cost at most $(2 + 2\alpha q_\ell)\omega_\ell + \alpha\omega_\ell$. Moreover, we have $k_2 = \sum_{j=h}^{\ell}(1 + q_j)$. The distance traversed from the h-th to the ℓ-th phases is not greater than:

$$\begin{aligned}
\sum_{j=h}^{\ell-1}(2+2\alpha q_j)\omega_j + (2+2\alpha q_\ell+\alpha)\omega_\ell &= \sum_{j=h}^{\ell}(2+2\alpha q_j)\omega_j + \alpha\omega_\ell, \\
&\leq 2\alpha k_2\omega_\ell + (2-2\alpha)(\ell-h+1)\omega_\ell + \alpha\omega_\ell, \qquad (8)\\
&\leq 2\alpha k_2\omega_\ell + (2-2\alpha)\mu_{\max}\omega_\ell + \alpha\omega_\ell, \qquad (9)\\
&\leq 2\alpha(k_2-\mu_{\max})\omega_\ell + 2\mu_{\max}\omega_\ell + \omega_\ell. \qquad (10)
\end{aligned}$$

Inequality (8) follows from $\omega_h \leq \cdots \leq \omega_\ell$. We obtain (9) from $\ell - h + 1 \leq \mu_{\max}$. Finally, $\alpha \leq 1$ implies Eq. (10).

Contrary to $P_1, \ldots, P_\ell$, some exploration paths $\widehat{P}$ may be such that the distance traversed on them is at most α multiplied by their own cost $\widehat{\omega}$. The distance traversed during the phases which are not associated with $P_1, \ldots, P_\ell$ is the cost of these exploration paths $\widehat{P}$ and their α-detours. As $\widehat{\omega} \leq \omega_{\mathrm{opt}}$, it is at most $2\alpha k_3\omega_{\mathrm{opt}}$. Applying Eq. (4), the competitive ratio of α-DETOUR admits the following upper-bound:

$$\begin{aligned}
\frac{\omega_{\alpha-\mathrm{DETOUR}}}{\omega_{\mathrm{opt}}} &\leq \frac{2k_1\omega_{h-1} + 2\alpha(k_2+k_3-\mu_{\max})\omega_{\mathrm{opt}} + 2\mu_{\max}\omega_{\mathrm{opt}} + \omega_{\mathrm{opt}}}{\omega_{\mathrm{opt}}}, \\
&\leq \frac{k_1\omega_\ell}{\alpha\omega_{\mathrm{opt}}} + 2\mu_{\max} + 2\alpha(k_2+k_3-\mu_{\max}) + 1, \qquad (11)\\
&\leq \frac{k_1}{\alpha} + 2\mu_{\max} + 2\alpha(k_2+k_3-\mu_{\max}) + 1. \qquad (12)
\end{aligned}$$

Inequality (11) follows from the partition $\{S_1, S_2\}$ imposing $2\alpha\omega_{h-1} < \omega_\ell$. □

Let $c_{\det}(k_1, k_2, k_3, \alpha)$ denote the value in (12). Parameters k_1, k_2, and k_3 depend on the road map (G, E_*), so only $\alpha \in (0, 1)$ can be tuned. Value $\alpha = \frac{\sqrt{2}}{2}$ minimizes $c_{\det}(k_1, k_2, k_3, \alpha)$ under the condition $k_1 + k_2 + k_3 = k$ for any $k > \mu_{\max}$. Formally,

$$\frac{\sqrt{2}}{2} = \underset{0 \leq \alpha \leq 1}{\mathrm{argmin}} \max_{\substack{k_1,k_2,k_3 \in \mathbb{N} \\ k_1+k_2+k_3=k}} c_{\det}(k_1, k_2, k_3, \alpha)$$

Corollary 1. *The competitive ratio of* DETOUR *is at most* $2\mu_{max} + \sqrt{2}(k - \mu_{max}) + 1$.

Proof. We set $\alpha = \frac{\sqrt{2}}{2}$ and $k_1 + k_2 + k_3 = k$ in Eq. (3). □

4.3 Discussion

In summary, strategy DETOUR is as competitive as REPOSITION/COMPARISON for the range $\mu_{\max} \geq k$ but more competitive for the range $1 \leq \mu_{\max} < k$. The slope of the competitive ratio of DETOUR when k varies is only $\sqrt{2}$ for $\mu_{\max} < k$. Figure 4 gives the shape of the competitive ratios of REPOSITION (in blue) and DETOUR (in red) as a function of k.

DETOUR strategy needs to identify the shortest (s,t)-paths and (pos,t)-paths at any moment of its execution. To achieve it, Dijkstra's algorithm [6] is computed once between two discoveries of blocked edges with t as the start point. Hence, the running time of DETOUR is $O(k(m + n\log n))$.

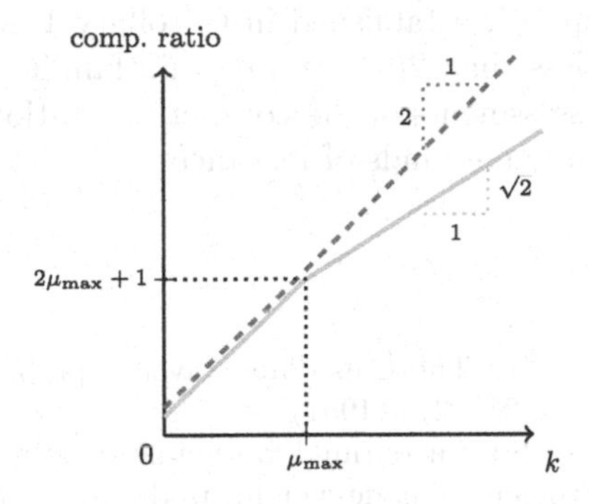

Fig. 4. Competitiveness of REPOSITION (blue) and DETOUR (red) versus k (Color figure online)

As for REPOSITION and COMPARISON, the execution of DETOUR strategy is independent of the value of k. Thus, it can be used when no upper bound on the number of blockages is known and its competitive ratio is $2\mu_{\max} + \sqrt{2}(|E_*| - \mu_{\max}) + 1$.

DETOUR strategy can be executed without knowing the value $\mu_{\max}$. Indeed, the competitive ratio of DETOUR depends on $\mu_{\max}$ but no decision is made based on $\mu_{\max}$ in Algorithm 1. In the next paragraph, we explain that value $\mu_{\max}$ cannot be computed in polynomial time.

Finding one of the largest minimal (s,t)-cuts $X_{\max}$, $|X_{\max}| = \mu_{\max}$ is a NP-hard problem, even for planar graphs [7]. A linear time algorithm computing $\mu_{\max}$ exists only for series-parallel graphs [5]. In summary, it is not possible to evaluate value $\mu_{\max}$ for any graph in polynomial time, assuming P$\neq$NP. Given an input graph G, the competitive ratio of DETOUR strategy on any road map (G, E_*) cannot be predicted fast. The only possibility is thus to execute DETOUR, which runs in polynomial time, on G directly and then to verify whether a gain of competitiveness is obtained.

5 Conclusion

Even if the global competitiveness of deterministic strategies for the k-CTP was fully treated by Westphal [10], families of graphs for which a competitive ratio better than $2k + 1$ can be achieved, remained to be identified. In this context, we designed DETOUR strategy to improve significantly the competitive ratio obtained on graphs satisfying $\mu_{\max} < k$. Its competitive ratio is $2\mu_{\max} + \sqrt{2}(k - \mu_{\max}) + 1$.

Some open questions emerge from this study. First, we wonder if a better strategy exists when $\mu_{\max} < k$. In other words, we do not know whether DETOUR is optimal for this family of graphs. Second, randomized strategies may offer the opportunity to decrease the ratio obtained with the deterministic framework, *i.e.* to go below the slope $\sqrt{2}$ established in Corollary 1. More generally, lots of strategies with a ratio less than $2k + 1$ on certain families of graphs may exist. We believe that local assessments of the competitive ratio can lead us to defeat strategy REPOSITION on many kinds of instances.

References

1. Bar-Noy, A., Schieber, B.: The Canadian traveller problem. In: Proceedings of ACM/SIAM SODA, pp. 261–270 (1991)
2. Bender, M., Westphal, S.: An optimal randomized online algorithm for the k-Canadian Traveller Problem on node-disjoint paths. J. Comb. Optim. **30**(1), 87–96 (2015)
3. Bergé, P., Hemery, J., Rimmel, A., Tomasik, J.: On the competitiveness of memoryless strategies for the k-Canadian Traveller Problem. In: Kim, D., Uma, R.N., Zelikovsky, A. (eds.) COCOA 2018. LNCS, vol. 11346, pp. 566–576. Springer, Cham (2018). https://doi.org/10.1007/978-3-030-04651-4_38
4. Borodin, A., El-Yaniv, R.: Online Computation and Competitive Analysis. Cambridge University Press, Cambridge (1998)
5. Chaourar, B.: A linear time algorithm for a variant of the MAX CUT problem in series parallel graphs. Adv. Oper. Res. (2017)
6. Dijkstra, E.W.: A note on two problems in connexion with graphs. Numerische Mathematik **1**(1), 269–271 (1959)
7. Haglin, D.J., Venkatesan, S.M.: Approximation and intractability results for the maximum cut problem and its variants. IEEE Trans. Comput. **40**(1), 110–113 (1991)
8. Papadimitriou, C., Yannakakis, M.: Shortest paths without a map. Theor. Comput. Sci. **84**(1), 127–150 (1991)
9. Shiri, D., Salman, F.S.: On the randomized online strategies for the k-Canadian traveler problem. J. Comb. Opt. **38**, 254–267 (2019)
10. Westphal, S.: A note on the k-Canadian Traveller Problem. Inf. Process. Lett. **106**(3), 87–89 (2008)
11. Xu, Y., Hu, M., Su, B., Zhu, B., Zhu, Z.: The Canadian traveller problem and its competitive analysis. J. Comb. Optim. **18**(2), 195–205 (2009)

Robust Online Algorithms for Certain Dynamic Packing Problems

Sebastian Berndt[1], Valentin Dreismann[2], Kilian Grage[1(✉)], Klaus Jansen[1], and Ingmar Knof[1]

[1] Department of Computer Science, Kiel University, Kiel, Germany
{seb,kig,kj}@informatik.uni-kiel.de, ingmar.knof@posteo.de
[2] University of Warwick, Coventry, UK
valentin.dreismann@posteo.de

Abstract. Online algorithms that allow a small amount of migration or recourse have been intensively studied in the last years. They are essential in the design of competitive algorithms for dynamic problems, where objects can also depart from the instance. In this work, we give a general framework to obtain so called robust online algorithms for a variety of dynamic problems: these online algorithms achieve an asymptotic competitive ratio of $\gamma + \epsilon$ with migration $O(1/\epsilon)$, where γ is the best known offline asymptotic approximation ratio. For our framework, we require only two ingredients: (i) the existence of an online algorithm for the static case (without departures) that provides a provably good solution compared to the total volume of the instance and (ii) that the optimal solution always exceeds this total volume. If these criteria are met, we can complement the online algorithm with any offline algorithm.

While these criteria are naturally fulfilled by many dynamic problems, they are especially suited for packing problems. In order to show the usefulness of our approach in this area, we improve upon the best known robust algorithms for the dynamic versions of generalizations of Strip Packing and Bin Packing, including the first robust algorithms for general d-dimensional Bin Packing and Vector Packing.

Keywords: Online algorithms · Dynamic algorithms · Competitive ratio · Recourse · Packing problems

1 Introduction

Online algorithms are a very natural way to deal with uncertain inputs. The main challenge for these algorithms is to produce a sequence of good solutions throughout the complete evolution of an instance. For a surprisingly large number of problems, one can obtain online algorithms that produce solutions within a

The authors thank Marten Maack and Malin Rau for initial discussions.
S. Berndt—Supported by DFG Project, "Robuste Online-Algorithmen für Scheduling- und Packungsprobleme", JA 612/19-1.
K. Grage—Supported by GIF-Project "Polynomial Migration for Online Scheduling".

E. Bampis and N. Megow (Eds.): WAOA 2019, LNCS 11926, pp. 43–59, 2020.
https://doi.org/10.1007/978-3-030-39479-0_4

constant factor of the optimal solutions. The worst-case ratio between an optimal solution and a solution produced by the online algorithm is called the *competitiveness* of the algorithm. For most algorithms with constant competitiveness, the algorithms rely heavily on the fact that the instances evolve in a *monotone* way: every object that becomes part of the instance will stay part of the instance forever. This monotonicity property is often not present in real-world applications, where the objects might be removed from the instance later on. These objects might be used at a different place, they might expire, or they are no longer relevant. Hence, in order to still give a performance guarantee, the online algorithms need to be able to modify parts of existing solutions. Clearly, such a modification might be costly and should thus be minimized. If no such boundary on the modification is given, one could easily use an offline approximation algorithm. A natural way to measure the amount of modification needed is the *migration factor*: it compares the total size of modified objects with the size of the newly inserted or removed object. An algorithm with a bounded migration factor roughly translates to the fact that the insertion or departure of a small object can only lead to small changes in the structure of the current solution. On the other hand, if a large (and thus impactful) object is inserted or removed, we are allowed to modify a larger part of the solution.

It is easy to see that online algorithms without migration can in general not achieve the same solution quality as offline algorithms. As an example, consider the well-studied Bin Packing problem. No online algorithm (without migration) can achieve a competitiveness smaller than 3/2 [27]. This lower bound was further improved by van Vliet to 1.54014 [26] and by Balogh et al. to 1.54037 [3]. On the other hand an offline algorithm with asymptotic approximation ratio of $1+\epsilon$ can be obtained in polynomial time [21]. The question how much these two values differ is one of the central questions in the field of online and approximation algorithms.

Clearly, an online algorithm that is allowed a certain amount of migration is able to gap between these extremes. If the amount of migration needed directly corresponds to the improvement of the solution guarantee, we call such an algorithm *robust*. A robust online algorithm would then have competitiveness $1+\epsilon$ and migration factor $f(1/\epsilon)$ for some function. In other words, we only need to increase the migration if we want to obtain a better solution. Such robust online algorithms thus have a continuous behavior between the performance of the best offline algorithm and the performance of the best online algorithm, depending on the choice of ϵ.

In this paper, we describe a general framework to construct robust online algorithms for certain dynamic packing problems. The only ingredients needed for our framework are (i) a good offline approximation algorithm and a (ii) suitably designed online algorithm for the *static* case that is related to the *volume* of an instance. To show the versatility of our approach, we improve upon existing robust algorithms and construct new robust algorithms.

1.1 Related Works

The migration factor model was introduced by Sanders, Sivadasan and Skutella [23]. They studied the classical problem of minimizing the makespan on parallel identical machines and made elegant use of sensitivity results of integer programming to obtain a robust $1 + \epsilon$-competitive algorithm. This spanned follow up works for many different problems:

- Skutella and Verschae [24] also studied the same problem and were able to obtain a robust $1+\epsilon$-competitive algorithm for the dynamic case, where jobs may depart. They also studied the dual version of the makespan minimization problem, where the minimum load should be maximized. This problem (called Machine Covering or Santa Claus) also admits a robust $1+\epsilon$-competitive algorithm even in the dynamic case. Furthermore, Skutella and Verschae proved that for all $\epsilon > 0$ there is no online algorithm for the Machine Covering problem that achieves competitive ratio $20/19 - \epsilon$ with worst-case migration $f(1/\epsilon)$ for any function f. Gálvez *et al.* [17] gave two simple and elegant competitive robust algorithms (one that is $1.7 + \epsilon$-competitive and one that is $4/3 + \epsilon$-competitive) with polynomial migration factor for the static version of the problem of makespan minimization.
- For the Bin Packing problem, Epstein and Levin [13] gave the first robust $1+\epsilon$-competitive algorithm for Bin Packing based on the same sensitivity results for integer programming. Jansen and Klein [18] designed new techniques in order to obtain a migration factor polynomial in $1/\epsilon$. These techniques were refined by Berndt *et al.* [8] to handle the dynamic version of the Bin Packing problem where items can also depart. In all of these works, the migration factor is worst-case and can thus not be saved up for later use. In contrast, Feldkord *et al.* [16] presented a nice robust $1 + \epsilon$-competitive algorithm with amortized migration factor that also works for the dynamic case. Berndt *et al.* showed that a worst-case migration factor of $\Omega(1/\epsilon)$ is needed [8]. This lower bound was shown to also hold for amortized migration by Feldkord *et al.* [16]. Feldkord *et al.* also studied a problem variant, where the migration needed for an item does not correspond to its size.
- Epstein and Levin [14] investigated a multidimensional extension of the Bin Packing problem, called Hypercube Packing, where a set of hypercubes is to be packed in the minimum number of unit-sized hypercubes. They present a robust $1 + \epsilon$-competitive algorithm with worst-case migration by adapting the offline algorithm of Bansal *et al.* [4].
- For the makespan problem, where jobs can be scheduled preemptively (i.e. they can be split, but parts are not allowed to run simultaneously), Epstein and Levin [15] designed an optimal online algorithm with migration factor $1+1/m$, where m is the number of machines. The migration factor used here is worst-case. They also showed that exact algorithms for the makespan minimization problem on uniform machines and for identical machines in the restricted assignment setting have worst-case migration factor at least $\Omega(m)$.
- Berndt *et al.* [7] studied the dual problem of Bin Packing called Bin Covering. They analyzed this problem both for worst-case migration and amortized

migration and for the static and the dynamic case to obtain lower bounds and matching algorithmic results (up to an additional additive term of ϵ).

The offline variants of the geometric packing problems have also been studied intensively. See e.g. the survey of Christensen *et al.* [10] and the references therein.

1.2 Our Results

We present a very general framework that allows to construct robust online algorithms that have amortized migration factor of $O(1/\epsilon)$ and achieve the same competitive ratio as the best known offline approximation algorithm (up to an additive error of ϵ). Note that the achieved migration factor is optimal for some of the problems we consider. See e.g. the work of Feldkord *et al.* [16] for a corresponding lower bound regarding the Bin Packing problem. The algorithms created by our framework can all deal with the dynamic versions of these problems, where items can also depart from the instance. The only requirement we have is the existence of an offline approximation algorithm and the existence of an online algorithm for the *static* case that is related to the *volume* of an instance. To present this general framework, we introduce the notion of *flexible online algorithms* and show that such algorithms can be combined with offline approximation algorithms under certain circumstances. We then show the versatility of our approach by looking at several well-studied problems including 2-D Strip Packing, 2-D Bin Packing, and Vector Packing. We give robust algorithms for the multi-dimensional variants of these problems, where we can also handle both the departure and the rotation of items. This improves and generalizes several known results (e.g. from [14,19]) and gives the first robust algorithms for all of the other problems. Additionally, our compact and clean framework gives much easier algorithms compared with the previously known algorithms. Due to space constraints, detailed descriptions and analysis of the online algorithms are omitted and can be found in the full version, which is attached in the appendix. Results on higher-dimensional Strip Packing can also be found there.

2 Online Algorithms for Dynamic Problems: A Framework

While the techniques presented in this paper also work for maximization problems, we will focus on minimization problems to improve the accessibility of our results.

Definition 1. *Let ITEMS* $\subseteq \{0,1\}^*$ *be a prefix-free set describing some items. A* minimization problem $\Pi = (\mathcal{I}, \text{SOL}, \text{COSTS})$ *consists of a set of* instances $\mathcal{I} \subseteq$ *ITEMS, a mapping* SOL *that maps an instance* $I \in \mathcal{I}$ *to a non-empty set of* feasible solution $\text{SOL}(I)$*, and a mapping* COSTS *that maps a solution* $S \in \text{SOL}(I)$ *to its* costs $\text{COSTS}(S) \in \mathbb{Q}_{\geq 0}$*. A solution* $S^* \in \text{SOL}(I)$ *is called* optimal*, if* $\text{COSTS}(S^*) \leq \text{COSTS}(S)$ *for all* $S \in \text{SOL}(I)$*. We denote the cost of any optimal solution* S^* *by* $\text{OPT}(I) := \text{COSTS}(S^*)$.

For the sake of simplicity, we sometimes treat Π and $\mathcal{I}$ interchangeably and also write $I \in \Pi$ to denote that I is an instance.

Throughout this paper, we make the following two natural assumptions that hold for all problems considered in this work:

Assumption 1

1. *For every instance I, every instance $I' \subseteq I$, and every solution $S \in \text{SOL}(I)$, the solution $S \restriction_{I'}$ induced by the items in I' is also feasible, i.e. $S \restriction_{I'} \in \text{SOL}(I')$. Hence, removing some items from a feasible solution results in a feasible solution of the remaining items.*
2. *For every instance I, and every instance $I' \subseteq I$, we have $\text{OPT}(I') \leq \text{OPT}(I)$. Hence, removing items from an instance can only decrease the optimum.*

In the classical *offline version* of $\Pi = (\mathcal{I}, \text{SOL}, \text{COSTS})$, we are given the complete instance $I \in \mathcal{I}$ all at once. An *α-approximation* for Π is a polynomial-time algorithm A such that for all instances $I \in \mathcal{I}$, we have $\mathsf{A}(I) \leq \alpha\text{OPT}(I) + c$ for some constant c not depending on the instance. Here, $\mathsf{A}(I)$ is the value of the solution produced by A.

In the *online version* of a minimization problem $\Pi = (\mathcal{I}, \text{SOL}, \text{COSTS})$, an *online instance* I is a sequence of instances $I_1, I_2, \ldots, I_{|I|} \in \mathcal{I}$ such that $|I_t \triangle I_{t+1}| = 1$ for all $t = 1, \ldots, |I| - 1$, where $\triangle$ is the symmetric difference of two sets. This means that we *insert* a new item i^* (if $I_{t+1} \setminus I_t = \{i^*\}$) or an item i^* *departs* (if $I_t \setminus I_{t+1} = \{i^*\}$). The set of online instances of Π is denoted as $\mathcal{I}^{\text{on}}$. An *online algorithm* A maintains a sequence of solutions $S_1, \ldots, S_t$, where $S_t \in \text{SOL}(I_t)$ and furthermore, $S_{t+1}(i) = S_t(i)$ for all $i \in I_{t+1} \cap I_t$. In the *static online version*, we only have insertions and thus $I_t \subsetneq I_{t+1}$. A non-static problem is called *dynamic*. An online algorithm A producing such a sequence of solutions $S_1, S_2, \ldots$ is *β-competitive*, if $\text{COSTS}(S_t) \leq \beta \cdot \text{OPT}(I_t) + c$ for all $t = 1, \ldots, |I|$ and some constant c not depending on the instance. This notion of competitiveness is sometimes also called *asymptotic competitiveness* in contrast to the notion of *absolute competitiveness*, where no additional additive term c is allowed. Whenever we talk about competitiveness, this is with regard to the notion of asymptotic competitiveness.

Migration. Achieving bounded competitiveness is usually impossible for dynamic problems, even for very simple problems such as Bin Packing where every item has the same size: Consider an instance with k^2 items of size $1/k$. Now, the adversary removes from each bin all items but one. If more than k bins are remaining, the adversary further removes items until only k bins are left. The optimal offline solution would pack these k items into a single bin, while the online algorithms uses k bins. This impossibility is simply due to the fact that an online algorithm is not allowed to change the position of an item once it is placed. This very strict restriction comes with a high cost in regards to the competitiveness. As many real-world applications are not static, but allow the departure of items, one must be more flexible. We will thus allow a small amount of repacking to be able to handle dynamic problems. The model of repacking we use is called the *amortized migration factor model.* In this setting, every item $i \in$ ITEMS comes with a *size* $v_i \in \mathbb{Q}_{\geq 0}$. Usually this size will be the space needed for an item, such as its total area, volume or some similar criteria like the side length of a hypercube. For a set of items $I \subseteq$ ITEMS, we denote by $\text{VOL}(I) = \sum_{i \in I} v_i$ the complete volume of I. If $I \in \mathcal{I}^{\text{on}}$ is an online instance and $\boldsymbol{S} = (S_1, \ldots, S_{|I|})$ is a sequence of solutions with $S_j \in \text{SOL}(I_j)$, the *migrated items* $M_t(\boldsymbol{S})$ at time t are defined as $M_t(\boldsymbol{S}) = \{i \in I_{t-1} \cap I_t \mid S_{t-1}(i) \neq S_t(i)\}$. The total *migration* $\mu(\boldsymbol{S}, t)$ used until time t is defined as $\mu(\boldsymbol{S}, t) := \sum_{j=1}^{t} \sum_{i \in M_j(\boldsymbol{S})} v_i$, i. e. as the sum of the sizes of the migrated items. Inserting an item i builds up a migration potential of v_i and migrating i has cost v_i. More formally, for $t = 1, \ldots, |I|$, let $A_t = \bigcup_{j \in \{0,1,\ldots,t-1\}} (I_{j+1} \setminus I_j)$ be the set of items that were inserted until time t and $D_t = \bigcup_{j \in \{0,1,\ldots,t-1\}} (I_j \setminus I_{j+1})$ be the set of items that departed until time t. Let A be an online algorithm that is allowed to migrate items. We say that A has *migration factor* β, if $\mu(\boldsymbol{S}, t) \leq \beta[\text{VOL}(A_t) + \text{VOL}(D_t)]$ for all $t = 1, \ldots, |I|$ and all $I \in \mathcal{I}^{\text{on}}$. Here $\boldsymbol{S}$ is the sequence of solutions produced by A. Note that our definition of migration is amortized, i. e. we can build a potential to use later on. This will essentially allow us to repack the complete instance from time to time. In contrast, in the notion of worst-case migration, the total size of all repacked items is at most $\beta \cdot v_t$ at each time t, i. e. one cannot save up migration for later use.

If A is $(\gamma + \epsilon)$-competitive for some constant γ and all $\epsilon > 0$, and additionally has migration factor bounded by $f(1/\epsilon)$ for some function f, we say that A is *robust.* This is due to the fact that the migration needed only depends on the desired quality of the solution.

The General Framework. Before we start looking at the specific problems, we will introduce our very general framework. The simple algorithm presented by Feldkord *et al.* [16] can be seen as a special case of our framework. By using the concept of amortized migration, we can save up repacking potential, to be used at a later time. Now the design of an algorithm boils down to three basic questions. How do we pack arriving items? How do we repack already packed items? And at what time do we repack items? The main idea behind our framework for packing problems is to use general algorithms for the first two problems: an online algorithm to pack arriving items and an offline algorithm for the repacking. The third problem is then solved by a generic combination of the two algorithms. In order for this approach to work, we need different criteria for both algorithms and the packing problem that we are trying to solve.

Definition 2. *Let Π be an minimization problem with sizes v_i. We call Π* space related, *if* $\text{OPT}(I) \geq \text{VOL}(I)$ *holds for all instances $I \in \Pi$. Here* $\text{VOL}(I) := \sum_{i \in I} v_i$ *is the total size of I.*

Intuitively, this definition captures the fact that Π is a packing problem that needs to pack items of a certain volume in some non-overlapping way. All problems in this paper will be space related. To make use of this relation, we also need online algorithms such that competitiveness not only holds for the optimum $\text{OPT}(I)$, but also for the volume $\text{VOL}(I)$. We therefore also formally introduce this necessity.

Definition 3. *Let Π be an online minimization problem with sizes v_i and* A *be an online algorithm for Π. We say that* A *is* space related with ratio *β, if $\text{COSTS}(S_t) \leq \beta \cdot \text{VOL}(I_t) + c$ for all online instances $I \in \mathcal{I}^{on}$, all time points $t = 1, 2, \ldots, |I|$ and some constant c. Here, S_t denotes the solution produced by* A *at time t.*

Trivially, a space related algorithm with ratio β for a space related problem implies β-competitiveness, like Next-Fit for the Bin Packing problem. As indicated above, we will combine such an online algorithm with another offline algorithm. To be able to efficiently combine these two algorithms the online algorithms need to be able to build flexibly on top of the solution of the offline algorithm.

More formally, we require that our online algorithm takes another optional argument $S \in \text{SOL}(I_t)$ describing an existing solution to the previous instance I_t to build upon.

Definition 4. *Let Π be an online minimization problem with sizes v_i and* A *be a space related online algorithm for Π with ratio β. Furthermore let $I \in \mathcal{I}^{on}$ be a static online instance (i. e. it does not need to handle departures), $t < t' \leq |I|$ two time points, and S be a solution of I_t. We say that* A *is* flexible, *if it also accepts S as another parameter and produces upon input $I_{t'}$ and S a solution S' with $S'(i) = S(i)$ for all $i \in I_t \cap I_{t'}$.*

Note that we define flexibility only with regard to static online instances where no items depart. One advantage of our framework is that we only need to design such online algorithms, but our combined algorithm will also be able to deal with departures. Remember that our online algorithm A is given some solution S to the previous items I_t. In order to be able to ignore the departure of items in the combined approach, we need to guarantee that A only introduces an error of $\beta[\text{VOL}(I_{t'}) - \text{VOL}(I_t)]$ when it packs instance $I_{t'}$.

Definition 5. *We say that* A *is* flexible with ratio *β, if* A *is flexible and $\mathsf{A}(I_{t'}, S) \leq \text{COSTS}(S) + \beta[\text{VOL}(I_{t'}) - \text{VOL}(I_t)] + c$ for some constant c, for all instances $I \in \mathcal{I}^{\text{on}}$, all $t \in \{1, \ldots, |I|\}$, and all $t' \in \{t+1, \ldots, |I|\}$.*

For the problems we will address later, Bin Packing and Strip Packing, we can simply solve the instance containing the newly arriving jobs separately and pack the new partial solution on top of the old one. Therefore this will be a property easily fulfilled by all online algorithms that we consider later. Note that a flexible algorithm with ratio β is also space related with ratio β, as we can simply choose $t = 1$ add a trivial solution S for this instance containing a single item.

Combining the Algorithms. In order to obtain a robust PTAS or a robust online algorithm that is $\gamma+\epsilon$-competitive, we will need a flexible online algorithm

Listing 1.1. Framework ALG($\mathsf{A}_{\text{on}}, \mathsf{A}_{\text{off}}$)

```
set V_total := 0; set V_changed := 0;
let S be an empty solution;
for each time t and arriving or departing item i
  if item i arrives do
    pack i according to online algorithm A_on and solution S;
  if item i departs do
    leave i in its position in the current solution S;
  set V_changed = V_changed + v_i;
  if V_changed > εV_total do
    compute offline solution S for I_t with A_off;
    Continue with new solution;
    set V_total = VOL(I_t); set V_changed = 0;
```

with a constant ratio. For the offline algorithm we will need an offline $\gamma + \epsilon$-approximation (in the case of $\gamma = 1$, this is simply a PTAS). Basically, our final algorithm will have the same ratios as the offline algorithm and migration factor $O(1/\epsilon)$. In order to achieve bounded competitiveness, we need to migrate items at some special time points. These time points will be determined by the total volume of items that have arrived or departed since the last such time point. At these special time points, we will use the offline algorithm to rebuild the solution completely. In between these points, we will only apply the online algorithm for the static case. We will therefore define phases such that during a phase we only apply the online algorithm.

Definition 6. *Let Π be an online minimization packing problem with sizes v_i and $I \in \mathcal{I}^{\text{on}}$ be an online instance.*

We partition $I_1, I_2, \ldots$ into phases *as follows: The start time of the first phase is 1. If τ is the start time of the current phase, and $t \geq \tau$ is some time point, we define the following values: (i) the complete volume of the instance at time τ is denoted by $V_\tau = \text{VOL}(I_\tau)$, (ii) the items inserted since τ are defined as $Ins_t = \bigcup_{i=\tau}^{t-1}(I_{i+1} \setminus I_i)$ and its volume is $A_t = \text{VOL}(Ins_t)$, and (iii) the items departed since τ are defined as $Dep_t = \bigcup_{i=\tau}^{t-1}(I_i \setminus I_{i+1})$ and its total volume is $R_t = \text{VOL}(Dep_t)$. The current phase ends at the earliest point of time $\tau' > \tau$ such that $A_{\tau'} + R_{\tau'} > \epsilon V_\tau$. The next phase then starts at time τ'.*

Basically we end a phase when the total size of items that were added or removed exceeds an ϵ factor of the total item size at the beginning of the phase. So our final algorithm basically packs inserted items using the online algorithm until a phase ends. Departed items stay at the same position. Whenever a phase ends, we compute an offline solution for the current instance and repack our solution completely. After that the next phase starts, where we again only pack with the online algorithm. Now we will prove that this combination of algorithms will be a robust $\gamma + \epsilon$ algorithm for $\gamma \in O(1)$ and fixed $\epsilon > 0$, when we combine two algorithms with the right properties.

Theorem 1. *Let Π be a minimization problem fulfilling Assumption 1. Furthermore let A_{off} be a $(\gamma+\epsilon)$-approximation algorithm for the offline version of Π for some constant $\gamma \in O(1)$, let $1/2 \geq \epsilon > 0$, and let A_{on} be a flexible algorithm with ratio β. Then the combination of these two algorithms, denoted with $\mathsf{ALG}(\mathsf{A}_{\text{on}}, \mathsf{A}_{\text{off}})$ (see Listing 1.1), is an $(\gamma + O(1)\beta\epsilon)$-competitive robust algorithm for Π with amortized migration factor $O(\frac{1}{\epsilon})$.*

The running time of ALG at time point t is at most $T_{\text{off}}(t) + T_{\text{on}}(t)$, where $T_{\text{off}}(t)$ (resp. $T_{\text{on}}(t)$) is the worst-case running time of A_{off} (resp. A_{on}) on an instance of t items.

Proof. *Migration factor:* Let I_τ be the instance at the start of a phase with volume $V_\tau = \text{VOL}(I_\tau)$ and let τ' be the ending time of this phase. As the phase ends at time τ', we have $A_{\tau'} + R_{\tau'} > \epsilon V_\tau$ (where $A_{\tau'}$ and $R_{\tau'}$ are defined as in Definition 6). As we only ever migrate items at the end of a phase, we only need to consider the amortized migration factor at these time points. We assign the volume of all items inserted and departed during the phase that ends at time τ' to this phase. Hence, a total migration potential of $A_{\tau'} + R_{\tau'}$ was build up in this phase. The total volume of the instance at this point is at most $V_\tau + A_{\tau'}$ and our amortized migration factor is thus $\frac{V_\tau + A_{\tau'}}{A_{\tau'}+R_{\tau'}} \leq \frac{V_\tau}{A_{\tau'}+R_{\tau'}} + 1 < \frac{V_\tau}{\epsilon V_\tau} + 1 = O(1/\epsilon)$. As the phases are disjoint, the total amortized migration factor is thus also bounded by $O(1/\epsilon)$.

Competitiveness: To show the competitiveness of ALG, assume that

$$\mathsf{A}_{\text{off}}(I_t) \leq (\gamma+\epsilon) \cdot \text{OPT}(I_t) + c_{\text{off}} \qquad (*_{\text{off}})$$

for some constant c_{off} and furthermore

$$\mathsf{A}_{\text{on}}(I_{t'}, S) \leq \text{COSTS}(S) + \beta[\text{VOL}(I_{t'}) - \text{VOL}(I_t)] + c_{\text{on}} \qquad (*_{\text{on}})$$

for some constant c_{on}.

Let $I \in \mathcal{I}^{\text{on}}$ be some online instance and $t \in \{1, \ldots, |I|\}$. We distinguish whether t is the end of a phase or in the middle of a phase. If $t = \tau$ is the end of a phase, we use the offline algorithm A_{off} and thus have $\mathsf{ALG}(I_t) = \mathsf{A}_{\text{off}}(I_t) \leq (\gamma+\epsilon) \cdot \text{OPT}(I_t) + c_{\text{off}}$, where the inequality follows from eq. $(*_{\text{off}})$. Now consider any point of time t during the phase starting at τ. Like above, let A_t, R_t denote the total volumes of arrived and removed items in this phase up to time t. Note that we have $A_t + R_t \leq \epsilon V_\tau$, since otherwise we would repack at time t.

Claim. We have $\text{OPT}(I_\tau) \leq \text{OPT}(I_t) + \beta\epsilon V_\tau + c_{\text{on}}$.

Proof. By Assumption 1, the value $\text{OPT}(I_t)$ is minimal if only departures happened. We can thus assume w. l. o. g. that up till time t some items were removed and no new items arrived. Consider an optimal solution S_t for the instance I_t, where items departed and let $D_t = I_\tau \setminus I_t$ be the set of departed items. Let $d_1, \ldots, d_k$ be some total ordering of D_t. We now construct a new online instance I' that somehow reverses the removal of D_t. The instance I' is of length $t + k$ with $I'_i = I_i$ for $i \leq t$. For $i > t$, we define $I'_i = I_{i-1} \cup \{d_i\}$. We now use the online algorithm A_{on} on this instance I' with solution S_t for instance I_t. At the end of instance I', eq. ($*_{\text{on}}$) implies that A_{on} gives a feasible solution to $I'_{t+k} = I_t \cup D_t = I_\tau$ with

$$\begin{aligned}\mathsf{A}_{\text{on}}(I'_{t+k}, S_t) \leq &\text{COSTS}(S_t) + \beta[\text{VOL}(I_{t+k}) - \text{VOL}(I_t)] + c_{\text{on}} = \\ &\text{OPT}(I_t) + \beta[\text{VOL}(I_\tau) - \text{VOL}(I_t)] + c_{\text{on}} = \text{OPT}(I_t) + \beta\text{VOL}(D_t) + c_{\text{on}} \leq \\ &\text{OPT}(I_t) + \beta\epsilon V_\tau + c_{\text{on}}.\end{aligned}$$

The last inequality follows from the fact that $R_t = \text{VOL}(D_t)$ by definition and $R_t \leq \epsilon V_\tau$ by assumption. As A_{on} produces a feasible solution for $I'_{t+k} = I_\tau$, we clearly have $\text{OPT}(I_\tau) \leq \mathsf{A}_{\text{on}}(I'_{t+k}, S_t)$ thus proving the claim.

Since the problem Π is space related, we can conclude that

$$\text{OPT}(I_t) \geq \text{VOL}(I_t) = V_\tau + A_t - R_t \geq V_\tau - R_t \geq V_\tau - \epsilon V_\tau \geq \frac{V_\tau}{2}. \qquad (*)$$

Hence $V_\tau \leq 2\text{OPT}(I_t)$.

Let S_τ be the solution produced by A_{off} at time τ that the online algorithm A_{on} is building upon. Note that we only remove the departed item at a time point where the offline algorithm is used. Hence, at time t, the online algorithm does not produce a solution for the instance I_t, where the departed items are already removed. The algorithm rather works on the instance that still contains all items that have departed since time τ. This instance is denoted by I'_t. We thus have

$$\begin{aligned}
&\mathsf{ALG}(I_t) = \mathsf{A}_{\text{on}}(I'_t, S_\tau) \leq && // \text{ eq. } (*_{\text{on}})\\
&\text{COSTS}(S_\tau) + \beta[\text{VOL}(I'_t) - \text{VOL}(I_\tau)] + c_{\text{on}} = && // \ S_\tau \text{ produced by } \mathsf{A}_{\text{off}}\\
&\mathsf{A}_{\text{off}}(I_\tau) + \beta[\text{VOL}(I'_t) - \text{VOL}(I_\tau)] + c_{\text{on}} \leq && // \text{ eq. } (*_{\text{off}})\\
&(\gamma + \epsilon) \cdot \text{OPT}(I_\tau) + c_{\text{off}} + \beta[\text{VOL}(I'_t) - \text{VOL}(I_\tau)] + c_{\text{on}} \leq && // \text{ Equation 2}\\
&(\gamma + \epsilon) \cdot [\text{OPT}(I_t) + \beta\epsilon V_\tau + c_{\text{on}}] + c_{\text{off}} + \beta[\text{VOL}(I'_t) - \text{VOL}(I_\tau)] + c_{\text{on}} = &&\\
&(\gamma + \epsilon) \cdot [\text{OPT}(I_t) + \beta\epsilon V_\tau + c_{\text{on}}] + c_{\text{off}} + \beta[A_t] + c_{\text{on}} \leq && // \ A_t + R_t \leq \epsilon V_\tau\\
&(\gamma + \epsilon) \cdot [\text{OPT}(I_t) + \beta\epsilon V_\tau + c_{\text{on}}] + c_{\text{off}} + \beta\epsilon V_\tau + c_{\text{on}} = &&\\
&(\gamma + \epsilon)\text{OPT}(I_t) + (\gamma + \epsilon + 1)\beta\epsilon V_\tau + (\gamma + \epsilon + 1)c_{\text{on}} + c_{\text{off}} \leq && // \text{ eq. } (*)\\
&(\gamma + \epsilon)\text{OPT}(I_t) + 2(\gamma + \epsilon + 1)\beta\epsilon\text{OPT}(I_t) + (\gamma + \epsilon + 1)c_{\text{on}} + c_{\text{off}} = &&\\
&(\gamma + \epsilon + 2(\gamma + \epsilon + 1)\beta\epsilon)\text{OPT}(I_t) + (\gamma + \epsilon + 1)c_{\text{on}} + c_{\text{off}}. &&
\end{aligned}$$

As γ, c_{on}, and c_{off} are constants, the last term can be written as

$$\begin{aligned}&(\gamma + \epsilon + 2(\gamma + \epsilon + 1)\beta\epsilon)\text{OPT}(I_t) + (\gamma + \epsilon + 1)c_{\text{on}} + c_{\text{off}} \leq \\ &(\gamma + O(1) \cdot \beta\epsilon))\text{OPT}(I_t) + O(1).\end{aligned}$$

The running time bound follows easily from the fact that we essentially only use A_{on} or A_{off} at any given time.

We further conclude that if β is constant, this framework gives us automatically an online algorithm with a competitive rate of $\gamma + \epsilon$, as we can scale down ϵ appropriately.

Note that we do not assume that any of the algorithms run in polynomial time. We could thus use an exact exponential-time algorithm for A_{off}. This allows us to conclude that all space-related minimization problems that have an exact exponential-time algorithm and a flexible online algorithm can achieve a competitive ratio of $1 + \epsilon$ with migration factor $O(1/\epsilon)$ in exponential time. This is a stark contrast to the setting without migration, where information-theoretic lower bounds prevent the existence of such algorithms. An example of such an information-theoretic lower bound is given by Balogh [3] for the bin packing problem.

We can further note that we considered the case of asymptotic algorithms. One can easily see, that this also works for algorithms with absolute ratios. If both the flexible and the offline algorithm have absolute ratio, the constants c_{on}, c_{off} are set to 0 and even the resulting combined algorithm has an absolute ratio.

3 2-Dimensional Strip Packing

In the *online Strip Packing* problem, we are given a two-dimensional strip of width 1 and infinite height. At time t, either a rectangle r_t with width $w(r_t) \leq 1$ and height $h(r_t) \leq 1$ is inserted and needs to be packed into this strip or a rectangle r_t is removed from the strip. A packing is valid if no two rectangles intersect. The size $v(r)$ of a rectangle r is defined as $v(r) = h(r) \cdot w(r)$. We first focus on the case that rectangles are not allowed to be rotated and will later see how to handle the rotations. In both cases, the goal is to minimize the height of the produced packing. This problem has been studied intensively in the online setting (see for example the works cited in [10]). Jansen *et al.* [19] studied the static case in the migration scenario, where rectangles can only arrive.

To use our framework, we need the following ingredients:

(i) We need to show that the Strip Packing problem is space related;
(ii) We need to construct a flexible online algorithm with ratio β;
(iii) We need to construct an offline approximation algorithm.

Concerning the first point, it is well-known that $\text{OPT}(I_t) \geq \text{VOL}(I_t)$, as the rectangles are not allowed to intersect and the width of the strip is exactly 1.

Remark 1. The Strip Packing problem is space related.

We will now present a flexible online algorithm with ratio $\beta = 4$. This algorithm is a simple adaption of the *shelf algorithms* presented by Baker and Schwarz [1]. In the notion of Csirik and Woeginger [12], this algorithm would be denoted as SHELF(FirstFit, 1/2). We first define several types of containers. A container c of type γ_0 has width 1 and height $h(c) = 1$. For $i \in \mathbb{N}_{\geq 1}$, a container of type γ_i has width 1 and height $h(c) = 2^{-i+1}$. For each $i \in \mathbb{Z}_{\geq 0}$, we will have at most one *active* container of type γ_i. For all other containers of this type – which we call *closed* – we will guarantee that at least 1/4 of their volume is used by items. We perform the following operation whenever a new rectangle r_t arrives:

- If $w(r_t) \geq 1/2$, check whether a container of type γ_0 exists. If not, open a new active container of type γ_0 and place r_t into it. If such a container c already exists and $h(r_t) + \sum_{r \in c} h(r) > 1$, declare c as closed and open a new active container of type γ_0. Otherwise ($h(r_t) + \sum_{r \in c} h(r) \leq 1$), put r_t on top of the top item in c.
- If $w(r_t) \leq 1/2$ and $h(r_t) \in (2^{-i}, 2^{-i+1}]$, check whether a container of type γ_i exist. If not, open a new active container of type γ_i and place r_t into it. If such a container c already exists and $w(r_t) + \sum_{r \in c} w(r) > 1$, declare c as closed and open a new active container of type γ_i. Otherwise ($w(r_t) + \sum_{r \in c} w(r) \leq 1$), put r_t right to the right-most item in c.

We have at most one active container of type γ_i for each $i \in \mathbb{Z}_{\geq 0}$: we only open a new active container if we simultaneously declare another container of the same type as closed. This observation directly leads to the following theorem about the competitiveness of the algorithm.

Theorem 2. *The presented algorithm* A_{SP} *is a flexible online algorithm for Strip Packing with ratio* 4.

We have now shown the first two ingredients for our framework: the problem is space related and we gave a suitable online algorithm. The final piece – an offline approximation algorithm – is given by the asymptotic fully polynomial time approximation scheme (AFPTAS) of Kenyon and Rémila [22], which is an $1 + \epsilon$-approximation. We can thus use Theorem 1 with $\gamma = 1 + \epsilon$ and $\beta = 4$ to conclude the following theorem.

Theorem 3. *There is a robust online algorithm for the dynamic Strip Packing problem that is* $1 + \epsilon$*-competitive and has amortized migration factor* $O(1/\epsilon)$.

Rotations. If rotations by 90° are allowed, the resulting problem is called Strip Packing With Rotations. For an instance I, we denote the height of a corresponding optimal packing by $\text{OPT}_R(I)$. As the volume of a rotated rectangle does not change, we have $\text{OPT}_R(I) \geq \text{VOL}(I)$. Similarly, the volume bound of Theorem 2 also remains true. We can thus conclude the following adaption of Theorem 2.

Theorem 4. *The presented algorithm* A_{SP} *is a flexible online algorithm for Strip Packing With Rotations with ratio* 4.

Instead of using the classical AFPTAS by Kenyon and Rémila [22], we use the AFPTAS of Jansen and van Stee [20] for the case that rotations are allowed. Using Theorem 1 with $\gamma = 1 + \epsilon$ and $\beta = 4$ gives the following theorem.

Theorem 5. *There is a robust online algorithm for the dynamic Strip Packing With Rotations problem that is* $1+\epsilon$*-competitive and has amortized migration factor* $O(1/\epsilon)$.

The best known online algorithm with migration known is due to Jansen *et al.* [19]. It also is $1+\epsilon$-competitive, but an amortized migration factor of $O(1/\epsilon^9 \log^2(1/\epsilon))$, only works for the static case (no rectangles are removed), and cannot handle rotations. The higher-dimensional case is treated in the full version.

4 Bin Packing

2-Dimensional Bin Packing. In 2-D Bin Packing, each item i is given by its height $h_i \leq 1$ and its width $w_i \leq 1$. The goal is to pack these items non-overlapping into as few unit-sized squares (called *bins*) as possible. As above, we will show the following: (i) We need to show that the 2-D Bin Packing problem is space related; (ii) We need to construct a flexible online algorithm with ratio β; (iii) We need to construct an offline approximation algorithm.

As the rectangles are not allowed to overlap and each bin has a total volume of 1, 2-D Bin Packing is space related.

Remark 2. The 2-D Bin Packing problem is space related.

We will now present a flexible online algorithm. This algorithm is a simple extension of the classical algorithm presented by Coppersmith and Raghavan [11]. We categorize items as follows: We call an item *vertical* if $w_i \leq h_i$ and *horizontal* if $w_i > h_i$. Note that squares with $h_i = w_i$ will be considered as vertical. Without loss of generality we will explain in the following how to pack vertical items. Horizontal items can and will be placed into separate bins with the same strategy, just altered for horizontal items. Note that we later also need to account for these horizontal bins.

We further assign each item i a *size class*. Item i is in size class $j \in \mathbb{N}_{\geq 1}$ if $1/2^{j-1} \geq h_i > 1/2^j$. Further we say an item i is *square-like*, if i is in size class j and furthermore $w_i > 1/2^j$. The general idea is that for every arriving item in size class j, we will assign a square slot of size $1/2^{j-1}$ in some bin. A square slot of this size is called a *slot of class* j. Note that an item of size class j always fits into a slot of class j, as we only handle vertical items with $w_i \leq h_i$ here. Our goal is to fill all opened slots with items until $1/4$ of the total area of the slot is covered. Square-like items will immediately fill such a slot to this extent. For the other items, we will *reserve* such slots for a size class j and stack items from left to right until the total width of all items in that slot exceeds $1/2^j$. Since the height of items assigned to this slot exceed $1/2^j$ as well, $1/4$ of the total slot will be covered at that point. A slot can have three states: it is either (i) *empty* and thus contains no item, (ii) *reserved* for class j and thus only contains items of class j or (iii) *closed* if at least $1/4$ of its total volume is filled with items.

In order to assign items to these slots, we will keep up to two open bins. The first open bin will hold items of size class 1 and we use the complete bin as a single reserved slot. The second open bin will receive items of size class ≥ 2. Initially, this bin is split into four empty slots of class 2. The online algorithm $\mathsf{A}_{2\text{-}D}$ now assigns a new item either to a non-closed slot of its class or, if no such slot exists, splits an empty slot of a larger class recursively. As every bin created by the algorithm contains at most three empty slots of a certain class, we can conclude the competitiveness of our algorithm.

Theorem 6. *The proposed algorithm $\mathsf{A}_{2\text{-}D}$ for 2-D Bin Packing is a flexible online algorithm with ratio $\frac{48}{5}$.*

Finally, we can use the approximation algorithm of Bansal and Khan [6] that is an 1.405-approximation for 2-D Bin Packing. We can thus use Theorem 1 with $\gamma = 1.405$ and $\beta = 48/5$ to conclude the following theorem.

Theorem 7. *There is a robust online algorithm for the dynamic 2-D Bin Packing problem that is $1.405 + \epsilon$-competitive and has amortized migration factor $O(1/\epsilon)$.*

To the best of our knowledge, this is the first robust online algorithm for dynamic 2-D Bin Packing. Note that the best known lower bound for the competitiveness of any online algorithm for online 2-D Bin Packing without migration is 1.856 due to Van Vliet [25].

Rotations. As for Strip Packing, allowing rotations of the rectangles by 90° gives rise to a problem called 2-D Bin Packing With Rotations. The corresponding optimal number of bins needed to pack instance I is denoted by $\text{OPT}_R(I)$. Rotations are invariant with regard to the volume of a rectangle and thus $\text{OPT}_R(I) \geq \text{VOL}(I)$. We can thus again use our online algorithm $\mathsf{A}_{2\text{-}D}$. The approximation algorithm of Bansal and Khan [6] used above can also handle the case of rotation and thus is an 1.405-approximation for 2-D Bin Packing With Rotations. We can thus use Theorem 1 with $\gamma = 1.405$ and $\beta = 48/5$ to conclude our theorem.

Theorem 8. *There is a robust online algorithm for the dynamic 2-D Bin Packing With Rotations problem that is $1.405 + \epsilon$-competitive and has amortized migration factor $O(1/\epsilon)$.*

To the best of our knowledge, this is the first robust online algorithm for dynamic 2-D Bin Packing With Rotations. Note that the best known lower bound for the competitiveness of any online algorithm for 2-D Bin Packing With Rotations without migration is at least 1.7515 due to Balogh *et al.* [2] improving the bound of 1.6707 due to Blitz *et al.* [9].

Higher Dimensions. Our presented algorithm can be simply adapted to the d-dimensional case to obtain the following result.

Theorem 9. *The proposed algorithm $\mathsf{A}_{d\text{-}D}$ for d-Dimensional Hyperrectangle Packing is a flexible online algorithm with ratio $\frac{2^{2d}-3\cdot 2^d+1}{2^{2d}(2^d-1)}$.*

Finally, we can use the offline APTAS with polynomial running time for constant d from Bansal *et al.* [4] for the special case of Hypercube Packing. Using Theorem 1 with $\gamma = 1 + \epsilon$ and $\beta = \frac{2^{2d} - 3 \cdot 2^d + 1}{2^{2d}(2^d - 1)}$ allows the conclusion of the following theorem.

Theorem 10. *For every constant $d \geq 2$, there is a robust online algorithm for the dynamic d-Dimensional Hypercube Packing problem that is $1 + \epsilon$-competitive and has amortized migration factor $O(1/\epsilon)$.*

In contrast, the best known online algorithm with migration for the d-Dimensional Hypercube Packing is due to Epstein and Levin [14]. It is also $1 + \epsilon$-competitive, but can only handle the static case and has migration factor $(1/\epsilon)^{\Omega(d)}$. Note however that they use worst-case migration, i. e. they are not allowed to repack the complete instance every once in a while but need to make slight adaptions carefully throughout the run of the algorithm.

5 Vector Packing

In the *online d-dimensional Vector Packing* problem, at time t either a vector $w_t \in (\mathbb{Q} \cap [0,1])^d$ is inserted and needs to be packed or is removed. The size $v(w_t)$ of such a vector $w_t = (w[1], \ldots, w[d])$ is defined as the average sum of its components, i. e. $v(w_t) = \sum_{j=1}^{d} w[j]/d$. The goal is to pack these vectors into as few as possible bins as possible. Here, a *bin* B is a subset of vector such that $\sum_{w \in B} w[j] \leq 1$ for $j = 1, \ldots, d$. As each bin can contain items of volume at most 1, it is easy to see that the problem is space related.

Remark 3. The d-dimensional Vector Packing problem is space related.

We will now present a flexible online algorithm with ratio $\beta = 2d$ that is a simple adaption of the well-known next fit online algorithm for bin packing. Every bin will have an index to guarantee a linear ordering. Whenever a vector w arrives, we first check whether w can be packed into an existing bin. If this is possible, we add w to such a bin with minimal index. If no such bin exists, we open a new bin containing w. If we are given a previous packing S, we simply ignore the previous bins and do not put any vector in them.

Theorem 11. *For every $d \geq 1$, the presented algorithm A_{VP} is a flexible online algorithm for d-dimensional Vector Packing with ratio $2d$.*

We have now shown the first two ingredients for our framework: the problem is space related and we gave a suitable online algorithm. The final piece – an offline approximation algorithm – is given by the algorithm of Bansal *et. al.* [5] which is a $\ln(d+1) + 0.807 + \epsilon$-approximation that runs in polynomial time for constant d. We can thus use Theorem 1 with $\gamma = \ln(d+1) + 0.807 + \epsilon$ and $\beta = 2d$ to conclude the following theorem. Note that if d is considered constant, so is β.

Theorem 12. *For every constant $d \geq 1$, there is a robust online algorithm for the dynamic d-dimensional Vector Packing problem that is $\ln(d+1) + 0.807 + \epsilon$-competitive and has amortized migration factor $O(1/\epsilon)$.*

References

1. Baker, B.S., Schwarz, J.S.: Shelf algorithms for two-dimensional packing problems. SIAM J. Comput. **12**(3), 508–525 (1983)
2. Balogh, J., Békési, J., Dósa, G., Epstein, L., Levin, A.: Lower bounds for several online variants of bin packing. In: Solis-Oba, R., Fleischer, R. (eds.) WAOA 2017. LNCS, vol. 10787, pp. 102–117. Springer, Cham (2018). https://doi.org/10.1007/978-3-319-89441-6_9
3. Balogh, J., Békési, J., Galambos, G.: New lower bounds for certain classes of bin packing algorithms. Theoret. Comput. Sci. **440–441**, 1–13 (2012). https://doi.org/10.1016/j.tcs.2012.04.017. http://www.sciencedirect.com/science/article/pii/S0304397512003611
4. Bansal, N., Correa, J.R., Kenyon, C., Sviridenko, M.: Bin packing in multiple dimensions: inapproximability results and approximation schemes. Math. Oper. Res. **31**(1), 31–49 (2006)
5. Bansal, N., Eliás, M., Khan, A.: Improved approximation for vector bin packing. In: Proceedings of SODA, pp. 1561–1579 (2016)
6. Bansal, N., Khan, A.: Improved approximation algorithm for two-dimensional bin packing. In: Proceedings of SODA, pp. 13–25 (2014)
7. Berndt, S., Epstein, L., Jansen, K., Levin, A., Maack, M., Rohwedder, L.: Online bin covering with limited migration. In: Proceedings of ESA (2019, Accepted)
8. Berndt, S., Jansen, K., Klein, K.-M.: Fully dynamic bin packing revisited. In: Proceedings of APPROX-RANDOM, pp. 135–151 (2015)
9. Blitz, D., Heydrich, S., van Stee, R., van Vliet, A., Woeginger, G.J.: Improved lower bounds for online hypercube and rectangle packing. CoRR, abs/1607.01229 (2016)
10. Christensen, H.I., Khan, A., Pokutta, S., Tetali, P.: Approximation and online algorithms for multidimensional bin packing: a survey. Comput. Sci. Rev. **24**, 63–79 (2017)
11. Coppersmith, D., Raghavan, P.: Multidimensional on-line bin packing: algorithms and worst-case analysis. Oper. Res. Lett. **8**(1), 17–20 (1989)
12. Csirik, J., Woeginger, G.J.: Shelf algorithms for on-line strip packing. Inf. Process. Lett. **63**(4), 171–175 (1997)
13. Epstein, L., Levin, A.: A robust APTAS for the classical bin packing problem. Math. Program. **119**(1), 33–49 (2009)
14. Epstein, L., Levin, A.: Robust approximation schemes for cube packing. SIAM J. Optim. **23**(2), 1310–1343 (2013)
15. Epstein, L., Levin, A.: Robust algorithms for preemptive scheduling. Algorithmica **69**(1), 26–57 (2014)
16. Feldkord, B., et al.: Fully-dynamic bin packing with little repacking. In: Proceedings of ICALP, pp. 51:1–51:24 (2018)
17. Gálvez, W., Soto, J.A., Verschae, J.: Symmetry exploitation for online machine covering with bounded migration. In: Proceedings of ESA, pp. 32:1–32:14 (2018)
18. Jansen, K., Klein, K.-M.: A robust AFPTAS for online bin packing with polynomial migration. In: Fomin, F.V., Freivalds, R., Kwiatkowska, M., Peleg, D. (eds.) ICALP 2013. LNCS, vol. 7965, pp. 589–600. Springer, Heidelberg (2013). https://doi.org/10.1007/978-3-642-39206-1_50
19. Jansen, K., Klein, K.-M., Kosche, M., Ladewig, L.: Online strip packing with polynomial migration. In: Proceedings of APPROX-RANDOM, pp. 13:1–13:18 (2017)

20. Jansen, K., van Stee, R.: On strip packing with rotations. In: Proceedings of STOC, pp. 755–761 (2005)
21. Karmarkar, N., Karp, R.M.: An efficient approximation scheme for the one-dimensional bin-packing problem. In: Proceedings of FOCS, pp. 312–320 (1982)
22. Kenyon, C., Rémila, E.: A near-optimal solution to a two-dimensional cutting stock problem. Math. Oper. Res. **25**(4), 645–656 (2000)
23. Sanders, P., Sivadasan, N., Skutella, M.: Online scheduling with bounded migration. Math. Oper. Res. **34**(2), 481–498 (2009)
24. Skutella, M., Verschae, J.: Robust polynomial-time approximation schemes for parallel machine scheduling with job arrivals and departures. Math. Oper. Res. **41**(3), 991–1021 (2016)
25. Van Vliet, A.: Lower and upper bounds for on-line bin packing and scheduling heuristics: Onder-en Bovengrenzen Voor On-line Bin Packing en Scheduling Heuristieken. Ph.D. thesis (1995)
26. van Vliet, A.: An improved lower bound for on-line bin packing algorithms. Inf. Process. Lett. **43**(5), 277–284 (1992). https://doi.org/10.1016/0020-0190(92)90223-I. http://www.sciencedirect.com/science/article/pii/002001909290223I
27. Yao, A.C.C.: New algorithms for bin packing. J. ACM **27**(2), 207–227 (1980)

Approximation Results for Makespan Minimization with Budgeted Uncertainty

Marin Bougeret[2](✉), Klaus Jansen[1], Michael Poss[2], and Lars Rohwedder[1]

[1] Department of Computer Science, Kiel University, 24098 Kiel, Germany
{kj,lro}@informatik.uni-kiel.de

[2] LIRMM, University of Montpellier, CNRS, Montpellier, France
{marin.bougeret,michael.poss}@lirmm.fr

Abstract. We study approximation algorithms for the problem of minimizing the makespan on a set of machines with uncertainty on the processing times of jobs. In the model we consider, which goes back to [3], once the schedule is defined an adversary can pick a scenario where deviation is added to some of the jobs' processing times. Given only the maximal cardinality of these jobs, and the magnitude of potential deviation for each job, the goal is to optimize the worst-case scenario. We consider both the cases of identical and unrelated machines. Our main result is an EPTAS for the case of identical machines. We also provide a 3-approximation algorithm and an inapproximability ratio of $2 - \epsilon$ for the case of unrelated machines.

Keywords: Makespan minimization · Robust Optimization · Approximation algorithms · EPTAS · Parallel machines · Unrelated machines

1 Introduction

Classical optimization models suppose perfect information over all parameters. This can lead to optimal solutions having poor performance when the actual parameters deviate, even by a small amount, from the predictions used in the optimization model. Different frameworks have been proposed to overcome this issue, among which Robust Optimization which tackles the uncertainty by providing a set of possible values for these parameters, and considering the worst outcome over that set. In this paper, we consider the problem of scheduling a set of jobs J on the set of machines M, so as to minimize the makespan, and considering that the processing times are uncertain. What is more, we consider the budgeted uncertainty model introduced by [3] where each processing time varies between its nominal value and the latter plus some deviation. Further, in

This work was partially supported by DFG Project, "Robuste Online-Algorithmen für Scheduling-und Packungsprobleme", JA 612/19-1, and ANR project ROBUST (ANR-16-CE40-0018).

E. Bampis and N. Megow (Eds.): WAOA 2019, LNCS 11926, pp. 60–71, 2020.
https://doi.org/10.1007/978-3-030-39479-0_5

any scenario, at most Γ of the uncertain parameters take the higher values, the other being at their nominal values.

Let us now formally define the Robust Scheduling on Unrelated Machines ($R|\mathcal{U}^{\Gamma}|C_{\max}$) problem. For any job $j \in J$ and machine $i \in M$, we denote by $\overline{p}_{ij} \geq 0$ the nominal processing time of j on i, and by $\hat{p}_{ij} \geq 0$ the (potential) deviation of j on i. A schedule σ is a function from $J \rightarrow M$. We write σ_i for the subset of jobs scheduled on machine i. Let $\mathcal{U}^{\Gamma} = \{\xi \in \{0,1\}^{|J|} : \|\xi\|_1 \leq \Gamma\}$ be the set of all possible scenarios where at most Γ jobs deviate. For any $\xi \in \mathcal{U}^{\Gamma}$, we set $p_{ij}^{\xi} = \overline{p}_{ij} + \xi_j \hat{p}_{ij}$ to be the actual processing time of j on i in scenario ξ.

Let us now formalize some common terms, but with dependence on scenario ξ. The load of machine i in scenario ξ is calculated as $\sum_{j \in \sigma_i} p_{ij}^{\xi}$. The makespan in scenario ξ is the maximum load in scenario ξ, i.e., $C_{\max}^{\xi}(\sigma) = \max_{i \in M} \sum_{j \in \sigma_i} p_{ij}^{\xi}$. Finally, $C_{\max}^{\Gamma}(\sigma) = \max_{\xi \in \mathcal{U}^{\Gamma}} C_{\max}^{\xi}(\sigma)$ denotes the objective function we consider in Robust Scheduling, where the adversary takes the worst scenario among $\mathcal{U}^{\Gamma}$.

Next, we will state important observations about the objective function. We first need to introduce the following notations. Given a set of jobs X_i scheduled on machine i, we define $\overline{p}(X_i) = \sum_{j \in X_i} \overline{p}_{ij}$, $\hat{p}(X_i) = \sum_{j \in X_i} \hat{p}_{ij}$, $\Gamma(X_i)$ as the set of the Γ jobs of X_i with the largest $\hat{p}_{ij}$ values (or $\Gamma(X_i) = \sigma_i$ when $|X_i| < \Gamma$) with ties broken arbitrarily. Finally, set $\hat{p}_{\Gamma}(X_i) = \hat{p}(\Gamma(X_i))$.

By definition we have $C_{\Gamma}(\sigma) = \max_{\xi \in \mathcal{U}^{\Gamma}} \max_{i \in M} \sum_{j \in \sigma_i} C_{\xi}(\sigma)$, and thus we can rewrite $C_{\Gamma}(\sigma) = \max_{i \in M} \max_{\xi \in \mathcal{U}^{\Gamma}} \sum_{j \in \sigma_i} C_{\xi}(\sigma) = \max_{i \in M} C_{\Gamma}(\sigma_i)$, where $C_{\Gamma}(\sigma_i) = \max_{\xi \in \mathcal{U}^{\Gamma}} \sum_{j \in \sigma_i} C_{\xi}(\sigma)$ is the worst-case makespan on machine i. The benefit of rewriting $C_{\Gamma}(\sigma)$ in this form is that it is now clear that $C_{\Gamma}(\sigma_i) = \overline{p}(\sigma_i) + \hat{p}_{\Gamma}(\sigma_i)$ as the worst scenario ξ for a fixed σ_i is obtained by picking the Γ jobs with highest $\hat{p}_{ij}$ and make them deviate. Thus, $R|\mathcal{U}^{\Gamma}|C_{\max}$ can also be thought as a "classical" scheduling problem (without adversary) where the makespan on a machine $C_{\Gamma}(\sigma_i)$ is simply the sum of all the nominal processing time of jobs of σ_i, plus only the Γ largest deviating values of jobs of σ_i. We are now ready to define $R|\mathcal{U}^{\Gamma}|C_{\max}$.

Problem 1. ROBUST SCHEDULING ON UNRELATED MACHINES ($R|\mathcal{U}^{\Gamma}|C_{\max}$)

- Input: $(J, M, \overline{p} \in \mathbb{Q}_{+}^{|M| \times |J|}, \hat{p} \in \mathbb{Q}_{+}^{|M| \times |J|})$ where J is the set of jobs, M the set of machines, $\overline{p}$ are the vectors of nominal processing times, and $\hat{p}$ the vectors of deviation
- Output: find a schedule $\sigma : J \rightarrow M$
- Objective function: min $C_{\Gamma}(\sigma) = \max_{\xi \in \mathcal{U}^{\Gamma}} \max_{i \in M} \sum_{j \in \sigma_i} [\overline{p}_{ij} + \xi_j \hat{p}_{ij}] = \max_{i \in M} C_{\Gamma}(\sigma_i)$, where $C_{\Gamma}(\sigma_i) = \overline{p}(\sigma_i) + \hat{p}_{\Gamma}(\sigma_i)$.

Following the classical three field notation, we denote by $R|\mathcal{U}^{\Gamma}|C_{\max}$ the previous problem. Notice that when all $\hat{p}_{ij} = 0$ the problem corresponds to the classical $R||C_{\max}$, for which we denote by $C(\sigma_i) = \sum_{j \in \sigma_i} \overline{p}_{ij}$ the makespan on machine i. We are also interested in a simplification of the above problem. This simplification is ROBUST SCHEDULING ON IDENTICAL MACHINES ($P|\mathcal{U}^{\Gamma}|C_{\max}$) where each has two processing times ($\overline{p}_j$ and $\hat{p}_j$), and we have $\overline{p}_{ij} = \overline{p}_j$ and $\hat{p}_{ij} = \hat{p}_j$ for any i.

Robust scheduling has been considered in the past, mostly for finite uncertainty sets without particular structure, see for instance [1,6,9,10,12]. More recently, [2,5,13] considered robust packing and scheduling with the budgeted uncertainty model $\mathcal{U}^\Gamma$ from [4]. Specifically, [5] (with authors in common) provided a 3-approximation algorithm and a $(1+\epsilon)$-approximation (PTAS) for $P|\mathcal{U}^\Gamma|C_{\max}$ but only for a constant Γ, as well as a randomized approximation algorithm for $R|\mathcal{U}^\Gamma|C_{\max}$ having an average ratio of $O(\log(m))$. They also considered problem $1|\mathcal{U}^\Gamma|\sum_j w_j C_{\max}$, proving that the problem is $\mathcal{NP}$-hard in the strong sense, and providing a polynomial-time algorithm when $w_j = 1$ for $j \in J$. Authors of [13] considered the robust one-machine problem for four commonly-used objective criteria: (weighted) total completion time, maximum lateness/tardiness, and number of late jobs. They showed that some of these problems are polynomially solvable and provide mixed-integer programming formulations for others. Their results considered $\mathcal{U}^\Gamma$ as well as two closely related uncertainty sets. Paper [2] (with also authors in common) considers robust bin-packing problem for $\mathcal{U}^\Gamma$ and one of the uncertainty sets considered by [13], and provided constant-factor approximations algorithms for the two problems.

In this paper we improve the results of [5] for $P|\mathcal{U}^\Gamma|C_{\max}$ and $R|\mathcal{U}^\Gamma|C_{\max}$. In Sect. 2 we show that any c-approximation for the classical $R||C_{\max}$ problem leads to a $(c+1)$-approximation for $R|\mathcal{U}^\Gamma|C_{\max}$, hence obtaining a 3-approximation algorithm for the latter problem, and a $(2+\epsilon)$-approximation for $P|\mathcal{U}^\Gamma|C_{\max}$. We point out that this result improves the *ad-hoc* 3-approximation of [5] for $P|\mathcal{U}^\Gamma|C_{\max}$, while having a simpler proof. In Sect. 3, we show through a reduction from the RESTRICTED ASSIGNMENT PROBLEM that there exists no $(2-\epsilon)$-approximation algorithm for $R|\mathcal{U}^\Gamma|C_{\max}$ unless $\mathcal{P} = \mathcal{NP}$. This implies that the best possible ratio (unless $\mathcal{P} = \mathcal{NP}$) for $R|\mathcal{U}^\Gamma|C_{\max}$ is somewhere between 2 and 3, contrasting with the classical $R||C_{\max}$ where the gap between $\frac{3}{2}$ and 2 is still open since [11].

In Sect. 4 we consider the $P|\mathcal{U}^\Gamma|C_{\max}$ problem and present the first step of our main result, namely a PTAS which is valid even when Γ is part of the input, i.e., not constant. Having Γ in the input (and not constant) requires a totally different technique from the one used in [5]. The algorithm is turned into an EPTAS in Sect. 5, i.e., a PTAS where the dependency of ϵ is not in the exponent of the encoding length.

2 A 3-Approximation for Unrelated Machines

Theorem 1. *Any polynomial c-approximation for $R||C_{\max}$ implies a polynomial $(c+1)$-approximation for $R|\mathcal{U}^\Gamma|C_{\max}$.*

Proof (Proof of Theorem 1). We design a dual approximation, i.e., given an instance I of $R|\mathcal{U}^\Gamma|C_{\max}$ and an threshold T, we either give a schedule σ of I with $C_\Gamma(\sigma) \leq (c+1)T$, or prove that $T < \mathrm{OPT}(I)$. Using a binary search on T this will imply a $(c+1)$-approximation algorithm.

For that, given an instance $I = (J, M, \overline{p}, \hat{p})$ of $R|\mathcal{U}^\Gamma|C_{\max}$, and T the current threshold, our objective is to define an instance $I' = (J, M, p)$ of the classical

$R||C_{\max}$ problem. The transformation of a solution for I' to a solution for I will be straightforward since the jobs and machines will be the same.

Given a machine i, let $B_i = \{j|\hat{p}_{ij} > \frac{T}{\Gamma}\}$ and $S_i = J \setminus B_i$. Define

$$p_{ij} := \begin{cases} \overline{p}_{ij} + \hat{p}_{ij} & \text{if } j \in B_i \\ \overline{p}_{ij} & \text{otherwise.} \end{cases} \tag{1}$$

Let us now prove that (1) if $\text{OPT}(I') > T$ then we have $\text{OPT}(I) > T$, and (2) every schedule σ with makespan $C^{I'}(\sigma)$ in I' has a makespan at most $C^{I'}(\sigma)+T$ in I ($C_\Gamma(\sigma) \leq C^{I'}(\sigma) + T$).

For (1), we prove that $\text{OPT}(I) \leq T$ implies that $\text{OPT}(I') \leq T$. Let σ be an optimal solution of I and i a machine. $C_\Gamma(\sigma) \leq T$ implies that $C_\Gamma(\sigma_i) \leq T$ for any i, and thus that $\overline{p}(\sigma_i) + \hat{p}_\Gamma(\sigma_i) \leq T$. Now, observe that $B_i \subseteq \Gamma(\sigma_i)$. Indeed, assume towards contradiction that there exists $j \in B_i \setminus \Gamma(\sigma_i)$. This implies that $|\Gamma(\sigma_i)| = \Gamma$. As by definition, any $j' \in \Gamma(\sigma_i)$ has $\hat{p}_{ij'} \geq \hat{p}_{ij} > \frac{T}{\Gamma}$, we get that $\hat{p}_\Gamma(\sigma_i) > T$, a contradiction. This implies $C^{I'}(\sigma_i) = \overline{p}(\sigma_i) + \hat{p}(B_i) \leq \overline{p}(\sigma_i) + \hat{p}_\Gamma(\sigma_i) \leq T$.

For (2), let σ be a solution of I'. Let $i \in M$. Observe that $\hat{p}(\Gamma(\sigma_i)) \leq \hat{p}(B_i)+T$ as $\Gamma(\sigma_i)$ contains at most Γ jobs in $\sigma_i \setminus B_i$, and these jobs have $\hat{p}_{ij} \leq \frac{T}{\Gamma}$. Thus, $C_\Gamma(\sigma_i) = \overline{p}(\sigma_i) + \hat{p}_\Gamma(\sigma_i) \leq \overline{p}(\sigma_i) + \hat{p}(B_i) + T = C^{I'}(\sigma_i) + T$.

Thus, given a T and I we create I' as above and run the c-approximation for $R||C_{\max}$ to get a solution σ. If $C^{I'}(\sigma) > cT$ then $\text{OPT}(I') > T$, implying $\text{OPT}(I) > T$, and thus we reject T. Otherwise, we consider σ as a solution for I, and $C_\Gamma(\sigma) \leq (c+1)T$. □

Using the well-known 2-approximation algorithm from [11], we obtain immediately the following.

Corollary 1. *There is a 3-approximation for* $R|\mathcal{U}^\Gamma|C_{\max}$.

Since by this reduction identical machines stay identical we also obtain the following using the EPTAS of [7] for the classical $Q||C_{\max}$ problem.

Corollary 2. *For every* $\epsilon > 0$ *there is a* $(2+\epsilon)$*-approximation for* $P|\mathcal{U}^\Gamma|C_{\max}$ *running in time* $2^{O(1/\epsilon \log(1/\epsilon)^4)} + poly(n)$.

3 A $2-\epsilon$ Inapproximability for Unrelated Machines

For the classical $R||C_{\max}$ problem, when all $p_{ij} \in \{1, \infty\}$, deciding if the optimal value is at most 1 is polynomially solvable as it can be reduced to finding a matching in a bipartite graph. The result below shows that answering the same question for $R|\mathcal{U}^\Gamma|C_{\max}$ is $\mathcal{NP}$-complete.

Theorem 2. *Given an instance I of* $R|\mathcal{U}^\Gamma|C_{\max}$, *it is NP-complete to decide if* $\text{OPT}(I) \leq 1$ *or* $\text{OPT}(I) \geq 2$, *and thus for any* $\epsilon > 0$ *is no* $(2-\epsilon)$*-approximation algorithm for* $R|\mathcal{U}^\Gamma|C_{\max}$ *unless* $\mathcal{P} = \mathcal{NP}$, *even for* $\Gamma = 1$ *and when each job can be scheduled on at most 3 machines.*

Proof. Let us define a reduction from 3-SAT to $R|\mathcal{U}^\Gamma|C_{\max}$ with $\Gamma = 1$. Let I_0 be an instance of 3-SAT with clauses $\{C_i, i \in [m_0]\}$ and variables $\{x_j, j \in [n_0]\}$. Each C_i is of the form $l_i^1 \vee l_i^2 \vee l_i^3$ where $l_i^k \in \{x_j, \bar{x}_j\}$ for some j. We define an instance I of $R|\mathcal{U}^\Gamma|C_{\max}$ with $m = 2n_0$ machines and $n = n_0 + m_0$ jobs as follows. To each variable x_j we associate two machines $\{j_f, j_t\}$. We create a set of n_0 variable jobs where for any $j \in [n_0]$, $\overline{p}_{j_f j} = \overline{p}_{j_t j} = 1$, $\overline{p}_{i' j} = \infty$ for any other i', and $\hat{p}_{ij} = 0$ for any $i \in [m]$. For any clause C_i, $i \in [m_0]$ we define M_i: the set of 3 machines corresponding to literals $\{l_i^k\}$ satisfying C_i. For example, if $C_7 = x_1 \vee \bar{x}_3 \vee x_5$ then $M_7 = \{1_t, 3_f, 5_t\}$. We now define a set of m_0 clause jobs as follows. For any $j \in [n_0+1, n_0+m_0]$, job j represents clause C_{j-n_0} with $\hat{p}_{ij} = 1$ iff $i \in M_{j-n_0}$, $\hat{p}_{i'j} = \infty$ for any other i', and $\overline{p}_{ij} = 0$ for any $i \in [m]$. For example, job $j = n_0 + 7$ is associated to C_7 where in particular $\hat{p}_{1_t j} = \hat{p}_{3_f j} = \hat{p}_{5_t j} = 1$. Notice that each clause job can be scheduled on at most 3 machines. Let us now verify that I_0 is satisfiable iff $\text{OPT}(I) = 1$.

$\Rightarrow$. Suppose I_0 is satisfied by assignment a. For any $j \in [n_0]$, we schedule j on j_t if x_j is set to false in a and on j_f otherwise. For any $j \in [n_0+1, n_0+m_0]$, we schedule job j on any machine $i \in M_{j-n_0}$ corresponding to a literal satisfying C_i in assignment a. Notice that in this schedule, a machine either receives exactly one variable job, implying a makespan of 1, or only clause jobs, also implying a makespan of 1 as $\Gamma = 1$.

$\Leftarrow$. Suppose that $\text{OPT}(I) = 1$ and let us define an assignment a. This implies that any variable job j is either scheduled on machine j_f, in which case we set x_j to true, or on machine j_t, in which case we set x_j to false. As $\text{OPT}(I) = 1$, and clause job $j \in [n_0 + 1, n_0 + m_0]$ is scheduled on a machine $i \in M_{j-n_0}$ that did not receive a variable job, implying that clause $j - n_0$ is satisfied by literal i. □

4 A PTAS for Identical Machines

Note that we can assume that $m < n$. If $m \geq n$, a trivial schedule with every job on a different machine is optimal. In some problems the encoding length may be much smaller than m, when m is only encoded in binary. However, here a polynomial time algorithm is allowed to have a polynomial dependency on m.

Recall that for the $P|\mathcal{U}^\Gamma|C_{\max}$ problem, given two n dimensional vectors $\hat{p}$ and $\overline{p}$ and the number of machine m, the objective is to create a schedule σ that minimizes $\max_{i \in M} C_\Gamma(\sigma_i)$. Recall also that $C_\Gamma(\sigma_i) = \overline{p}(\sigma_i) + \hat{p}_\Gamma(\sigma_i)$, where $\overline{p}(\sigma_i) = \sum_{j \in \sigma_i} \overline{p}_j$, and $\hat{p}_\Gamma(\sigma_i)$ is the sum of the $\hat{p}_j$ values of the Γ largest jobs (w.r.t. $\hat{p}_j$) of σ_i (or the sum of all $\hat{p}_j$ values if $|\sigma_i| \leq \Gamma$). To obtain a PTAS for $P|\mathcal{U}^\Gamma|C_{\max}$, we will reduce to the following problem, which admits an EPTAS (see [8]).

Problem 2. Unrelated Machines with few Machine Types

- Input: n jobs and a set M of m machines with processing times $p_{ij} \geq 0$ for job j on machine i. Moreover, there is a constant k and machine types $T_1 \dot\cup \cdots \dot\cup T_k = \{1, \ldots, m\}$, such that every machine within a type behaves the same. Formally, for every k', every $i, i' \in T_{k'}$ and every $j \leq n$ it holds that $p_{ij} = p_{i'j}$

- Output: find a schedule $\sigma : J \rightarrow M$
- Objective function: minimize makespan $C(\sigma) = \max_{i \in M} C(\sigma_i)$, where $C(\sigma_i) = \sum_{j \in \sigma_i} p_{ij}$

Notice that the EPTAS of [8] for this problem provides an $(1+\epsilon)$-approximation running in time $f(|I|, \epsilon, k) = 2^{O(k \log(k) \frac{1}{\epsilon} \log^4(\frac{1}{\epsilon}))} + poly(|I|)$.

We also introduce the following decision problem.

Problem 3. UNRELATED MACHINES WITH FEW MACHINE TYPES AND CAPACITIES

- Input: as above, but in addition every machine i has a capacity $c_i \in (0, 1]$. Moreover, capacities are the same among a type (for any $k' \in [k]$, for any $i, i' \in T_{k'}$, $c_i = c_{i'}$)
- Output: decide if there is a schedule where $C(\sigma_i) \leq c_i$ for any i.

Notice that the EPTAS for Problem 2 allows to approximately decide Problem 3 in the following sense.

Lemma 1. *There is an algorithm that for any $\epsilon > 0$, either outputs a schedule with $C(\sigma_i) \leq (1+\epsilon) \cdot c_i$ for any i, or reject the instance, proving that there is no schedule with $C(\sigma_i) \leq c_i$ for any i. This algorithm runs in time $f(|I|, \epsilon, k)$ where f is the complexity of the above EPTAS to get a $(1+\epsilon)$-approximation.*

Proof. Let A be the EPTAS of [8] for Problem 2. Given a input I of Problem 3 we define an input I' of Problem 2 in the following way. For every $j \leq n$, scale p_{ij} to p_{ij}/c_i. Then, if $A(I') \leq (1+\epsilon)$, we can convert the solution found by A into a solution for I of makespan at most $(1+\epsilon) \cdot c_i$ for any i. Otherwise, as A is a $(1+\epsilon)$-approximation, it implies that $\text{OPT}(I') > 1$, and thus that no solution can have makespan at most c_i for any i. □

Let us now describe the PTAS for $P|\mathcal{U}^\Gamma|C_{\max}$. Our objective is to provide a $(1+O(\epsilon))$ dual approximation for $P|\mathcal{U}^\Gamma|C_{\max}$. The constant factor with ϵ can be ignored, since we can divide ϵ with this constant in the preprocessing.

1. Guess the makespan and scale OPT to 1. Let I be an input of $P|\mathcal{U}^\Gamma|C_{\max}$, and T be a positive value (representing the current threshold). We start by redefining I by scaling $p_j := \frac{p_j}{T}$. Our objective is now to produce a schedule σ with $C_\Gamma(\sigma) \leq 1+\epsilon$, or to prove that $\text{OPT}(I) > 1$.

2. Rounding deviations. Let us now define I^1 (having vectors $\overline{p}^1$ and $\hat{p}^1$) in the following way. For any j, if $\hat{p}_j < \epsilon/\Gamma$ then we set $\hat{p}^1_j \leftarrow 0$. Intuitively, this will only result in an error of at most $\Gamma \cdot \epsilon/\Gamma$ on every machine. Otherwise ($\hat{p}_j \geq \epsilon/\Gamma$), we define $\hat{p}^1_j$ by rounding $\hat{p}_j$ to the closest smaller value of the form $\epsilon/\Gamma \cdot (1+\epsilon)^i$.

Observation 1. *In I^1 there are at most $O(1/\epsilon \log(\Gamma/\epsilon))$ deviation values, and at most $O(1/\epsilon \log(1/\epsilon))$ deviation values in the interval $[\epsilon/\Gamma, 1/\Gamma]$.*

In the following, we will denote by $C_\Gamma^{I'}(\sigma)$ the cost of σ for instance I'.

Observation 2. *If* $\mathrm{OPT}(I) \leq 1$ *then* $\mathrm{OPT}(I^1) \leq 1$. *If we get solution* σ^1 *of* I^1, *then* $C_\Gamma^I(\sigma^1) \leq (1+\epsilon)C_\Gamma^{I^1}(\sigma^1) + \epsilon$

It only remains now to either produce a good solution of I^1 (of cost at most $1 + O(\epsilon)$), or prove that $\mathrm{OPT}(I^1) > 1$.

3. Machine thresholds. Given any solution σ of I^1 such that $C_\Gamma^{I^1}(\sigma) \leq 1$, we can associate to σ an outline $t = o(\sigma)$ which is defined as follows. For any machine i with more that Γ jobs, the threshold value t_i is such that any job on i with $\hat{p}_j > t_i$ deviates (belongs to $\Gamma(\sigma_i)$) and none of the jobs with $\hat{p}_j < t_i$ deviate. Notice that among jobs with $\hat{p}_j = t_i$, some may deviate, but not necessarily all. For any machine i with at most Γ jobs, we define $t_i = 0$, implying again that any job with $\hat{p}_j > t_i$ deviates on i. Notice that in both cases we have $\hat{p}_\Gamma(\sigma_i) \geq \Gamma \cdot t_i$. Notice also that $C_\Gamma^{I^1}(\sigma) \leq 1$ implies $t_i \leq \frac{1}{\Gamma}$. Indeed, if we had $t_i > \frac{1}{\Gamma}$, there would be Γ deviating jobs with $\hat{p}_j > t_i$, implying $C_\Gamma^{I^1}(\sigma_i) > 1$, a contradiction. Let us denote by Δ the set of all possible values of a t_i. According to Observation 1 we have $|\Delta| = O(1/\epsilon \log(1/\epsilon))$. Let $\mathcal{P} = \Delta^m$ be the set of all outlines (of solutions of cost at most 1).

Lemma 2. *Consider a solution* σ^{1*} *of* I^1 *such that* $C_\Gamma(\sigma^{1*}) \leq 1$, *and let* $t^* = o(\sigma^{1*})$. *Then, we can guess in* $m^{O(1/\epsilon \log(1/\epsilon))}$ *time the vector* t^* *(or a permutation thereof).*

Proof. As $t^* \in T$, all the t_i^* have a value in $\{0\} \cup [\frac{\epsilon}{\Gamma}, \frac{1}{\Gamma}]$. Thus, as deviating values are rounded in I^1, there are only a constant number of possible threshold value and we can guess them. For every possible threshold, we guess how many machines in the optimal solution have it. □

Thus, we can now assume that we know the vector t^*.

4. Constructing an instance with few machine types and capacities. To give an insight of the correct reduction defined below, let us first see what happen if we define an instance $I^2(t^*)$ of $R||C_{\max}$ as follows. For simplicity, we also assume that there are no job with $\hat{p}_j = t_i^*$ on each machine i in the previously considered optimal solution of I^1. For any machine i and job j, define the processing time in $I^2(t^*)$ as $p_{ij} = \overline{p}_j + \hat{p}_j$ if $\hat{p}_j \geq t_i^*$, and $p_{ij} = \overline{p}_j$ otherwise. Then, consider the following implications.

1. if $OPT(I^1) \leq 1$, then $\mathrm{OPT}(I^2(t^*)) \leq 1$
2. for any solution σ' of $I^2(t^*)$, $C_\Gamma^{I^1}(\sigma') \leq C^{I^2}(\sigma')$ (implying that if there exists σ' with $C^{I^2}(\sigma') \leq 1 + \epsilon$, then we will have our solution for I^1 of cost $1 + \epsilon$)

While Property (1) holds, this is not the case for Property (2). Indeed, suppose that in σ' there is a machine i such that for all jobs j scheduled on i, $\hat{p}_j < t_i^*$. This implies that $C(\sigma_i) = \sum_{j \in \sigma_i} \overline{p}_j$. However, if we look now at σ' in I^1, we get

$C_\Gamma^{I^1}(\sigma_i) = C^{I^2}(\sigma_i) + \hat{p}(\Gamma(\sigma_i))$, which is greater than the claimed value. To solve this problem we have to remember in $R||C_{\max}$ that there will be a space of size at most $\Gamma \cdot t_i$ which will be occupied by deviations.

Let us now turn to the correct version.

Definition 1. *For any $t \in \mathcal{P}$, we define the following input $I^2(t)$ of Problem 3. We set the machine capacity to*

$$c_i := 1 - \Gamma \cdot t_i + \epsilon.$$

The addition of ϵ is only a technicality to ensure that all c_i are non-zero. Note that if there are less than Γ jobs on i, then t_i must be 0 and therefore $c_i = 1 + \epsilon$. For every job j set

$$p_{ij} := \begin{cases} \overline{p}_j + \hat{p}_j - t_i & \text{if } \hat{p}_j \geq t_i, \\ \overline{p}_j & \text{if } \hat{p}_j < t_i. \end{cases}$$

Note that at $\hat{p}_j = t_i$, the values of both cases are equal. Notice also that in $I^2(t)$ there are only $|\Delta|$ different machine types.

Lemma 3. *If $\mathrm{OPT}(I^1) \leq 1$ and t is the outline of an optimal solution σ^2, for any i, $C^{I^2(t)}(\sigma_i^2) \leq c_i$.*

Proof. Let us consider jobs σ_i^2 scheduled on machine i. If $t_i = 0$, then

$$\sum_{j \in \sigma_i^2} p_{ij} = \sum_{j \in \sigma_i^2} \overline{p}_j + \hat{p}_j \leq 1 < c_i.$$

Assume now $t_i > 0$, implying that $|\Gamma(\sigma_i^2)| \geq \Gamma$. By choice of t_i, every job $j \in \Gamma(\sigma_i^2)$ has $\hat{p}_j \geq t_i$ and every $j \in \sigma_i^2 \setminus \Gamma(\sigma_i^2)$ has $\hat{p}_j \leq t_i$. This implies

$$\sum_{j \in \sigma_i^2} p_{ij} = \sum_{j \in \Gamma(\sigma_i^2)} p_{ij} + \sum_{j \in \sigma_i^2 \setminus \Gamma(\sigma_i^2)} p_{ij} = \sum_{j \in \Gamma(\sigma_i^2)} [\overline{p}_j + \hat{p}_j - t_i] + \sum_{j \in \sigma_i^2 \setminus \Gamma(\sigma_i^2)} \overline{p}_j \leq 1 \quad \Gamma \cdot t_i < c_i.$$

□

Lemma 4. *For any $t \in \mathcal{P}$, if there is a solution σ^2 of $I^2(t)$ such that $C^{I^2(t)}(\sigma_i^2) \leq (1 + \epsilon) \cdot c_i$ for any i, then $C_\Gamma^{I^1}(\sigma^2) \leq (1 + \epsilon)^2$.*

Proof. Let i be a machine. Then for every $j \in \sigma_i^2$,

$$p_{ij} = \begin{cases} \overline{p}_j + \hat{p}_j - t_i \geq \overline{p}_j & \text{if } \hat{p}_j \geq t_i, \\ \overline{p}_j & \text{if } \hat{p}_j < t_i. \end{cases}$$

Furthermore, for every $j \in \Gamma(\sigma_i^2)$,

$$p_{ij} = \begin{cases} \overline{p}_j + \hat{p}_j - t_i & \text{if } \hat{p}_j \geq t_i, \\ \overline{p}_j > \overline{p}_j + \hat{p}_j - t_i & \text{if } \hat{p}_j < t_i. \end{cases}$$

This implies,

$$\sum_{j\in\Gamma(\sigma_i^2)}[\overline{p}_j+\hat{p}_j]+\sum_{j\in\sigma_i^2\setminus\Gamma(\sigma_i^2)}\overline{p}_j\le\Gamma\cdot t_i+\sum_{j\in\Gamma(\sigma_i^2)}[\overline{p}_j+\hat{p}_j-t_i]+\sum_{j\in\sigma_i^2\setminus\Gamma(\sigma_i^2)}\overline{p}_j$$

$$\le\Gamma\cdot t_i+\sum_{j\in\Gamma(\sigma_i^2)}p_{ij}+\sum_{j\in\sigma_i^2\setminus\Gamma(\sigma_i^2)}p_{ij}=\Gamma\cdot t_i+\underbrace{\sum_{j\in\sigma_i^2}p_{ij}}_{\le(1+\epsilon)\cdot c_i}\le\Gamma\cdot t_i+(1+\epsilon)\cdot(1-\Gamma\cdot t_i+\epsilon)\le(1+\epsilon)^2.$$

□

Theorem 3. *There is a $(1+\epsilon)$-approximation algorithm for $P|\mathcal{U}^\Gamma|C_{\max}$ running in time $O(m^{O(1/\epsilon\log(1/\epsilon))}\times f(|I|,\epsilon,O(1/\epsilon\log(1/\epsilon)))$ where f is the function of Lemma 1.*

Proof. Given I input of $P|\mathcal{U}^\Gamma|C_{\max}$ and a threshold T, we run algorithm A of Lemma 1 on $I^2(t)$ for any $t\in\mathcal{P}$ with a precision ϵ. If A rejects all the $I^2(t)$ then we can reject T according to Observation 2 and Lemma 3. Otherwise, there exists t_0 such that $A(I^2(t_0))$ outputs a schedule σ^2 where $C^{I^2(t_0)}(\sigma^2)\le(1+\epsilon)\cdot c_i$ for any i, implying $C^I_\Gamma(\sigma^2)\le(1+\epsilon)C^{I^1}_\Gamma(\sigma^2)+\epsilon\le(1+\epsilon)^3+\epsilon\le1+5\epsilon$ according to Observation 2 and Lemma 4 (for sufficiently small ϵ). Finally, the running time is as claimed due to the bound of $\mathcal{P}$ in Lemma 2. □

5 EPTAS for Identical Machines

The approach for an EPTAS is similar to the PTAS above. We would like to remove the bottleneck from the previous section, which is the guessing the thresholds. In the PTAS we notice that even if the thresholds were chosen incorrectly, but we find a solution to the derived problem, we can get a good solution for the initial problem. Informally, we will now still create an instance of Problem 3, but we only guess approximately the number of machines for each threshold.

We start by defining I^1 as in the previous section. Given any solution σ^1 of I^1 such that $C^{I^1}_\Gamma(\sigma^1)\le1$, we can associate to σ a restricted outline $\overline{m}=\overline{o}(\sigma)$ where $\overline{m}$ is defined as follows. Let $t=o(\sigma)$. For any threshold value $l\in\Delta$, let $m_l=|\{i|t_i=l\}|$ be the number of machines with threshold l in σ^1. We define $\overline{m}_l\in\{0,1,2,4,8,\dots,2^{\lfloor\log(m)\rfloor}\}$ such that $\overline{m}_l\le m_l<2\overline{m}_l$. Let $\overline{\mathcal{P}}=\{\overline{m}\in\{0,1,2,4,8,\dots,2^{\lfloor\log(m)\rfloor}\}^\Delta$ such that $\frac{m}{2}\le\sum_l\overline{m}_l\le m\}$ be the set of restricted outlines (of solutions of cost at most 1).

Lemma 5. *Consider a solution σ^{1*} of I^1 such that $C_\Gamma(\sigma^{1*})\le1$, and let $\overline{m}^*=\overline{o}(\sigma^{1*})$. Then, we can guess in time $2^{O(1/\epsilon\log^2(1/\epsilon))}+m^{O(1)}$ the vector $\overline{m}^*$.*

Proof. Clearly it suffices to iterate over all values $\overline{m}_i\in\{0,1,2,4,8,\dots,2^{\lfloor\log(m)\rfloor}\}$, i.e., $O(\log(m))$ many. Guessing this number for every threshold value in Δ takes $\log^{O(1/\epsilon\log(1/\epsilon))}(m)$ time. Consider first the case when $\log(m)/\log\log(m)\le1/\epsilon\log(1/\epsilon)$. For sufficiently large m it holds that $\log^{1/2}(m)\le\log(m)/\log\log(m)\le1/\epsilon\log(1/\epsilon)$. Hence,

$$\log^{O(1/\epsilon\log(1/\epsilon))}(m) = (\log^{1/2}(m))^{2\cdot O(1/\epsilon\log(1/\epsilon))} \le (1/\epsilon\log(1/\epsilon))^{O(1/\epsilon\log(1/\epsilon))} \le 2^{O(1/\epsilon\log^2(1/\epsilon))}.$$

If on the other hand $\log(m)/\log\log(m) \ge 1/\epsilon\log(1/\epsilon)$, then

$$\log^{O(1/\epsilon\log(1/\epsilon))}(m) \le \log^{O(\log(m)/\log\log(m))}(m) = 2^{O(\log(m)/\log\log(m)\cdot\log\log(m))} = m^{O(1)}.$$

We conclude,

$$\log^{O(1/\epsilon\log(1/\epsilon))}(m) \le 2^{O(1/\epsilon\log^2(1/\epsilon))} + m^{O(1)}.$$

From all the guesses, we report fail whenever $\sum_i m_i < m/2$ or $\sum_i m_i > m$. □

For any $\overline{m} \in \overline{\mathcal{P}}$, we define the following input $I^2(\overline{m})$ of Problem 3. We first create for any l a set M_l of $\overline{m}_l$ machines where for each machine $i \in M_l$ the capacity and the p_{ij} are defined as in Definition 1 for threshold $t_i = l$. Then, we create another set M'_l of $\overline{m}_l$ machines (that we call cloned machines) with the same capacity and the same p_{ij} values. Let $m' = \sum \overline{m}_l$. Notice that the total number of machines is $2m'$, with $m \le 2m' < 2m$. Thus, we have to ensure that not too many machines are used in total. For that purpose we add a set of $2m' - m$ dummy jobs D, where all $j \in D$ have $p_{ij} = \infty$ on the original machines $i \in M_l$ and $p_{ij} = c_i$ on every cloned machine $i \in M'_l$. Notice that the number of types is now $2|\Delta|$, which is still small enough to get an EPTAS. Let us call the non-dummy jobs *regular* jobs.

Lemma 6. *If* $\mathrm{OPT}(I^1) \le 1$ *and* $\overline{m}$ *is the restricted outline of an optimal solution, then there exists a solution* σ^2 *of* $I^2(\overline{m})$ *such that for any* i, $C^{I^2(\overline{m})}(\sigma^2_i) \le c_i$.

Proof. Let $m^*_l = |\{i|t_i = l\}|$ be the number of machines with threshold l in the considered optimal solution of I^1. Let $l \in \Delta$ be a threshold value. We first schedule $2\overline{m}_l - m^*_l$ many dummy jobs on cloned machines of M'_l. This will cover all dummy jobs, since

$$\sum_l [2\overline{m}_l - m^*_l] = 2\sum_l \overline{m}_l - \sum_l m^*_l = 2m' - m.$$

We will now schedule all remaining jobs on the empty machines. For every threshold value l we have $2\overline{m}_l - (2\overline{m}_l - m^*_l) = m^*_l$ many empty machines. In other words, we are left with an instance with the exact same number of machines for each threshold as in the optimal solution and with the original jobs. As argued in Lemma 3, we get the desired claim. □

Lemma 7. *For any* $\overline{m} \in \overline{\mathcal{P}}$, *if there is a solution* σ^2 *of* $I^2(\overline{m})$ *such that* $C^{I^2(\overline{m})}(\sigma^2_i) \le c_i + \epsilon$ *for any* i, *then we can deduce a solution* σ^3 *for* I^1 *with* $C^{I^1}_{\Gamma}(\sigma^3) \le (1 + 2\epsilon)^2$.

Proof. We will first normalize σ^2. Since dummy jobs have $p_{ij} = c_i$ on cloned machines, in a $(1+\epsilon)$-approximation there can only be one per machine (assuming that $\epsilon < 1$). Indeed, there may still be a load of $\epsilon \cdot c_i$ from other jobs on the same machine. We want to ensure that every machine either has a dummy job or some regular load, but not both. For every threshold value $l \in \Delta$, there can be at most $\overline{m}_l$ machines in M'_l that have a dummy job. For any such machine in M'_l, we remove all the regular jobs (of total load of at most $\epsilon \cdot c_i$) from it and move them to one of the original machines in M_l, without using the same machine in M_l twice. Since for any $i \in M_l$ we had $C^{I^2(\overline{m})}(\sigma_i^2) \leq (1+\epsilon)c_i$ before moving the jobs, and since regular jobs have the same processing time on machines M_l and M'_l, after moving the jobs we get $C^{I^2(\overline{m})}(\sigma_i^2) \leq (1+2\epsilon)c_i$ for any $i \in M_l$. We now have a solution violating the capacities by at most $2\epsilon \cdot c_i$ such that a machine with a dummy job has no other jobs.

We now forget about all dummy jobs and the machines they are on. What we are left with is a set of m machines (with some thresholds t) such that for any i we have $C^{I^2(\overline{m})}(\sigma_i^2) \leq (1+2\epsilon)c_i$. By Lemma 4 we get the desired result. □

All in all, we were able to reduce the number of instances created to only $2^{O(1/\epsilon \log^2(1/\epsilon))} + m^{O(1)}$ many and removed the bottleneck from the PTAS this way. As in Theorem 3, given an instance of $P|\mathcal{U}^\Gamma|C_{\max}$ we will use the algorithm of Lemma 1 on $I^2(m)$ for any $m \in \overline{\mathcal{P}}$. This leads to the following result.

Theorem 4. *There is a $(1+\epsilon)$-approximation algorithm for $P|\mathcal{U}^\Gamma|C_{\max}$ running in time $O(2^{O(1/\epsilon \log^2(1/\epsilon))} + m^{O(1)}) \times f(|I|, \epsilon, O(1/\epsilon \log(1/\epsilon)))$ where f is the function of Lemma 1.*

References

1. Aloulou, M.A., Croce, F.D.: Complexity of single machine scheduling problems under scenario-based uncertainty. Oper. Res. Lett. **36**(3), 338–342 (2008)
2. Basu Roy, A., Bougeret, M., Goldberg, N., Poss, M.: Approximating robust bin packing with budgeted uncertainty. In: Friggstad, Z., Sack, J.-R., Salavatipour, M.R. (eds.) WADS 2019. LNCS, vol. 11646, pp. 71–84. Springer, Cham (2019). https://doi.org/10.1007/978-3-030-24766-9_6
3. Bertsimas, D., Sim, M.: Robust discrete optimization and network flows. Math. Program. **98**(1–3), 49–71 (2003)
4. Bertsimas, D., Sim, M.: The price of robustness. Oper. Res. **52**(1), 35–53 (2004)
5. Bougeret, M., Pessoa, A.A., Poss, M.: Robust scheduling with budgeted uncertainty. Discrete Appl. Math. **261**, 93–107 (2019)
6. Daniels, R.L., Kouvelis, P.: Robust scheduling to hedge against processing time uncertainty in single-stage production. Manag. Sci. **41**(2), 363–376 (1995)
7. Jansen, K., Klein, K., Verschae, J.: Closing the gap for makespan scheduling via sparsification techniques. In: 43rd International Colloquium on Automata, Languages, and Programming, ICALP 2016, 11–15 July 2016, Rome, Italy, pp. 72:1–72:13 (2016)
8. Jansen, K., Maack, M.: An EPTAS for scheduling on unrelated machines of few different types. Algorithms and Data Structures. LNCS, vol. 10389, pp. 497–508. Springer, Cham (2017). https://doi.org/10.1007/978-3-319-62127-2_42

9. Kasperski, A., Kurpisz, A., Zielinski, P.: Approximating a two-machine flow shop scheduling under discrete scenario uncertainty. Eur. J. Oper. Res. **217**(1), 36–43 (2012)
10. Kasperski, A., Kurpisz, A., Zieliński, P.: Parallel machine scheduling under uncertainty. In: Greco, S., Bouchon-Meunier, B., Coletti, G., Fedrizzi, M., Matarazzo, B., Yager, R.R. (eds.) IPMU 2012. CCIS, vol. 300, pp. 74–83. Springer, Heidelberg (2012). https://doi.org/10.1007/978-3-642-31724-8_9
11. Lenstra, J.K., Shmoys, D.B., Tardos, E.: Approximation algorithms for scheduling unrelated parallel machines. Math. Program. **46**(1–3), 259–271 (1990)
12. Mastrolilli, M., Mutsanas, N., Svensson, O.: Approximating single machine scheduling with scenarios. In: Goel, A., Jansen, K., Rolim, J.D.P., Rubinfeld, R. (eds.) APPROX/RANDOM -2008. LNCS, vol. 5171, pp. 153–164. Springer, Heidelberg (2008). https://doi.org/10.1007/978-3-540-85363-3_13
13. Tadayon, B., Smith, J.C.: Algorithms and complexity analysis for robust single-machine scheduling problems. J. Scheduling **18**(6), 575–592 (2015)

Streaming Algorithms for Bin Packing and Vector Scheduling

Graham Cormode and Pavel Veselý[(✉)]

Department of Computer Science, University of Warwick, Coventry, UK
{G.Cormode,Pavel.Vesely}@warwick.ac.uk

Abstract. Problems involving the efficient arrangement of simple objects, as captured by bin packing and makespan scheduling, are fundamental tasks in combinatorial optimization. These are well understood in the traditional online and offline cases, but have been less well-studied when the volume of the input is truly massive, and cannot even be read into memory. This is captured by the streaming model of computation, where the aim is to approximate the cost of the solution in one pass over the data, using small space. As a result, streaming algorithms produce concise input summaries that approximately preserve the optimum value.

We design the first efficient streaming algorithms for these fundamental problems in combinatorial optimization. For Bin Packing, we provide a streaming asymptotic $1+\varepsilon$-approximation with $\widetilde{\mathcal{O}}\left(\frac{1}{\varepsilon}\right)$ memory, where $\widetilde{\mathcal{O}}$ hides logarithmic factors. Moreover, such a space bound is essentially optimal. Our algorithm implies a streaming $d+\varepsilon$-approximation for Vector Bin Packing in d dimensions, running in space $\widetilde{\mathcal{O}}\left(\frac{d}{\varepsilon}\right)$. For the related Vector Scheduling problem, we show how to construct an input summary in space $\widetilde{\mathcal{O}}(d^2 \cdot m/\varepsilon^2)$ that preserves the optimum value up to a factor of $2-\frac{1}{m}+\varepsilon$, where m is the number of identical machines.

Keywords: Streaming algorithms · Bin Packing · Scheduling

1 Introduction

The streaming model captures many scenarios when we must process very large volumes of data, which cannot fit into the working memory. The algorithm makes one or more passes over the data with a limited memory, but does not have random access to the data. Thus, it needs to extract a concise summary of the huge input, which can be used to approximately answer the problem under consideration. The main aim is to provide a good trade-off between the space used for processing the input stream (and hence, the summary size) and the accuracy of the (best possible) answer computed from the summary. Other relevant parameters are the time and space needed to make the estimate, and the number of passes, ideally equal to one.

The work is supported by European Research Council grant ERC-2014-CoG 647557.

E. Bampis and N. Megow (Eds.): WAOA 2019, LNCS 11926, pp. 72–88, 2020.
https://doi.org/10.1007/978-3-030-39479-0_6

While there have been many effective streaming algorithms designed for a range of problems in statistics, optimization, and graph algorithms (see surveys by Muthukrishnan [37] and McGregor [36]), there has been little attention paid to the core problems of packing and scheduling. These are fundamental abstractions, which form the basis of many generalizations and extensions [13,14]. In this work, we present the first efficient algorithms for packing and scheduling that work in the streaming model.

A first conceptual challenge is to resolve what form of answer is desirable in this setting. If items in the input are too many to store, then it is also unfeasible to require a streaming algorithm to provide an explicit description of how each item is to be handled. Rather, our objective is for the algorithm to provide the cost of the solution, in the form of the number of bins or the duration of the schedule. Moreover, many of our algorithms can provide a concise *description* of the solution, which describes in outline how the jobs are treated in the design.

A second issue is that the problems we consider, even in their simplest form, are NP-hard. The additional constraints of streaming computation do not erase the computational challenge. In some cases, our algorithms proceed by adopting and extending known polynomial-time approximation schemes for the offline versions of the problems, while in other cases, we come up with new approaches. The streaming model effectively emphasizes the question of how compactly can the input be summarized to allow subsequent approximation of the problem of interest. Our main results show that in fact the inputs for many of our problems of interest can be "compressed" to very small intermediate descriptions which suffice to extract near-optimal solutions for the original input. This implies that they can be solved in scenarios which are storage or communication constrained.

We proceed by formalizing the streaming model, after which we summarize our results. We continue by presenting related work, and contrast with the online setting.

1.1 Problems and Streaming Model

Bin Packing. The BIN PACKING problem is defined as follows: The input consists of N items with sizes $s_1, \ldots, s_N$ (each between 0 and 1), which need to be packed into bins of unit capacity. That is, we seek a partition of the set of items $\{1, \ldots, N\}$ into subsets $B_1, \ldots, B_m$, called bins, such that for any bin B_i, it holds that $\sum_{j \in B_i} s_j \leq 1$. The goal is to minimize the number m of bins used.

We also consider the natural generalization to VECTOR BIN PACKING, where the input consists of d-dimensional vectors, with the value of each coordinate between 0 and 1 (i.e., the scalar items s_i are replaced with vectors $\mathbf{v^i}$). The vectors need to be packed into d-dimensional bins with unit capacity in each dimension, we thus require that $\| \sum_{\mathbf{v} \in B_i} \mathbf{v} \|_\infty \leq 1$ (where the infinity norm $\|\mathbf{v}\|_\infty = \max_i \mathbf{v}_i$).

Scheduling. The MAKESPAN SCHEDULING problem is closely related to BIN PACKING but, instead of filling bins with bounded capacity, we try to balance the loads assigned to a fixed number of bins. Now we refer to the input as comprising a set of *jobs*, with each job j defined by its processing time p_j. Our goal is to assign each job on one of m identical machines to minimize the *makespan*, which is the maximum load over all machines.

In Vector Scheduling, a job is described not only by its processing time, but also by, say, memory or bandwidth requirements. The input is thus a set of jobs, each job j characterized by a vector $\mathbf{v^j}$. The goal is to assign each job into one of m identical machines such that the maximum load over all machines and dimensions is minimized.

Streaming Model. In the streaming scenario, the algorithm receives the input as a sequence of items, called the input stream. We do not assume that the stream is ordered in any particular way (e.g., randomly or by item sizes), so our algorithms must work for arbitrarily ordered streams. The items arrive one by one and upon receiving each item, the algorithm updates its memory state. A streaming algorithm is required to use space sublinear in the length of the stream, ideally just polylog(N), while it processes the stream. After the last item arrives, the algorithm computes its estimate of the optimal value, and the space or time used during this final computation is not restricted.

For many natural optimization problems outputting some explicit solution of the problem is not possible owing to the memory restriction (as the algorithm can store only a small subset of items). Thus the goal is to find a good approximation of the *value* of an offline optimal solution. Since our model does not assume that item sizes are integers, we express the space complexity not in bits, but in words (or memory cells), where each word can store any number from the input; a linear combination of numbers from the input; or any integer with $\mathcal{O}(\log N)$ bits (for counters, pointers, etc.).

1.2 Our Results

Bin Packing. In Sect. 3, we present a streaming algorithm for Bin Packing, which outputs an asymptotic $1+\varepsilon$-approximation of OPT, the optimal number of bins, using $\mathcal{O}\left(\frac{1}{\varepsilon}\cdot\log\frac{1}{\varepsilon}\cdot\log\mathsf{OPT}\right)$ memory.[1] This means that the algorithm uses at most $(1+\varepsilon)\cdot\mathsf{OPT}+o(\mathsf{OPT})$ bins, and in our case, the additive $o(\mathsf{OPT})$ term is bounded by the space used. The novelty of our contribution is to combine a data structure that approximately tracks all quantiles in a numeric stream [25] with techniques for approximation schemes [17,32]. We show that we can improve upon the $\log\mathsf{OPT}$ factor in the space complexity if randomization is allowed or if item sizes are drawn from a bounded-size set of real numbers. On the other hand, we argue that our result is close to optimal, up to a factor of $\mathcal{O}\left(\log\frac{1}{\varepsilon}\right)$, if item sizes are accessed only by comparisons (including comparisons with some fixed constants). Thus, one cannot get an estimate with at most $\mathsf{OPT}+o(\mathsf{OPT})$ bins by a streaming algorithm, unlike in the offline setting [27]. The hardness

[1] We remark that some online algorithms can be implemented in the streaming model, as described in Sect. 2.1, but they give worse approximation guarantees.

emerges from the space complexity of the quantiles problem in the streaming model.

For VECTOR BIN PACKING, we design a streaming asymptotic $d + \varepsilon$-approximation algorithm running in space $\mathcal{O}\left(\frac{d}{\varepsilon} \cdot \log \frac{d}{\varepsilon} \cdot \log \mathsf{OPT}\right)$. This is done by a reduction to the 1-dimensional case and using the aforementioned streaming algorithm; the details are deferred to the full version of the paper. We remark that if vectors are rounded into a sublinear number of types, then better than d-approximation is not possible [7].

Scheduling. For MAKESPAN SCHEDULING, one can obtain a straightforward streaming $1+\varepsilon$-approximation[2] with space of only $\mathcal{O}(\frac{1}{\varepsilon} \cdot \log \frac{1}{\varepsilon})$ by rounding sizes of suitably large jobs to powers of $1+\varepsilon$ and counting the total size of small jobs. In a higher dimension, it is also possible to get a streaming $1+\varepsilon$-approximation, by the rounding introduced by Bansal *et al.* [8]. However, the memory required for this algorithm is exponential in d, precisely of size $\mathcal{O}\left(\left(\frac{1}{\varepsilon} \log \frac{d}{\varepsilon}\right)^d\right)$, and thus only practical when d is a very small constant. Moreover, such a huge amount of memory is needed even if the number m of machines (and hence, of big jobs) is small as the algorithm rounds small jobs into exponentially many types.

In case m and d make this feasible, we design a new streaming $\left(2 - \frac{1}{m} + \varepsilon\right)$-approximation with $\mathcal{O}\left(\frac{1}{\varepsilon^2} \cdot d^2 \cdot m \cdot \log \frac{d}{\varepsilon}\right)$ memory, which implies a 2-approximation streaming algorithm running in space $\mathcal{O}(d^2 \cdot m^3 \cdot \log dm)$. We thus obtain a much better approximation than for VECTOR BIN PACKING with a reasonable amount of memory (although to compute the actual makespan from our input summary, it takes time doubly exponential in d [8]). Our algorithm is not based on rounding, as in the aforementioned algorithms, but on combining small jobs into containers, and the approximation guarantee of this approach is at least $2 - \frac{1}{m}$. We describe the algorithm in Sect. 4.

2 Related Work

We give an overview of related work in offline, online, and sublinear algorithms, and highlight the differences between online and streaming algorithms. Recent surveys of Christensen *et al.* [13] and Coffman *et al.* [14] have a more comprehensive overview.

2.1 Bin Packing

Offline Approximation Algorithms. BIN PACKING is an NP-complete problem and indeed it is NP-hard even to decide whether two bins are sufficient or at least three bins are necessary. This follows by a simple reduction from the PARTITION problem and presents the strongest inapproximability to date. Most work in

[2] Unlike for BIN PACKING, an additive constant or even an additive $o(\mathsf{OPT})$ term does not help in the definition of the approximation ratio, since we can scale every number on input by any $\alpha > 0$ and OPT scales by α as well.

the offline model focused on providing *asymptotic* R-approximation algorithms, which use at most $R \cdot \mathsf{OPT} + o(\mathsf{OPT})$ bins. In the following, when we refer to an approximation for BIN PACKING we implicitly mean the asymptotic approximation. The first *polynomial-time approximation scheme* (PTAS), that is, a $1+\varepsilon$-approximation for any $\varepsilon > 0$, was given by Fernandez de la Vega and Lueker [17]. Karmarkar and Karp [32] provided an algorithm which returns a solution with $\mathsf{OPT}+\mathcal{O}(\log^2 \mathsf{OPT})$ bins. Recently, Hoberg and Rothvoß [27] proved it is possible to find a solution with $\mathsf{OPT} + \mathcal{O}(\log \mathsf{OPT})$ bins in polynomial time.

The input for BIN PACKING can be described by N numbers, corresponding to item sizes. While in general these sizes may be distinct, in some cases the input description can be compressed significantly by specifying the number of items of each size in the input. Namely, in the HIGH-MULTIPLICITY BIN PACKING problem, the input is a set of pairs $(a_1, s_1), \dots, (a_\sigma, s_\sigma)$, where for $i = 1, \dots, \sigma$, a_i is the number of items of size s_i (and all s_i's are distinct). Thus, σ encodes the number of item sizes, and hence the size of the description. The goal is again to pack these items into bins, using as few bins as possible. For constant number of sizes, σ, Goemans and Rothvoß [23] recently gave an exact algorithm for the case of rational item sizes running in time $(\log \Delta)^{2^{\mathcal{O}(\sigma)}}$, where Δ is the largest multiplicity of an item or the largest denominator of an item size, whichever is the greater.

While these algorithms provide satisfying theoretical guarantees, simple heuristics are often adopted in practice to provide a "good-enough" performance. FIRST FIT [31], which puts each incoming item into the first bin where it fits and opens a new bin only when the item does not fit anywhere else achieves 1.7-approximation [16]. For the high-multiplicity variant, using an LP-based Gilmore-Gomory cutting stock heuristic [21,22] gives a good running time in practice [2] and produces a solution with at most $\mathsf{OPT}+\sigma$ bins. However, neither of these algorithms adapts well to the streaming setting with possibly distinct item sizes. For example, FIRST FIT has to remember the remaining capacity of each open bin, which in general can require space proportional to OPT.

VECTOR BIN PACKING proves to be substantially harder to approximate, even in a constant dimension. For fixed d, Bansal, Eliáš, and Khan [7] showed an approximation factor of $\approx 0.807+\ln(d+1)+\varepsilon$. For general d, a relatively simple algorithm based on an LP relaxation, due to Chekuri and Khanna [11], remains the best known, with an approximation guarantee of $1 + \varepsilon d + \mathcal{O}(\log \frac{1}{\varepsilon})$. The problem is APX-hard even for $d = 2$ [39], and cannot be approximated within a factor better than $d^{1-\varepsilon}$ for any fixed $\varepsilon > 0$ [13] if d is arbitrarily large. Hence, our streaming $d+\varepsilon$-approximation for VECTOR BIN PACKING asymptotically achieves the offline lower bound.

Sampling-Based Algorithms. Sublinear-time approximation schemes constitute a model related to, but distinct from, streaming algorithms. Batu, Berenbrink, and Sohler [9] provide an algorithm that takes $\widetilde{\mathcal{O}}\left(\sqrt{N} \cdot \mathrm{poly}(\frac{1}{\varepsilon})\right)$ weighted samples, meaning that the probability of sampling an item is proportional to its size. It outputs an asymptotic $1+\varepsilon$-approximation of OPT. If uniform samples are also available, then sampling $\widetilde{\mathcal{O}}\left(N^{1/3} \cdot \text{poly}(\frac{1}{\varepsilon})\right)$ items is sufficient. These results are tight, up to a poly$(\frac{1}{\varepsilon}, \log N)$ factor. Later, Beigel and Fu [10] focused on uniform sampling of items, proving that $\widetilde{\Theta}(N/\mathsf{SIZE})$ samples are sufficient and necessary, where SIZE is the total size of all items. Their approach implies a streaming approximation scheme by uniform sampling of the substream of big items. However, the space complexity in terms of $\frac{1}{\varepsilon}$ is not stated in the paper, but we calculate this to be $\Omega\left(\varepsilon^{-c}\right)$ for a constant $c \geq 10$. Moreover, $\Omega(\frac{1}{\varepsilon^2})$ samples are clearly needed to estimate the number of items with size close to 1. Note that our approach is deterministic and substantially different than taking a random sample from the stream.

Online Algorithms. Online and streaming algorithms are similar in the sense that they are required to process items one by one. However, an online algorithm must make all its decisions immediately—it must fix the placement of each incoming item on arrival. A streaming algorithm can postpone such decisions to the very end, but is required to keep its memory small, whereas an online algorithm may remember all items that have arrived so far. Hence, online algorithms apply in the streaming setting only when they have small space cost, including the space needed to store the solution constructed so far. The approximation ratio of online algorithms is quantified by the *competitive ratio.*

For Bin Packing, the best possible competitive ratio is substantially worse than what we can achieve offline or even in the streaming setting. Balogh *et al.* [5] designed an asymptotically 1.5783-competitive algorithm, while the current lower bound on the asymptotic competitive ratio is 1.5403 [6]. This (relatively complicated) online algorithm is based on the Harmonic algorithm [34], which for some integer K classifies items into size groups $(0, \frac{1}{K}], (\frac{1}{K}, \frac{1}{K-1}], \dots, (\frac{1}{2}, 1]$. It packs each group separately by Next Fit, keeping just one bin open, which is closed whenever the next item does not fit. Thus Harmonic can run in memory of size K and be implemented in the streaming model, unlike most other online algorithms which require maintaining the levels of all bins opened so far. Its competitive ratio tends to approximately 1.691 as K goes to infinity. Surprisingly, this is also the best possible ratio if only a bounded number of bins is allowed to be open for an online algorithm [34], which can be seen as the intersection of online and streaming models.

For Vector Bin Packing, the best known competitive ratio of $d + 0.7$ [19] is achieved by First Fit. A lower bound of $\Omega(d^{1-\varepsilon})$ on the competitive ratio was shown by Azar *et al.* [3]. It is thus currently unknown whether or not online algorithms outperform streaming algorithms in the vector setting.

2.2 Scheduling

Offline Approximation Algorithms. Makespan Scheduling is strongly NP-complete [20], which in particular rules out the possibility of a PTAS with time complexity poly$(\frac{1}{\varepsilon}, n)$. After a sequence of improvements, Jansen, Klein, and Verschae [30] gave a PTAS with time complexity $2^{\widetilde{\mathcal{O}}(1/\varepsilon)} + \mathcal{O}(n \log n)$, which is essentially tight under the Exponential Time Hypothesis (ETH) [12].

For constant dimension d, VECTOR SCHEDULING also admits a PTAS, as shown by Chekuri and Khanna [11]. However, the running time is of order $n^{(1/\varepsilon)^{\tilde{\mathcal{O}}(d)}}$. The approximation scheme for a fixed d was improved to an efficient PTAS, namely to an algorithm running in time $2^{(1/\varepsilon)^{\tilde{\mathcal{O}}(d)}} + \mathcal{O}(dn)$, by Bansal *et al.* [8], who also showed that the running time cannot be significantly improved under ETH. In contrast our streaming poly(d, m)-space algorithm computes an input summary maintaining 2-approximation of the original input. This respects the lower bound, since to compute the actual makespan from the summary, we still need to execute an offline algorithm, with running time doubly exponential in d. The state-of-the-art approximation ratio for large d is $\mathcal{O}(\log d/(\log\log d))$ [26,29], while α-approximation is not possible in polynomial time for any constant $\alpha > 1$ and arbitrary d, unless NP = ZPP.

Online Algorithms. For the scalar problem, the optimal competitive ratio is known to lie in the interval (1.88, 1.9201) [1,18,24,28], which is substantially worse than what can be done by a simple streaming $1 + \varepsilon$-approximation in space $\mathcal{O}(\frac{1}{\varepsilon} \cdot \log \frac{1}{\varepsilon})$. Interestingly, for VECTOR SCHEDULING, the algorithm by Im *et al.* [29] with ratio $\mathcal{O}(\log d/(\log\log d))$ actually works in the online setting as well and needs space $\mathcal{O}(d \cdot m)$ only during its execution (if the solution itself is not stored), which makes it possible to implement it in the streaming setting. This online ratio cannot be improved as there is a lower bound of $\Omega(\log d/(\log\log d))$ [4,29], whereas in the streaming setting we can achieve a 2-approximation with a reasonable memory (or even $1 + \varepsilon$-approximation for a fixed d). If all jobs have sufficiently small size, we improve the analysis in [29] and show that the online algorithm achieves $1 + \varepsilon$-approximation; see Sect. 4.

3 Bin Packing

Notation. For an instance I, let $N(I)$ be the number of items in I, let $\mathsf{SIZE}(I)$ be the total size of all items in I, and let $\mathsf{OPT}(I)$ be the number of bins used in an optimal solution for I. Clearly, $\mathsf{SIZE}(I) \le \mathsf{OPT}(I)$. For a bin B, let $s(B)$ be the total size of items in B. For a given $\varepsilon > 0$, we use $\tilde{\mathcal{O}}(f(\frac{1}{\varepsilon}))$ to hide factors logarithmic in $\frac{1}{\varepsilon}$ and $\mathsf{OPT}(I)$, i.e., to denote $\mathcal{O}\big(f(\frac{1}{\varepsilon}) \cdot \text{polylog}\,\frac{1}{\varepsilon} \cdot \text{polylog}\,\mathsf{OPT}(I)\big)$.

Overview. We first briefly describe the approximation scheme of Fernandez de la Vega and Lueker [17], whose structure we follow in outline. Let I be an instance of BIN PACKING. Given a precision requirement $\varepsilon > 0$, we say that an item is *small* if its size is at most ε; otherwise, it is *big*. Note that there are at most $\frac{1}{\varepsilon}\mathsf{SIZE}(I)$ big items. The rounding scheme in [17], called "linear grouping", works as follows: We sort the big items by size non-increasingly and divide them into groups of $k = \lfloor \varepsilon \cdot \mathsf{SIZE}(I) \rfloor$ items (the first group thus contains the k biggest items). In each group, we round up the sizes of all items to the size of the biggest item in that group. It follows that the number of groups and thus the number of distinct item sizes (after rounding) is bounded by $\lceil \frac{1}{\varepsilon^2} \rceil$. Let I_{R} be the instance of HIGH-MULTIPLICITY BIN PACKING consisting of the big items with rounded sizes. It can be shown that $\mathsf{OPT}(I_B) \leq \mathsf{OPT}(I_R) \leq (1+\varepsilon) \cdot \mathsf{OPT}(I_B)$, where I_B is the set of big items in I (we detail a similar argument in Sect. 3.1). Due to the bounded number of distinct item sizes, we can find a close-to-optimal solution for I_R efficiently. We then translate this solution into a packing for I_B in the natural way. Finally, small items are filled greedily (e.g., by First Fit) and it can be shown that the resulting complete solution for I is a $1+\mathcal{O}(\varepsilon)$-approximation.

Karmarkar and Karp [32] proposed an improved rounding scheme, called "geometric grouping". It is based on the observation that item sizes close to 1 should be approximated substantially better than item sizes close to ε. We present a version of such a rounding scheme in Sect. 3.1.

Our algorithm follows a similar outline with two stages (rounding and finding a solution for the rounded instance), but working in the streaming model brings two challenges: First, in the rounding stage, we need to process the stream of items and output a rounded high-multiplicity instance with few item sizes that are not too small, while keeping only a small number of items in the memory. Second, the rounding of big items needs to be done carefully so that not much space is "wasted", since in the case when the total size of small items is relatively large, we argue that our solution is close to optimal by showing that the bins are nearly full on average.

Input Summary Properties. More precisely, we fix some $\varepsilon > 0$ that is used to control the approximation guarantee. During the first stage, our algorithm has one variable which accumulates the total size of all small items in the input stream, i.e., those of size at most ε. Let I_B be the substream consisting of all big items. We process I_B and output a rounded high-multiplicity instance I_R with the following properties:

(P1) There are at most σ item sizes in I_R, all of them larger than ε, and the memory required for processing I_B is $\mathcal{O}(\sigma)$.
(P2) The i-th biggest item in I_R is at least as large as the i-th biggest item in I_B (and the number of items in I_R is the same as in I_B). This immediately implies that any packing of I_R can be used as a packing of I_B (in the same number of bins), so $\mathsf{OPT}(I_B) \leq \mathsf{OPT}(I_R)$, and moreover, $\mathsf{SIZE}(I_B) \leq \mathsf{SIZE}(I_R)$.
(P3) $\mathsf{OPT}(I_R) \leq (1+\varepsilon) \cdot \mathsf{OPT}(I_B) + \mathcal{O}(\log \frac{1}{\varepsilon})$.
(P4) $\mathsf{SIZE}(I_R) \leq (1+\varepsilon) \cdot \mathsf{SIZE}(I_B)$.

In words, (P2) means that we are rounding item sizes up and, together with (P3), it implies that the optimal solution for the rounded instance approximates $\mathsf{OPT}(I_B)$ well. The last property is used in the case when the total size of small items constitutes a large fraction of the total size of all items. Note that $\mathsf{SIZE}(I_R) - \mathsf{SIZE}(I_B)$ can be thought of as bin space "wasted" by rounding.

Observe that the succinctness of the rounded instance depends on σ. First, we show a streaming algorithm for rounding with $\sigma = \widetilde{\mathcal{O}}(\frac{1}{\varepsilon^2})$. Then we improve upon it and give an algorithm with $\sigma = \widetilde{\mathcal{O}}(\frac{1}{\varepsilon})$, which is essentially the best possible, while guaranteeing an error of $\varepsilon \cdot \mathsf{OPT}(I_B)$ introduced by rounding (elaborated on in Sect. 3.2). More precisely, we show the following:

Lemma 1. *Given a steam I_B of big items, there is a deterministic streaming algorithm that outputs a* High-Multiplicity Bin Packing *instance satisfying (P1)–(P4) with* $\sigma = \mathcal{O}\left(\frac{1}{\varepsilon} \cdot \log \frac{1}{\varepsilon} \cdot \log OPT(I_B)\right)$.

Before describing the rounding itself, we explain how to use it to calculate an accurate estimate of the number of bins.

Calculating a Bound on the Number of Bins After Rounding. First, we obtain a solution $\mathcal{S}$ of the rounded instance I_R. For instance, we may round the solution of the linear program introduced by Gilmore and Gomory [21,22], and get a solution with at most $\mathsf{OPT}(I_R) + \sigma$ bins. Or, if item sizes are rational numbers, we may compute an optimal solution for I_R by the algorithm of Goemans and Rothvoß [23]; however, the former approach appears to be more efficient and more general. In the following, we thus assume that $\mathcal{S}$ uses at most $\mathsf{OPT}(I_R) + \sigma$ bins.

We now calculate a bound on the number of bins in the original instance. Let W be the total free space in the bins of $\mathcal{S}$ that can be used for small items. To be precise, W equals the sum over all bins B in $\mathcal{S}$ of $\max(0, 1 - \varepsilon - s(B))$. Note that the capacity of bins is capped at $1 - \varepsilon$, because it may happen that all small items are of size ε while the packing leaves space of just under ε in any bin. Then we would not be able to pack small items into these bins. Reducing the capacity by ε removes this issue. On the other hand, if a small item does not fit into a bin, then the remaining space in the bin is smaller than ε.

Let s be the total size of all small items in the input stream. If $s \leq W$, then all small items surely fit into the free space of bins in $\mathcal{S}$ (and can be assigned there greedily by First Fit). Consequently, we output that the number of bins needed for the stream of items is at most $|\mathcal{S}|$, i.e., the number of bins in solution $\mathcal{S}$ for I_R. Otherwise, we need to place small items of total size at most $s' = s - W$ into new bins and it is easy to see that opening at most $\lceil s'/(1-\varepsilon) \rceil \leq (1+\mathcal{O}(\varepsilon)) \cdot s' + 1$ bins for these small items suffices. Hence, in the case $s > W$, we output that $|\mathcal{S}| + \lceil s'/(1-\varepsilon) \rceil$ bins are sufficient to pack all items in the stream.

It holds that the number of bins that we output in either case is a good approximation of the optimal number of bins, provided that $\mathcal{S}$ is a good solution for I_R. The proof is deferred to the full version of the paper.

Lemma 2. *Let I be given as a stream of items. Suppose that $0 < \varepsilon \leq \frac{1}{3}$, that the rounded instance I_R, created from I, satisfies properties (P1)–(P4), and that the solution $\mathcal{S}$ of I_R uses at most $OPT(I_R) + \sigma$ bins. Let $ALG(I)$ be the number of bins that our algorithm outputs. Then, it holds that* $OPT(I) \leq ALG(I) \leq (1 + 3\varepsilon) \cdot OPT(I) + \sigma + \mathcal{O}\left(\log \frac{1}{\varepsilon}\right)$.

3.1 Processing the Stream and Rounding

The streaming algorithm of the rounding stage makes use of the deterministic quantile summary of Greenwald and Khanna [25]. Given a precision $\delta > 0$ and an input stream of numbers $s_1, \ldots, s_N$, their algorithm computes a data structure

$Q(\delta)$ which is able to answer a quantile query with precision δN. Namely, for any $0 \leq \phi \leq 1$, it returns an element s of the input stream such that the rank of s is $[(\phi-\delta)N, (\phi+\delta)N]$, where the rank of s is the position of s in the non-increasing ordering of the input stream.[3] The data structure stores an ordered sequence of tuples, each consisting of an input number s_i and valid lower and upper bounds on the true rank of s_i in the input sequence.[4] The first and last stored items correspond to the maximum and minimum numbers in the stream, respectively. Note that the lower and upper bounds on the rank of any stored number differ by at most $\lfloor 2\delta N \rfloor$ and upper (or lower) bounds on the rank of two consecutive stored numbers differ by at most $\lfloor 2\delta N \rfloor$ as well. The space requirement of $Q(\delta)$ is $\mathcal{O}(\frac{1}{\delta} \cdot \log \delta N)$, however, in practice the space used is observed to scale linearly with $\frac{1}{\delta}$ [35]. (Note that an offline optimal data structure for δ-approximate quantiles uses space $\mathcal{O}\left(\frac{1}{\delta}\right)$.) We use data structure $Q(\delta)$ to construct our algorithm for processing the stream I_{B} of big items.

Simple Rounding Algorithm. We begin by describing a simpler solution with $\delta = \frac{1}{4}\varepsilon^2$, resulting in a rounded instance with $\widetilde{\mathcal{O}}(\frac{1}{\varepsilon^2})$ item sizes. Subsequently, we introduce a more involved solution with smaller space cost. The algorithm uses a quantile summary structure to determine the rounding scheme. Given a (big) item s_i from the input, we insert it into $Q(\delta)$. After processing all items, we extract from $Q(\delta)$ the set of stored input items (i.e., their sizes) together with upper bounds on their rank (where the largest size has highest rank 1, and the smallest size has least rank N_{B}). Note that the number N_{B} of big items in I_{B} is less than $\frac{1}{\varepsilon}\mathsf{SIZE}(I_{\mathrm{B}}) \leq \frac{1}{\varepsilon}\mathsf{OPT}(I_{\mathrm{B}})$ as each is of size more than ε. Let q be the number of items (or tuples) extracted from $Q(\delta)$; we get that $q = \mathcal{O}(\frac{1}{\delta} \cdot \log \delta N_{\mathrm{B}}) = \mathcal{O}(\frac{1}{\varepsilon^2} \cdot \log(\varepsilon \cdot \mathsf{OPT}(I_{\mathrm{B}})))$. Let $(a_1, u_1 = 1), (a_2, u_2), \ldots, (a_q, u_q = N_{\mathrm{B}})$ be the output pairs of an item size and the bound on its rank, sorted so that $a_1 \geq a_2 \geq \cdots \geq a_q$. We define the rounded instance I_{R} with at most q item sizes as follows: I_{R} contains $(u_{j+1} - u_j)$ items of size a_j for each $j = 1, \ldots, q-1$, plus one item of size a_q. (See Fig. 1.)

We show that the desired properties (P1)–(P4) hold with $\sigma = q$. Property (P1) follows easily from the definition of I_{R} and the design of data structure $Q(\delta)$. Note that the number of items is preserved. To show (P2), suppose for a contradiction that the i-th biggest item in I_{B} is bigger than the i-th biggest item in I_{R}, whose size is a_j for $j = 1, \ldots, q-1$, i.e., $i \in [u_j, u_{j+1})$ (note that $j < q$ as a_q is the smallest item in I_{B} and is present only once in I_{R}). We get that the rank of item a_j in I_{B} is strictly more than i, and as $i \geq u_j$, we get a contradiction with the fact that u_j is a valid upper bound on the rank of a_j in I_{B}.

Next, we give bounds for $\mathsf{OPT}(I_{\mathrm{R}})$ and $\mathsf{SIZE}(I_{\mathrm{R}})$, which are required by properties (P3) and (P4). We pack the $\lfloor 4\delta N_{\mathrm{B}} \rfloor$ biggest items in I_{R} separately into

[3] Note that if s appears more times in the stream, its rank is an interval rather than a single number. Also, unlike in [25], we order numbers non-increasingly, which is more convenient for Bin Packing.

[4] More precisely, valid lower and upper bounds on the rank of s_i can be computed easily from the set of tuples.

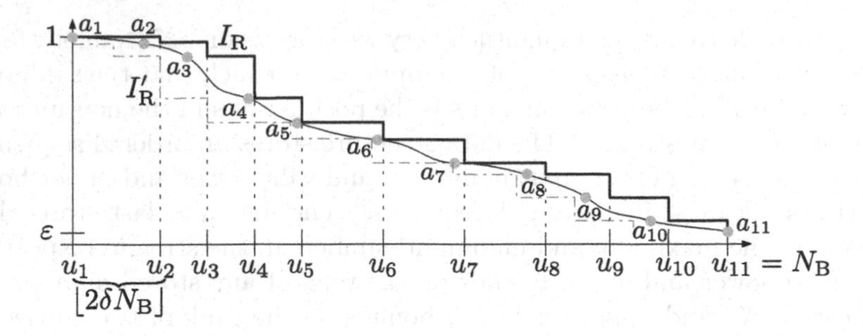

Fig. 1. An illustration of the original distribution of sizes of big items in I_B, depicted by a smooth curve, and the distribution of item sizes in the rounded instance I_R, depicted by a bold "staircase" function. The distribution of I'_R (which is I_R without the $\lfloor 4\delta N_B \rfloor$ biggest items) is depicted a (blue) dash dotted line. Selected items $a_i, \ldots, a_q$, with $q = 11$, are illustrated by (red) dots, and the upper bounds $u_1, \ldots, u_q$ on the ranks appear on the x axis. (Color figure online)

"extra" bins. Using the choice of $\delta = \frac{1}{4}\varepsilon^2$ and $N_B \leq \frac{1}{\varepsilon}\mathsf{SIZE}(I_B)$, we bound the number of these items and thus extra bins by $4\delta N_B \leq \varepsilon \cdot \mathsf{SIZE}(I_B) \leq \varepsilon \cdot \mathsf{OPT}(I_B)$. Let I'_R be the remaining items in I_R. We claim that the i-th biggest item b_i in I_B is bigger than the i-th biggest item in I'_R with size equal to a_j for $j = 1, \ldots, q$. For a contradiction, suppose that $b_i < a_j$, which implies that the rank r_j of a_j in I_B is less than i. Note that $j < q$ as a_q is the smallest item in I_B. Since we packed the $\lfloor 4\delta N_B \rfloor$ biggest items from I_R separately, one of the positions of a_j in the ordering of I_R is $i + \lfloor 4\delta N_B \rfloor$ and so we have $i + \lfloor 4\delta N_B \rfloor < u_{j+1} \leq u_j + \lfloor 2\delta N_B \rfloor$, where the first inequality holds by the construction of I_R and the second inequality is by the design of data structure $Q(\delta)$. It follows that $i < u_j - \lfloor 2\delta N_B \rfloor$. Combining this with $r_j < i$, we obtain that the rank of a_j in I_B is less than $u_j - \lfloor 2\delta N_B \rfloor$, which contradicts that $u_j - \lfloor 2\delta N_B \rfloor$ is a valid lower bound on the rank of a_j.

The claim implies $\mathsf{OPT}(I'_R) \leq \mathsf{OPT}(I_B)$ and $\mathsf{SIZE}(I'_R) \leq \mathsf{SIZE}(I_B)$. We thus get that $\mathsf{OPT}(I_R) \leq \mathsf{OPT}(I'_R) + \lfloor 4\delta N_B \rfloor \leq \mathsf{OPT}(I_B) + \varepsilon \cdot \mathsf{OPT}(I_B)$, proving property (P3). Similarly, $\mathsf{SIZE}(I_R) \leq \mathsf{SIZE}(I'_R) + \lfloor 4\delta N_B \rfloor \leq \mathsf{SIZE}(I_B) + \varepsilon \cdot \mathsf{SIZE}(I_B)$, showing (P4).

Better Rounding Algorithm. Our improved rounding algorithm reduces the number of sizes in the rounded instance (and also the memory requirement) from $\widetilde{\mathcal{O}}(\frac{1}{\varepsilon^2})$ to $\widetilde{\mathcal{O}}(\frac{1}{\varepsilon})$. It is based on the observation that the number of items of sizes close to ε can be approximated with much lower accuracy than the number of items with sizes close to 1, without affecting the quality of the overall approximation. This was observed already by Karmarkar and Karp [32].

The full description of rounding, which also gives the proof of Lemma 1, is deferred to the full version of the paper. Here, we give a brief overview. Big items are split into groups based on size such that for an integer $j \geq 1$, the j-th group contains items with sizes in $(2^{-j-1}, 2^{-j}]$. Thus, there are $\lceil \log_2 \frac{1}{\varepsilon} \rceil$ groups. For each group j, we use a separate data structure $Q_j := Q(\delta)$ with $\delta = \frac{1}{8}\varepsilon$.

After all items arrive, we extract stored items from each data structure Q_j and create the rounded instance for each group as in the previous section. Then, the input summary is just the union of the rounded instances over all groups. We show that properties (P1)–(P4) hold for the input summary in a similar way as for the simple rounding algorithm, also using the following observation: Let N_j be the number of big items in group j. Then $\mathsf{SIZE}(I_{\mathrm{B}}) > \sum_j N_j \cdot 2^{-j-1}$. This holds as any item in group j has size exceeding 2^{-j-1}.

3.2 Bin Packing and Quantile Summaries

In the previous section, the deterministic quantile summary data structure from [25] allows us to obtain a streaming approximation scheme for Bin Packing. We argue that this connection runs deeper.

We start with two scenarios for which there exist better quantile summaries, thus implying a better space bound for achieving a streaming $1+\varepsilon$-approximation for Bin Packing in a similar way as in Sect. 3.1. First, if all big item sizes belong to a universe $U \subset (\varepsilon, 1]$, known in advance, then it can be better to use the quantile summary of Shrivastava *et al.* [38], which provides a guarantee of $\mathcal{O}(\frac{1}{\delta} \cdot \log |U|)$ on the space complexity, where δ is the precision requirement. Second, if we allow the algorithm to use randomization and fail with probability γ, we can employ the optimal randomized quantile summary of Karnin, Lang, and Liberty [33], which, for a given precision δ and failure probability η, uses space $\mathcal{O}(\frac{1}{\delta} \cdot \log\log \frac{1}{\eta})$ and does not provide a δ-approximate quantile for some quantile query with probability at most η.

More intriguingly, the connection between quantile summaries and Bin Packing also goes in the other direction. Namely, we show that a streaming $1+\varepsilon$-approximation algorithm for Bin Packing with space bounded by $S(\varepsilon, \mathsf{OPT})$ (or $S(\varepsilon, N)$) implies a data structure of size $S(\varepsilon, N)$ for the following Estimating Rank problem: Create a summary of a stream of N numbers which is able to provide a δ-approximate rank of any query q, i.e., the number of items in the stream which are larger than q, up to an additive error of $\pm\delta N$. A summary for Estimating Rank is essentially a quantile summary and we can actually use it to find an approximate quantile by doing a binary search over possible item names. However, this approach does *not* guarantee that the item name returned will correspond to one of the items present in the stream.

The reduction from Estimating Rank to Bin Packing is deferred to the full version. In [15] we show a space lower bound of $\Omega(\frac{1}{\varepsilon} \cdot \log \varepsilon N)$ for comparison-based data structures for Estimating Rank (and for quantile summaries as well).

Theorem 1 (Theorem 13 in [15]). *For any* $0 < \varepsilon < \frac{1}{16}$, *there is no deterministic comparison-based data structure for* Estimating Rank *which stores* $o\left(\frac{1}{\varepsilon} \cdot \log \varepsilon N\right)$ *items on any input stream of length* N.

We conclude that there is no comparison-based streaming algorithm for Bin Packing which stores $o(\frac{1}{\varepsilon} \cdot \log \mathsf{OPT})$ items on any input stream ($N = \mathcal{O}(\mathsf{OPT})$

in our reduction). Note that our algorithm is comparison-based if we employ the comparison-based quantile summary of Greenwald and Khanna [25], except that it needs to determine the size group for each item, which can be done by comparisons with 2^{-j} for integer values of j. Nevertheless, comparisons with a fixed set of constants does not affect the reduction from ESTIMATING RANK, thus the lower bound of $\Omega\left(\frac{1}{\varepsilon} \cdot \log \mathsf{OPT}\right)$ applies to our algorithm as well. This yields near optimality of our approach, up to a factor of $\mathcal{O}\left(\log \frac{1}{\varepsilon}\right)$.

4 Vector Scheduling

We provide a novel approach for creating an input summary for VECTOR SCHEDULING, based on combining small items into containers. Our streaming algorithm stores all big jobs and all containers, created from small items, that are relatively big as well. Thus, there is a bounded number of big jobs and containers, and the space used is also bounded. We show that this simple summarization preserves the optimal makespan up to a factor of $2 - \frac{1}{m} + \varepsilon$ for any $0 < \varepsilon \leq 1$. Take $m \geq 2$, since for $m = 1$ there is a trivial streaming algorithm that just sums up the vectors of all jobs to get the optimal makespan. We assume that the algorithm knows (an upper bound on) m in advance.

Algorithm Description. For $0 < \varepsilon \leq 1$ and $m \geq 2$, the algorithms works as follows: For each $k = 1, \ldots, d$, it keeps track of the total load of all jobs in dimension k, denoted L_k. Note that the optimal makespan satisfies $\mathsf{OPT} \geq \max_k \frac{1}{m} \cdot L_k$ (an alternative lower bound on OPT is the maximum ℓ_∞ norm of a job seen so far, but our algorithm does not use this). For brevity, let $\mathsf{LB} = \max_k \frac{1}{m} \cdot L_k$.

Let $\gamma = \Theta\left(\varepsilon^2 / \log \frac{d^2}{\varepsilon}\right)$; the constant hidden in Θ follows from the analysis. We say that a job with vector $\mathbf{v}$ is *big* if $\|\mathbf{v}\|_\infty > \gamma \cdot \mathsf{LB}$; otherwise it is *small*. The algorithm stores all big jobs (i.e., the full vector of each big job), while it aggregates small jobs into containers, and does not store any small job directly. A *container* is simply a vector $\mathbf{c}$ that equals the sum of vectors for small jobs assigned to this container, and we ensure that $\|\mathbf{c}\|_\infty \leq 2\gamma \cdot \mathsf{LB}$. Furthermore, container $\mathbf{c}$ is *closed* if $\|\mathbf{c}\|_\infty > \gamma \cdot \mathsf{LB}$, otherwise, it is *open*. As two open containers can be combined into one (open or closed) container, we maintain only one open container. We execute a variant of the NEXT FIT algorithm to pack the containers, adding an incoming small job into the open container, where it always fits as any small vector $\mathbf{v}$ satisfies $\|\mathbf{v}\|_\infty \leq \gamma \cdot \mathsf{LB}$. All containers are retained in the memory.

When a new job vector $\mathbf{v}$ arrives, we update the values of L_k for $k = 1, \ldots, d$ (by adding $\mathbf{v}_k$) and also of LB. If LB increases, any previously big job $\mathbf{u}$ that has become small (w.r.t. new LB), is considered to be an open container. Moreover, it may happen that a previously closed container $\mathbf{c}$ becomes open again, i.e., $\|\mathbf{c}\|_\infty \leq \gamma \cdot \mathsf{LB}$. If we indeed have more open containers, we keep aggregating arbitrary two open containers as long as we have at least two of them. Finally, if the new job $\mathbf{v}$ is small, we add it in an open container (if there is no open container, we first open a new, empty one). This completes the description of the algorithm. (We remark that for packing the containers, we may also use another, more efficient algorithm, such as FIRST FIT, which however makes no difference in the approximation guarantee.)

Properties of the Input Summary. After all jobs are processed, we can assume that $\mathsf{LB} = \max_k \frac{1}{m} \cdot L_k = 1$, which implies that $\mathsf{OPT} \geq 1$. This is without loss of generality by scaling every quantity by $1/\mathsf{LB}$. Since any big job and any closed container, each characterized by a vector $\mathbf{v}$, satisfy $\|\mathbf{v}\|_\infty > \gamma$, it holds that there are at most $\frac{1}{\gamma} \cdot d \cdot m$ big jobs and closed containers. As at most one container remains open in the end and any job or container is described by d numbers, the space cost is $\mathcal{O}\left(\frac{1}{\gamma} \cdot d^2 \cdot m\right) = \mathcal{O}\left(\frac{1}{\varepsilon^2} \cdot d^2 \cdot m \cdot \log \frac{d}{\varepsilon}\right)$.

We now analyze the maximum approximation factor that can be lost by this summarization. Let I_{R} be the resulting instance formed by big jobs and containers with small items (i.e., the input summary), and let I be the original instance, consisting of jobs in the input stream. We prove that $\mathsf{OPT}(I_{\mathrm{R}})$ and $\mathsf{OPT}(I)$ are close together, up to a factor of $2 - \frac{1}{m} + \varepsilon$. Note, however, that we still need to execute an offline algorithm to get (an approximation of) $\mathsf{OPT}(I_{\mathrm{R}})$, which is not an explicit part of the summary.

The crucial part of the proof is to show that containers for small items can be assigned to machines so that the loads of all machines are nearly balanced in every dimension, especially in the case when containers constitute a large fraction of the total load of all jobs. Let L_k^{C} be the total load of containers in dimension k (equal to the total load of small jobs). Let $I_{\mathrm{C}} \subseteq I_{\mathrm{R}}$ be the instance consisting of all containers in I_{R}.

Lemma 3. *Supposing that* $\max_k \frac{1}{m} \cdot L_k = 1$, *the following holds:*

(i) There is a solution for instance I_{C} *with load at most* $\max(\frac{1}{2}, \frac{1}{m} \cdot L_k^{\mathrm{C}}) + 2\varepsilon + 4\gamma$ *in each dimension* k *on every machine.*
(ii) $\mathsf{OPT}(I) \leq \mathsf{OPT}(I_{\mathrm{R}}) \leq (2 - \frac{1}{m} + 3\varepsilon) \cdot \mathsf{OPT}(I)$.

The full proof is deferred to the full version; here we give its brief outline. To show (i), we obtain the solution with the desired load from the randomized online algorithm by Im *et al.* [29]. Although this algorithm has ratio $\mathcal{O}(\log d / \log \log d)$ on general instances, we show that it behaves substantially better when jobs are small enough, namely, that it creates a nearly balanced assignment as claimed in (i). Item (ii) follows from an arbitrary combination of an optimal solution for big jobs in I_{R} only (i.e., excluding containers) and the solution ensured by (i) for containers.

It remains open whether or not the above algorithm with $\gamma = \Theta(\varepsilon)$ also gives $(2 - \frac{1}{m} + \varepsilon)$-approximation, which would imply a better space bound of $\mathcal{O}(\frac{1}{\varepsilon} \cdot d^2 \cdot m)$. The approximation guarantee of this approach cannot be improved, however, which we demonstrate by an example in the full version of the paper. More importantly, it would be interesting to know whether or not there is a streaming $\mathcal{O}(1)$-approximation with space $\text{poly}(\frac{1}{\varepsilon}, d, \log m)$, that is, polylogarithmic in m, or even independent of m. Recall that rounding from [8] achieves space independent of m, but exponential in d.

Acknowledgments. The authors wish to thank Michael Shekelyan for fruitful discussions.

References

1. Albers, S.: Better bounds for online scheduling. SIAM J. Comput. **29**(2), 459–473 (1999)
2. Applegate, D., Buriol, L.S., Dillard, B.L., Johnson, D.S., Shor, P.W.: The cutting-stock approach to bin packing: theory and experiments. In: ALENEX, vol. 3, pp. 1–15 (2003)
3. Azar, Y., Cohen, I.R., Kamara, S., Shepherd, B.: Tight bounds for online vector bin packing. In: Proceedings of the 25th Annual ACM Symposium on Theory of Computing, STOC 2013, pp. 961–970. ACM (2013)
4. Azar, Y., Cohen, I.R., Panigrahi, D.: Randomized algorithms for online vector load balancing. In: Proceedings of the 29th Annual ACM-SIAM Symposium on Discrete Algorithms, SODA 2018, pp. 980–991. SIAM (2018)
5. Balogh, J., Békési, J., Dósa, G., Epstein, L., Levin, A.: A new and improved algorithm for online bin packing. In: 26th Annual European Symposium on Algorithms (ESA 2018), LIPIcs, vol. 112, pp. 5:1–5:14. Schloss Dagstuhl-Leibniz-Zentrum fuer Informatik (2018)
6. Balogh, J., Békési, J., Galambos, G.: New lower bounds for certain classes of bin packing algorithms. Theoret. Comput. Sci. **440–441**, 1–13 (2012)
7. Bansal, N., Eliáš, M., Khan, A.: Improved approximation for vector bin packing. In: Proceedings of the 27th Annual ACM-SIAM Symposium on Discrete Algorithms, SODA 2016, pp. 1561–1579. SIAM (2016)
8. Bansal, N., Oosterwijk, T., Vredeveld, T., van der Zwaan, R.: Approximating vector scheduling: almost matching upper and lower bounds. Algorithmica **76**(4), 1077–1096 (2016)
9. Batu, T., Berenbrink, P., Sohler, C.: A sublinear-time approximation scheme for bin packing. Theoret. Comput. Sci. **410**(47–49), 5082–5092 (2009)
10. Beigel, R., Fu, B.: A dense hierarchy of sublinear time approximation schemes for bin packing. In: Snoeyink, J., Lu, P., Su, K., Wang, L. (eds.) AAIM/FAW -2012. LNCS, vol. 7285, pp. 172–181. Springer, Heidelberg (2012). https://doi.org/10.1007/978-3-642-29700-7_16
11. Chekuri, C., Khanna, S.: On multidimensional packing problems. SIAM J. Comput. **33**(4), 837–851 (2004)
12. Chen, L., Jansen, K., Zhang, G.: On the optimality of approximation schemes for the classical scheduling problem. In: Proceedings of the 25th Annual ACM-SIAM Symposium on Discrete Algorithms, SODA 2014, pp. 657–668. SIAM (2014)
13. Christensen, H.I., Khan, A., Pokutta, S., Tetali, P.: Approximation and online algorithms for multidimensional bin packing: a survey. Comput. Sci. Rev. **24**, 63–79 (2017)
14. Coffman, E.G., Csirik, J., Galambos, G., Martello, S., Vigo, D.: Bin packing approximation algorithms: survey and classification. In: Pardalos, P.M., Du, D.-Z., Graham, R.L. (eds.) Handbook of Combinatorial Optimization, pp. 455–531. Springer, New York (2013). https://doi.org/10.1007/978-1-4419-7997-1_35

15. Cormode, G., Veselý, P.: Tight lower bound for comparison-based quantile summaries. arXiv e-prints, page arXiv:1905.03838, May 2019
16. Dósa, G., Sgall, J.: First fit bin packing: a tight analysis. In: 30th International Symposium on Theoretical Aspects of Computer Science (STACS 2013), LIPIcs, vol. 20, pp. 538–549. Schloss Dagstuhl-Leibniz-Zentrum fuer Informatik (2013)
17. Fernandez de la Vega, W., Lueker, G.S.: Bin packing can be solved within $1 + \varepsilon$ in linear time. Combinatorica **1**(4), 349–355 (1981)
18. Fleischer, R., Wahl, M.: On-line scheduling revisited. J. Scheduling **3**(6), 343–353 (2000)
19. Garey, M.R., Graham, R.L., Johnson, D.S., Yao, A.C.-C.: Resource constrained scheduling as generalized bin packing. J. Comb. Theory Ser. A **21**(3), 257–298 (1976)
20. Garey, M.R., Johnson, D.S.: Computers and Intractability: A Guide to the Theory of NP-Completeness. WH Freeman, New York (1979)
21. Gilmore, P.C., Gomory, R.E.: A linear programming approach to the cutting-stock problem. Oper. Res. **9**(6), 849–859 (1961)
22. Gilmore, P.C., Gomory, R.E.: A linear programming approach to the cutting stock problem–part II. Oper. Res. **11**(6), 863–888 (1963)
23. Goemans, M.X., Rothvoß, T.: Polynomiality for bin packing with a constant number of item types. In: Proceedings of the 25th Annual ACM-SIAM Symposium on Discrete Algorithms, SODA 2014, pp. 830–839. SIAM (2014)
24. Gormley, T., Reingold, N., Torng, E., Westbrook, J.: Generating adversaries for request-answer games. In: Proceedings of the 11th ACM-SIAM Symposium on Discrete Algorithms, SODA 2000, pp. 564–565. SIAM (2000)
25. Greenwald, M., Khanna, S.: Space-efficient online computation of quantile summaries. In: Proceedings of the ACM SIGMOD International Conference on Management of Data, SIGMOD 2001, pp. 58–66, November 2001
26. Harris, D.G., Srinivasan, A.: The Moser-Tardos framework with partial resampling. In: 2013 IEEE 54th Annual Symposium on Foundations of Computer Science, FOCS 2013, pp. 469–478, October 2013
27. Hoberg, R., Rothvoss, T.: A logarithmic additive integrality gap for bin packing. In: Proceedings of the 28th Annual ACM-SIAM Symposium on Discrete Algorithms, SODA 2017, pp. 2616–2625. SIAM (2017)
28. Rudin III, J.F.: Improved bounds for the on-line scheduling problem. Ph.D. thesis, The University of Texas at Dallas (2001)
29. Im, S., Kell, N., Kulkarni, J., Panigrahi, D.: Tight bounds for online vector scheduling. SIAM J. Comput. **48**(1), 93–121 (2019)
30. Jansen, K., Klein, K.-M., Verschae, J.: Closing the gap for makespan scheduling via sparsification techniques. In: 43rd International Colloquium on Automata, Languages, and Programming (ICALP 2016), LIPIcs vol. 55, pp. 72:1–72:13. Schloss Dagstuhl-Leibniz-Zentrum fuer Informatik (2016)
31. Johnson, D.S.: Fast algorithms for bin packing. J. Comput. Syst. Sci. **8**, 272–314 (1974)
32. Karmarkar, N., Karp, R.M.: An efficient approximation scheme for the one-dimensional bin-packing problem. In: 23rd Annual Symposium on Foundations of Computer Science, SFCS 1982, pp. 312–320, November 1982
33. Karnin, Z., Lang, K., Liberty, E.: Optimal quantile approximation in streams. In: 2016 IEEE 57th Annual Symposium on Foundations of Computer Science (FOCS), pp. 71–78, October 2016
34. Lee, C.C., Lee, D.T.: A simple on-line bin-packing algorithm. J. ACM **32**, 562–572 (1985)

35. Ge Luo, L., Wang, K.Y., Cormode, G.: Quantiles over data streams: experimental comparisons, new analyses, and further improvements. VLDB J. **25**(4), 449–472 (2016)
36. McGregor, A.: Graph stream algorithms: a survey. SIGMOD Rec. **43**(1), 9–20 (2014)
37. Muthukrishnan, S.: Data streams: algorithms and applications. Found. Trends® Theoret. Comput. Sci. **1**(2), 117–236 (2005)
38. Shrivastava, N., Buragohain, C., Agrawal, D., Suri, S.: Medians and beyond: new aggregation techniques for sensor networks. In: Proceedings of the 2nd International Conference on Embedded Networked Sensor Systems, SenSys 2004, pp. 239–249. ACM (2004)
39. Woeginger, G.J.: There is no asymptotic PTAS for two-dimensional vector packing. Inf. Process. Lett. **64**(6), 293–297 (1997)

An Improved Upper Bound for the Ring Loading Problem

Karl Däubel(✉)

Institut für Mathematik, Technische Universität Berlin, Berlin, Germany
daeubel@math.tu-berlin.de

Abstract. The *Ring Loading Problem* emerged in the 1990s to model an important special case of telecommunication networks (SONET rings) which gained attention from practitioners and theorists alike. Given an undirected cycle on n nodes together with non-negative demands between any pair of nodes, the *Ring Loading Problem* asks for an unsplittable routing of the demands such that the maximum cumulated demand on any edge is minimized. Let L be the value of such a solution. In the relaxed version of the problem, each demand can be split into two parts where the first part is routed clockwise while the second part is routed counter-clockwise. Denote with L^* the maximum load of a minimum split routing solution. In a landmark paper, Schrijver, Seymour and Winkler [22] showed that $L \leq L^* + \frac{3}{2}D$, where D is the maximum demand value. They also found (implicitly) an instance of the *Ring Loading Problem* with $L = L^* + \frac{101}{100}D$. Recently, Skutella [25] improved these bounds by showing that $L \leq L^* + \frac{19}{14}D$, and there exists an instance with $L = L^* + \frac{11}{10}D$. We contribute to this line of research by showing that $L \leq L^* + \frac{13}{10}D$. We also take a first step towards lower and upper bounds for small instances.

Keywords: Ring Loading Problem · SONET ring · Load balancing · Unsplittable flow

1 Introduction

Given an undirected cycle on n nodes together with non-negative demands between any pair of nodes, the *Ring Loading Problem* asks for an unsplittable routing of the demands such that the maximum cumulated demand on any edge is minimal. Formally, we are given a graph $G = (V, E)$ with nodes $V = [n] := \{1, \dots, n\}$, edges $\{i, i+1\}$ for each $i \in V$, where we assume throughout the paper that $\{n, n+1\} := \{n, 1\}$, and demands for each pair of nodes $i < j$ of value $d_{i,j} \geq 0$. By a slight abuse of notation, we refer to both the demand from i to j and its value as $d_{i,j}$. An unsplittable solution decides for each demand whether it should be routed clockwise, sending all of its value along the path $\{i, i+1, \dots, j\}$, or counter-clockwise, sending all of its value along the

A preprint of this paper with full proofs is available at [4].

E. Bampis and N. Megow (Eds.): WAOA 2019, LNCS 11926, pp. 89–105, 2020.
https://doi.org/10.1007/978-3-030-39479-0_7

path $\{i, i-1, \ldots, 1, n, \ldots, j\}$. The *load* of an edge, for a given solution, is the sum of all demand values that are routed on paths that use the edge. We call the maximum load on any edge of the ring the *load* of the solution. The problem is to find an unsplittable routing that minimizes the load. We denote with L the load of such an optimal unsplittable solution. See Fig. 1 for an example.

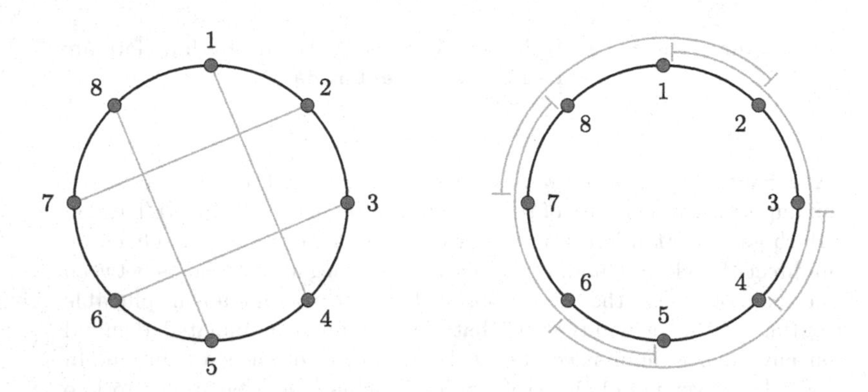

Fig. 1. An instance of the *Ring Loading Problem* on 8 nodes and 4 non-zero demands with $d_{1,4} = d_{2,7} = d_{3,6} = d_{5,8} = 1$ (left) together with an optimum unsplittable routing of load 2 (right).

The problem was introduced by Cosares and Saniee [3] to mathematically model survivable networks with respect to the emerging standard of synchronous optical networks (SONET). The underlying structure to this technology, the SONET ring, is a set of network nodes and links that are arranged in a cycle. In this way, even in the event of a link failure, most of the traffic could be recovered. See [1,11,27] for further resources on technical details. To the best of our knowledge, Cosares and Saniee also established the name *Ring Loading Problem.* They further showed via a reduction of the *Partition Problem* that the problem is NP-hard and provided an algorithm that returns an unsplittable solution with load at most $2L$. Using a result from Schrijver et al. [22], Khanna [12] showed that there exists a PTAS, i.e. a class of poly-time algorithms that return a solution with load at most $(1+\varepsilon)\,L$, for each fixed $\varepsilon > 0$. If all non-zero demands have the same value, Frank [7] showed that the *Ring Loading Problem* can be solved in polynomial time.

Although a PTAS for the *Ring Loading Problem* exists, there remain unsolved problems that connect unsplittable solutions to a relaxed version of the *Ring Loading Problem.* To this end, consider the *Ring Loading Problem* where demands are allowed to be routed splittably, i.e. a demand can be routed partly clockwise while the remaining part is routed counter-clockwise. The definition of the load of an edge and the load of a solution generalize naturally to the relaxed version. We denote with L^* the optimum load of a split solution. The relaxed version of the *Ring Loading Problem* has a linear programming formulation [3] and can thus be solved in polynomial time. Further effort was put into finding more efficient algorithms (see [5,18,19,22,26,28]). It was also shown in [5,19,22] that $L \leq 2L^*$, and this bound is tight ([19,22]).

In a landmark paper, Schrijver, Seymour and Winkler [22] proved in this context that $L \leq L^* + \frac{3}{2}D$, where we denote with $D := \max_{i<j} d_{i,j}$ the maximum demand value. They furthermore gave the "guarantee" that $L \leq L^* + D$, which was later restated as conjecture in the survey on multicommodity flows by Shepherd [23]. More recently, Skutella [25] improved the upper bound by showing that $L \leq L^* + \frac{19}{14}D$. He also found an instance of the *Ring Loading Problem* with $L = L^* + \frac{11}{10}D$, disproving the long-standing conjecture by Schrijver et al. and Shepherd. Skutella furthermore conjectured that $L \leq L^* + \frac{11}{10}D$.

Interestingly, Schrijver et al. gave an instance of the *Ring Loading Problem* together with a split routing that cannot be turned into an unsplittable routing without increasing the load on some edge by at least $\frac{101}{100}D$, whereas Skutella writes that this "does not imply a gap strictly larger than D between the optimum values of split and unsplittable routings". We show in Lemma 7 that this implication *does* hold, and that Schrijver, Seymour and Winkler therefore (implicitly) found a counterexample to their own conjecture.

Our Contributions. The following theorem is the main contribution of this work.

Theorem 1. *Any split routing solution to the* Ring Loading Problem *can be turned into an unsplittable routing while increasing the load on any edge by at most* $\frac{13}{10}D$. *In particular, we have* $L \leq L^* + \frac{13}{10}D$.

In order to prove the theorem, we first define a general framework that unifies structural results of split routings introduced by Skutella. We then apply this framework in a new way to obtain the better upper bound. The algorithm implicitly given in the proof of Theorem 1 runs in linear time. This result is the first progress towards closing the remaining additive gap since Skutella.

As all previous lower bound examples are of relative small size, it is interesting to settle these cases conclusively. We take a step into this direction by showing upper and lower bounds for small instances. The upper bounds are deduced from a mixed integer linear program that verifies for a given instance size that no worse examples can exist. Although the lower bounds also follow from this formulation, we provide further examples to enrich the view on instances where the difference $L - L^*$ is large with respect to D. In fact, we give an infinite family of instances with $L > L^* + D$.

A summary of previous results on lower and upper bounds together with new advancements is shown in Fig. 2 on the right vertical line, while on the left results are given with respect to $\delta \in \left[0, \frac{1}{2}\right]$ that, loosely speaking, parametrizes instances of the *Ring Loading Problem* and guarantees that all demands are either small or large with respect to δ and D, i.e. all demand values lie in $[0, \delta D] \cup [(1-\delta)D, D]$. A formal definition is given in Sect. 2.

Just as Schrijver et al. and Skutella before, we mention a nice combinatorial implication of our result. Schrijver et al. define β to be the infimum of all reals α

such that the following combinatorial statement holds: For all positive integers m and nonnegative reals $u_1, \ldots, u_m$ and $v_1, \ldots, v_m$ with $u_i + v_i \leq 1$, there exist $z_1, \ldots, z_m$ such that for every k, $z_k \in \{v_k, -u_k\}$ and

$$\left| \sum_{i=1}^{k} z_i - \sum_{i=k+1}^{m} z_i \right| \leq \alpha.$$

Schrijver et al. prove that $\beta \in \left[\frac{101}{100}, \frac{3}{2}\right]$. Skutella reduces the size of the interval to $\beta \in \left[\frac{11}{10}, \frac{19}{14}\right]$. As a result of our work, we obtain $\beta \in \left[\frac{11}{10}, \frac{13}{10}\right]$.

Further Related Work. In the *Ring Loading Problem* with integer demand splitting, each demand is allowed to be split into two integer parts which are routed in different directions along the ring. The objective is to find an integer split routing that minimizes the load. Let L' be the load of an optimal integer split routing solution. Lee et al. [15] showed an algorithm that returns an integer split routing solution with load at most $L' + 1$. Schrijver et al. [22] found an optimal solution in pseudo-polynomial time. Vachani et al. [26] provided an $O\left(n^3\right)$ algorithm. In [17] Myung presented an algorithm with runtime $O\left(nk\right)$ where k is the number of non-zero demands. Wang [28] proved the existence of an $O\left(k + t_S\right)$ algorithm where t_S is the time for sorting k nodes.

More recently, the weighted *Ring Loading Problem* was introduced where each edge has a weight associated with it, and the *weighted load* of an edge is the product of its weight and the smallest integer greater or equal than its load. In the case where demand splitting is allowed, Nong et al. [20] gave an $O\left(n^2 k\right)$ algorithm. If integer demand splitting is allowed, the authors present a pseudo-polynomial time algorithm. Later, Nong et al. [21] present an $O\left(n^3 k\right)$ algorithm. If the demands have to be send unsplittably, Nong et al. [20] prove the existence of a PTAS.

In a broader context, the *Ring Loading Problem* is a special case of unsplittable multicommodity flows. We mention the case of single source unsplittable flows, as similarities between theorems and conjectures for these problems exist (see [6,8,16,24]). We also refer to the survey of Shepherd [23].

A more geometric problem with similarities to the *Ring Loading Problem* is the dynamic storage allocation problem. In this problem, axis-parallel rectangles in the positive quadrant of the plane are given. The horizontal position is fixed while vertical shifts are allowed. These rectangles have to be placed pairwise disjoint such that the supremum of the y-coordinates is minimized. Each rectangle can be seen as demand whose value is its height. Multiple approximation algorithms for the problem exist, e.g. [2,9,10,13,14].

Outline. In Sect. 2, we introduce some notation and provide useful results from Schrijver et al. and Skutella that we need. We then continue in Sect. 3 with the proof of Theorem 1. In Sect. 4, we turn our attention to upper and lower bounds for small instances. We wrap everything up with our conclusions in Sect. 5.

2 Preliminaries

In this section, we introduce further notation and mention results already presented in [22,25]. We start with a preprocessing step to reduce the size and complexity of an instance to the *Ring Loading Problem.*

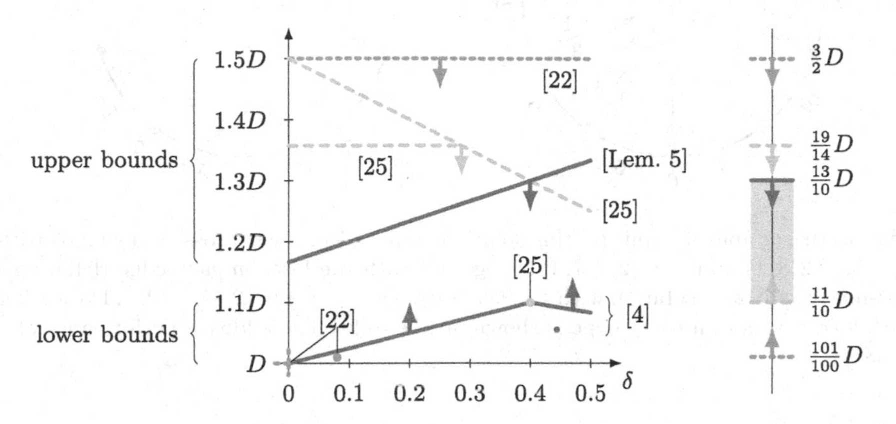

Fig. 2. Summary of known results dependent on δ (left) and independent of δ (right). The currently best bounds are due to Theorem 1 together with the lower bound in [25].

Two demands $d_{i,j}$ and $d_{k,l}$ are *parallel* if there exists a path from i to j and a path from k to l that are edge-disjoint, otherwise they are *crossing.* Note that the demands $d_{i,j}$ and $d_{i,k}$ are parallel.

As Theorem 1 only argues about the load increase on all edges for split routing solutions, we can ignore and delete all demands that are routed unsplittably. The following observation shows that we can assume that there are not too many remaining demands.

Observation 1 ([22]). *Given a split routing of two parallel demands d_1 and d_2. The routing can be altered such that at most one demand is routed splittably, without increasing the load on any edge.*

If we apply Observation 1 to an arbitrary split routing and delete afterwards all demands that are routed unsplittably, we can concentrate on instances with pairwise crossing demands, implying in particular that every node is end point of at most one demand. If a node is not the end point of a demand, the load on its adjacent edges have the same value, allowing us to delete the node and merge the edges.

After this process we are left with a ring on $n = 2m$ nodes, demands $d_i := d_{i,i+m} > 0$ for $i \in [m]$ and a split routing. We denote for all $i \in [m]$ with $u_i > 0$ the amount of flow from demand d_i routed clockwise and likewise with $v_i > 0$ the remainder of flow routed counter-clockwise. Note that $u_i + v_i = d_i$, $i \in [m]$. From now on we refer to an instance with this structure as *split routing solution.* An example is given in Fig. 3 on the left.

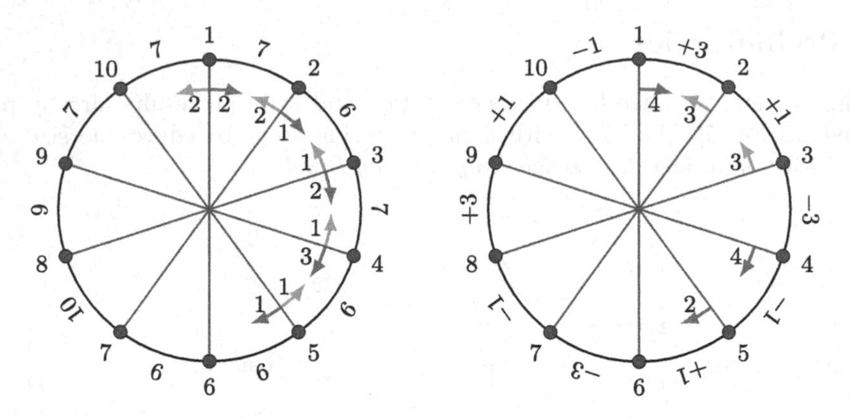

Fig. 3. An example of a split routing solution on $m = 5$ pairwise crossing demands with $u = (2,1,2,3,1)$ and $v = (2,2,1,1,1)$ together with the load on each edge (left). One possible unsplittable solution with $z = (v_1, -u_2, -u_3, v_4, v_5) = (2,-1,-2,1,1)$ together with load changes on every edge is shown on the right. The additive performance of z is 3.

The following definition describes for a given $\delta \in \left[0, \frac{1}{2}\right]$ all split routing solutions with small and large demands only (with respect to δ and D) and ensures the existence of a demand on the boundary to medium demands. Formally we call these split routing solutions δ-instances:

Definition 1. Let $\delta \in \left[0, \frac{1}{2}\right]$. We call a split routing solution a *δ-instance*, if $d_i \in \{\delta D, (1-\delta) D\}$ for all $i \in \arg\min_{j\in[m]} \left(\left|\frac{1}{2}D - d_j\right|\right)$.

A $\frac{1}{2}$-instance for example has a demand of value $\frac{1}{2}D$, whereas a 0-instance only has demands of value D. An important property of δ-instances is that $d_i \in [0, \delta D] \cup [(1-\delta) D, D]$ for all $i \in [m]$. All demands are therefore either small or large with respect to δ and D.

In order to turn a given split routing solution into an unsplittable solution, we have to decide for each demand d_i whether u_i units of flow are rerouted to use the counter-clockwise direction, or whether v_i units of flow are rerouted to use the clockwise direction. We encode this decision using $z = (z_1, \ldots, z_m)$, with $z_i \in \{v_i, -u_i\}$ for all $i \in [m]$, where $z_i = v_i$ means that we send the demand completely in clockwise direction, whereas $z_i = -u_i$ means that we completely send the demand in counter-clockwise direction. In either case the z_i values model exactly the increase of load on the clockwise edges from i to $i+m$, and the decrease of load on the counter-clockwise edges. For $k \in [m]$ the load on an edge $\{k, k+1\}$ changes by

$$\sum_{i=1}^{k} z_i - \sum_{i=k+1}^{m} z_i,$$

while the load on the opposite edge $\{k+m, k+m+1\}$ changes by the negative amount. The maximum increase of load on any edge is therefore

$$\max_{k\in[m]} \left| \sum_{i=1}^{k} z_i - \sum_{i=k+1}^{m} z_i \right| .$$

As described by Skutella [25], we refer to this quantity as the *additive performance* of z. In Fig. 3 an example of the load change and the additive performance is given.

Let $x \in \mathbb{R}$ be fixed, we define $p_z(k) := x + \sum_{i=1}^{k} z_i$, for $k \in \{0,\ldots,m\}$. We refer to p_z as a *pattern* starting at $x = p_z(0)$ and ending at $y = p_z(m)$. We denote with $a := \min_{k\in\{0,\ldots,m\}} p_z(k)$ and $b := \max_{k\in\{0,\ldots,m\}} p_z(k)$ the minimum and maximum of pattern p_z, respectively. We refer to $[a, b]$ as strip and say that the pattern p_z lives on the strip $[a, b]$ of width $b - a$. As $p_z(k) - p_z(k-1) = z_i$, when we refer to a pattern p_z we also refer to the corresponding unsplittable solution. As the choice of x might vary, multiple patterns correspond to a single unsplittable solution. A pattern can be visualized as seen in Fig. 4.

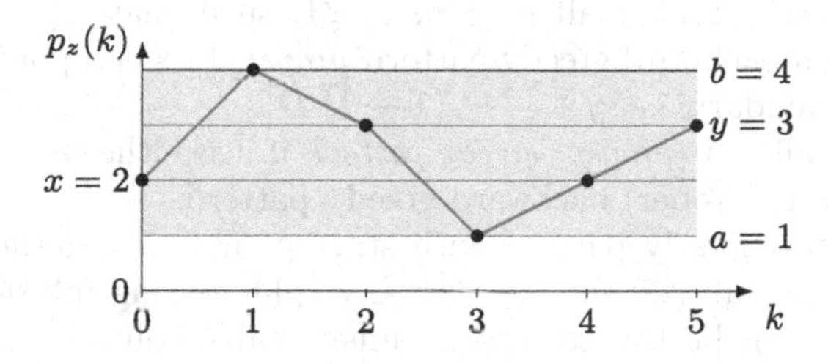

Fig. 4. An example of a pattern p_z that corresponds to the split routing given in Fig. 3 with $z = (2, -1, -2, 1, 1)$ and start point $x = 2$, end point $y = 3$, minimum value $a = 1$ and maximum value $b = 4$. The additive performance of the pattern due to Observation 2 is $\max\{2b - x - y, x + y - 2a\} = 3$.

Observation 2 ([25]). *Given an unsplittable solution z with corresponding pattern p_z with start point x, end point y living on a strip of $[a, b]$, then the additive performance of pattern p_z is*

$$\max_{k\in[m]} \left| \sum_{i=1}^{k} z_i - \sum_{i=k+1}^{m} z_i \right| = \max\{2b - x - y, x + y - 2a\} . \tag{1}$$

Observation 3 ([25]). *Let $\varepsilon > 0$. Given an unsplittable solution z with corresponding pattern p_z with start point x and end point y living on a strip of $[a, b]$, the additive performance of pattern p_z is at most $b-a+\varepsilon$ if and only if the pattern starts at x and ends at $y \in [y_{opt} - \varepsilon, y_{opt} + \varepsilon] \cap [a, b]$ with $y_{opt} := a + b - x$.*

Given a strip of width D, say $[0, D]$. Let $x \in [0, D]$, we denote throughout with $\bar{x} := D - x$ the reflection of x across $\frac{1}{2}D$. We can construct a pattern with start point x living on a strip $[a, b] \subseteq [0, D]$ by applying iteratively the following observation.

Observation 4. *Let $d = e + f$ with $e, f \in \mathbb{R}_{\geq 0}$. If I is an interval of size at least d and $x \in I$, then $x + e \in I$ or $x - f \in I$ (or both).*

Formally, we construct the pattern p_z by setting $p_z(0) = x$ for some $x \in [0, D]$ and choose $z_k \in \{-u_k, v_k\}$ iteratively such that $p_z(k) = p_z(k-1) + z_k \in [0, D]$ for all $k = 1, \ldots, m$, which always works by Observation 4. If this decision is not unique, we set z_k such that $\left|\frac{1}{2}D - p_z(k)\right|$ is minimal, i.e. $p_z(k)$ is as close as possible to the middle $\frac{1}{2}D$ of the interval $[0, D]$. Remaining ties are broken arbitrarily. A pattern that is constructed with respect to this procedure is called a *forward greedy pattern*. For technical reasons, we call a forward greedy pattern p_z *proper* if its start point is far enough away from the boundary, i.e. $x \in \left[\frac{\delta}{4}D, \left(1 - \frac{\delta}{4}\right) D\right]$. This requirement is used in Lemma 2.

We obtain a *backward greedy pattern* p_z by applying this procedure backwards. We define $p_z(m) = y$, for some $y \in [0, D]$, and iteratively choose $p_z(k-1) = p_z(k) - z_k$, for all $k = m, \ldots, 1$, such that $\left|p_z(k-1) - \frac{1}{2}D\right|$ is minimal. We call a backward greedy pattern *proper* if its end point is far enough away from the boundary, i.e. $y \in \left[\frac{\delta}{4}D, \left(1 - \frac{\delta}{4}\right) D\right]$.

A pattern is called a *(proper) greedy pattern* if it is either a (proper) forward greedy pattern or a (proper) backward greedy pattern.

Using a forward greedy pattern with start point $\frac{1}{2}D$ together with Observation 3, Schrijver et al. [22] showed that any split routing solution to the *Ring Loading Problem* can be turned into an unsplittable solution while increasing the load on any edge by at most $\frac{3}{2}D$.

Although the following structural properties of (greedy) patterns are crucial for our results, we refer the reader for complete proofs to [25].

Definition 2 ([25]). Let $\varepsilon \geq 0$. Two patterns p_z and $p_{z'}$ are said to be ε-close if $|p_z(k) - p_{z'}(k)| \leq \varepsilon$ for some $k \in \{0, 1, \ldots, m\}$.

The following lemma combines two ε-close patterns to a single pattern while preserving crucial properties.

Lemma 1 ([25]). *Consider a fixed split routing solution. Let $p_{z'}$ be a pattern with start point x' living on strip $[a', b']$, and $p_{z''}$ a pattern with end point y'' living on strip $[a'', b'']$. If the two patterns are ε-close for some $\varepsilon \geq 0$, then there is a pattern p_z living on a sub-strip of*

$$\left[\min\{a', a''\} - \frac{1}{2}\varepsilon, \max\{b', b''\} + \frac{1}{2}\varepsilon\right]$$

with start point x and end point y such that $x + y = x' + y''$.

This lemma describes situations where $\frac{\delta}{2}D$-close patterns exist.

Lemma 2 ([25]). *Consider three proper greedy patterns $p_{z_a}, p_{z_b}, p_{z_c}$, all three living on sub-strips of $[0, D]$. If the sorting of the patterns by their end points is not a cyclic permutation of the sorting of their start points, then (at least) two of the three patterns are $\frac{1}{2}\delta D$-close.*

At the end of the section reconsider Fig. 2. Both previous and new results are shown with respect to δ on the left and the consequences for all instances independent of δ on the right.

3 Improved Upper Bound

In this section we prove Theorem 1. We start by defining a general framework that allows us to use Lemmas 1 and 2 in a very unified manner. The following definition is at the heart of this framework (see Fig. 5).

Definition 3. Given a greedy pattern p_{z_a} living on a sub-strip of $[0, D]$ with start point x_a and end point y_a, we call a forward greedy pattern p_{z_b} *induced* by p_{z_a}, if it lives on a sub-strip of $[0, D]$ with start point $x_b := \frac{2}{3}\bar{y}_a + \frac{1}{3}x_a$. Likewise, we call a backward greedy pattern p_{z_c} *induced* by p_{z_a}, if it lives on a sub-strip of $[0, D]$ with end point $y_c := \frac{2}{3}\bar{x}_a + \frac{1}{3}y_a$.

If a greedy pattern p_{z_a} and its induced patterns are proper, the following lemma ensures the existence of a pattern with an additive performance that only depends on the start and end points of p_{z_a} together with δ. It is therefore possible to pick a single pattern, check if its induced patterns are proper, and obtain a strong bound on the additive performance.

Lemma 3. *For a δ-instance with $\delta \in \left[0, \frac{1}{2}\right]$, let p_{z_a} be a greedy pattern living on a sub-strip of $[0, D]$ with start point x_a and end point y_a. Denote with p_{z_b} a forward greedy pattern induced by p_{z_a} and with p_{z_c} a backward greedy pattern induced by p_{z_a}. If all three greedy patterns are proper, then there exists a pattern with additive performance at most*

$$\max\left\{\frac{4}{3}D - \frac{1}{3}(x_a + y_a) + \frac{\delta}{2}D, \frac{2}{3}D + \frac{1}{3}(x_a + y_a) + \frac{\delta}{2}D\right\}.$$

Proof. We first show that the sorting of the start points is not a cyclic permutation of the sorting of the end points. This allows us to use Lemma 2 that guarantees the existence of two patterns that are $\frac{\delta}{2}D$-close. We then conclude the lemma by showing that if any two of the three patterns are $\frac{\delta}{2}D$-close, that there exists a pattern with the required additive performance. In Fig. 5 is an illustration of the procedure.

By the definition of p_{z_b} as forward greedy pattern induced by p_{z_a} and p_{z_c} as backward greedy pattern induced by p_{z_a}, we know that $x_b = \frac{2}{3}\bar{y}_a + \frac{1}{3}x_a$ and $y_c = \frac{2}{3}\bar{x}_a + \frac{1}{3}y_a$. The definitions are such that the interval between x_a and $\bar{y}_a$

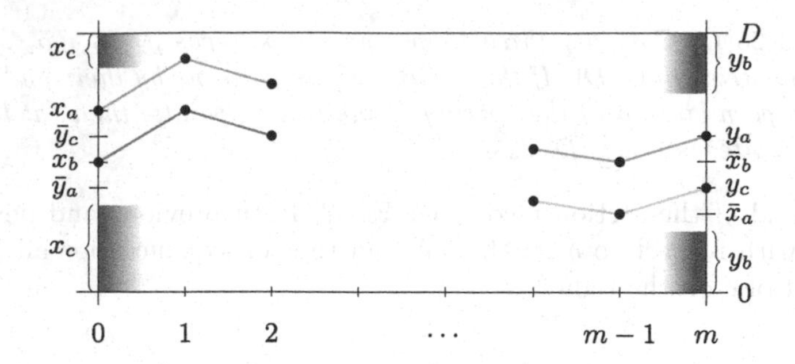

Fig. 5. An illustration of a greedy pattern p_{z_a} with induced forward greedy pattern p_{z_b} and induced backward greedy pattern p_{z_c} as defined in Definition 3.

is divided into three equal parts by the points x_b and $\bar{y}_c$. A straightforward computation shows that

$$\bar{y}_c = \frac{x_a + x_b}{2}, \qquad \bar{x}_b = \frac{y_a + y_c}{2}, \tag{2}$$

i.e. the optimal start point for pattern p_{z_c} is in the middle between x_a and x_b, and the optimal end point of p_{x_b} is in the middle between y_a and y_c. Because $|x_a - \bar{y}_a| = |\bar{x}_a - y_a|$, and the symmetric definitions of x_b and y_c, it also holds that $|x_a - x_b| = |y_a - y_c|$.

For the sake of brevity, we define $\varepsilon := \max\left\{\frac{1}{3}D - \frac{1}{3}(x_a + y_a) + \frac{\delta}{2}D, \frac{1}{3}(x_a + y_a) - \frac{1}{3}D + \frac{\delta}{2}D\right\}$. In fact, we want to show that there exists a pattern with additive performance at most $D + \varepsilon$.

We now show that $|x_a - x_b| \leq 2\varepsilon$, which also implies that $|y_a - y_c| \leq 2\varepsilon$. By definition of $x_b = \frac{2}{3}\bar{y}_a + \frac{1}{3}x_a = \frac{2}{3}D - \frac{2}{3}y_a + \frac{1}{3}x_a$, we can conclude that

$$\begin{aligned}
|x_a - x_b| &= \max\{x_b - x_a, x_a - x_b\} \\
&= \max\left\{\frac{2}{3}D - \frac{2}{3}(x_a + y_a), \frac{2}{3}(x_a + y_a) - \frac{2}{3}D\right\} \\
&\leq 2\max\left\{\frac{1}{3}D - \frac{1}{3}(x_a + y_a) + \frac{\delta}{2}D, \frac{1}{3}(x_a + y_a) - \frac{1}{3}D + \frac{\delta}{2}D\right\} \\
&= 2\varepsilon,
\end{aligned} \tag{3}$$

where the inequality follows from the fact that $\delta \geq 0$. With Eqs. (2) and (3), it follows that

$$|\bar{y}_c - x_b| = |\bar{y}_c - x_a| = |\bar{x}_b - y_a| = |\bar{x}_b - y_c| \leq \varepsilon. \tag{4}$$

If for the start point of p_{z_c} holds $x_c \in [\bar{y}_c - \varepsilon, \bar{y}_c + \varepsilon]$, we know by Observation 3 that the additive performance of p_{z_c} is at most $D + \varepsilon$, from which the lemma follows. We can thus assume that $x_c \in [0, \bar{y}_c - \varepsilon] \cup [\bar{y}_c + \varepsilon, D]$. Using Eq. (4), we can therefore conclude that either $x_c \leq \bar{y}_c - \varepsilon \leq \min\{x_a, x_b\}$ or $\max\{x_a, x_b\} \leq \bar{y}_c + \varepsilon \leq x_c$.

Equivalently, if for the end point of p_{z_b} holds $y_b \in [\bar{x}_b - \varepsilon, \bar{x}_b + \varepsilon]$, we know by Observation 3 that the additive performance of p_{z_b} is at most $D + \varepsilon$, from which the lemma follows. We can thus assume that $y_b \in [0, \bar{x}_b - \varepsilon] \cup [\bar{x}_b + \varepsilon, D]$. Using Eq. (4), we can therefore conclude that either $y_b \leq \bar{x}_b - \varepsilon \leq \min\{y_a, y_c\}$ or $\max\{y_a, y_c\} \leq \bar{x}_b + \varepsilon \leq y_b$.

Assume first that $\bar{y}_a \leq x_a$ (as shown in Fig. 5), then $\bar{y}_a \leq x_b \leq \bar{y}_c \leq x_a$ and $\bar{x}_a \leq y_c \leq \bar{x}_b \leq y_a$, which implies that either $x_c \leq x_b \leq x_a$ or $x_b \leq x_a \leq x_c$ and that either $y_b \leq y_c \leq y_a$ or $y_c \leq y_a \leq y_b$. In either case, the sorting of the patterns by their start points is not a cyclic permutation of the patterns by their end points.

Assume now that $x_a \leq \bar{y}_a$, then $x_a \leq \bar{y}_c \leq x_b \leq \bar{y}_a$ and $y_a \leq \bar{x}_b \leq y_c \leq \bar{x}_a$, which implies that either $x_c \leq x_a \leq x_b$ or $x_a \leq x_b \leq x_c$ and that either $y_b \leq y_a \leq y_c$ or $y_a \leq y_c \leq y_b$. In either case, the sorting of the patterns by their start points is not a cyclic permutation of the patterns by their end points.

As p_{z_a}, p_{z_b} and p_{z_c} are proper greedy patterns, we can apply Lemma 2, ensuring the existence of two patterns that are $\frac{\delta}{2}D$-close. We conclude the proof by showing that the closeness of any two patterns guarantees the existence of a pattern with the claimed additive performance.

(i) Assume that p_{z_a} and p_{z_b} are $\frac{\delta}{2}D$-close. Then Lemma 1 assures the existence of a pattern with start point x and end point y such that $x + y = x_b + y_a = \frac{2}{3}D + \frac{1}{3}(x_a + y_a)$ on a sub-strip of $\left[-\frac{\delta}{4}D, D + \frac{\delta}{4}D\right]$. Using Observation 2, a straightforward calculation shows that this pattern has additive performance at most

$$\max\left\{\frac{4}{3}D - \frac{1}{3}(x_a + y_a) + \frac{\delta}{2}D, \frac{2}{3}D + \frac{1}{3}(x_a + y_a) + \frac{\delta}{2}D\right\}.$$

(ii) Assume that p_{z_a} and p_{z_c} are $\frac{1}{2}\delta D$-close. Then Lemma 1 assures the existence of a pattern with start point x and end point y such that $x + y = x_a + y_c = \frac{2}{3}D + \frac{1}{3}(x_a + y_a)$ on a sub-strip of $\left[-\frac{\delta}{4}D, D + \frac{\delta}{4}D\right]$. Using Observation 2, a straightforward calculation shows that this pattern has additive performance at most

$$\max\left\{\frac{4}{3}D - \frac{1}{3}(x_a + y_a) + \frac{\delta}{2}D, \frac{2}{3}D + \frac{1}{3}(x_a + y_a) + \frac{\delta}{2}D\right\}.$$

(iii) Assume that p_{z_b} and p_{z_c} are $\frac{1}{2}\delta D$-close. Then Lemma 1 assures that there exists a pattern with start point x and end point y such that $x + y = x_b + y_c = \frac{4}{3}D - \frac{1}{3}(x_a + y_a)$ on a sub-strip of $\left[-\frac{\delta}{4}D, D + \frac{\delta}{4}D\right]$. Using Observation 2, a straightforward calculation shows that this pattern has additive performance at most

$$\max\left\{\frac{2}{3}D + \frac{1}{3}(x_a + y_a) + \frac{\delta}{2}D, \frac{4}{3}D - \frac{1}{3}(x_a + y_a) + \frac{\delta}{2}D\right\}.$$

In either case, the lemma follows.

If both the start point and the end point of a greedy pattern p_{z_a} are far enough away from the boundary, the next lemma ensures that its induced patterns are proper. Note that this is a stronger requirement on p_{z_a} than being a proper greedy pattern.

Lemma 4. *For a δ-instance with $\delta \in \left[0, \frac{1}{2}\right]$, let p_{z_a} be a greedy pattern living on a sub-strip of $[0, D]$ with start point x_a and end point y_a. Denote with p_{z_b} a forward greedy pattern induced by p_{z_a} and with p_{z_c} a backward greedy pattern induced by p_{z_a}. If $x_a, y_a \in \left[\frac{\delta}{4}D, \left(1 - \frac{\delta}{4}\right) D\right]$, then p_{z_a}, p_{z_b} and p_{z_c} are proper.*

Proof. By definition, p_{z_a} is proper. For $x \in [0, D]$ holds that $\bar{x} \in \left[\frac{\delta}{4}D, \left(1 - \frac{\delta}{4}\right) D\right]$ if and only if $x \in \left[\frac{\delta}{4}D, \left(1 - \frac{\delta}{4}\right) D\right]$. By assumption, we therefore also know that $\bar{x}_a$ and $\bar{y}_a$ are far enough away from the boundary. By the definition of induced patterns, we know that $x_b = \frac{2}{3}\bar{y}_a + \frac{1}{3}x_a$, which implies in particular that $\min\{x_a, \bar{y}_a\} \leq x_b \leq \max\{x_a, \bar{y}_a\}$. The start point x_b of the induced forward greedy pattern is consequently far enough away from the boundary. Using the same argumentation for the definition of $y_c = \frac{2}{3}\bar{x}_a + \frac{1}{3}y_a$, the lemma follows.

A crucial part of the proof of Theorem 1, and the main contribution of this work is the following auxiliary lemma.

Lemma 5. *For a δ-instance with $\delta \in \left[0, \frac{1}{2}\right]$ there exists a pattern with additive performance at most $\left(\frac{7}{6} + \frac{\delta}{3}\right) D$.*

Proof. We start the proof by modifying the instance such that the special demand of value either δD or $(1 - \delta) D$ is the last demand. These two cases will be treated separately. In either case, we then use the nice structure of the newly created instance to find a greedy pattern that can be used with Lemma 3.

Let d_i be the demand that minimizes $\left|\frac{1}{2}D - d_i\right|$ over all $i \in [m]$. By the definition of a δ-instance, we know that $d_i \in \{\delta D, (1 - \delta) D\}$.

We now rotate the instance such that the specially chosen demand d_i has index m, and is thus the last demand of the instance. By a slight abuse of notation we will refer to this newly created instance again as instance. Recall that now the demand d_m has the property that $d_m \in \{\delta D, (1 - \delta) D\}$.

The following procedure is similar to the one described by Skutella [25] when dealing with demands of medium size. An example is depicted in Fig. 6. We first delete the last demand m to obtain a smaller instance. We define a backward greedy pattern ending at $\frac{1}{2}(D + d_m) - v_m$ and starting at some $x_a \in [0, D]$. This backward greedy pattern can be extended in two possible ways to create a pattern that includes demand m, once with end point $y_a^1 := \frac{1}{2}(D - d_m)$ and once with end point $y_a^2 := \frac{1}{2}(D + d_m)$. A crucial observation is that both possible extensions produce a valid backward greedy pattern for the original instance. Depending on the particular start point x_a, we choose in which way the pattern will be extended: If $x_a \leq \frac{1}{2}D$, we extend the pattern with end point y_a^2, otherwise we extend the pattern with y_a^1. For the rest of the proof, we may assume that x_a is at most $\frac{1}{2}D$ and the pattern is therefore extended with end point $y_a := y_a^2$. This assumption can be made, as the following construction is highly symmetric

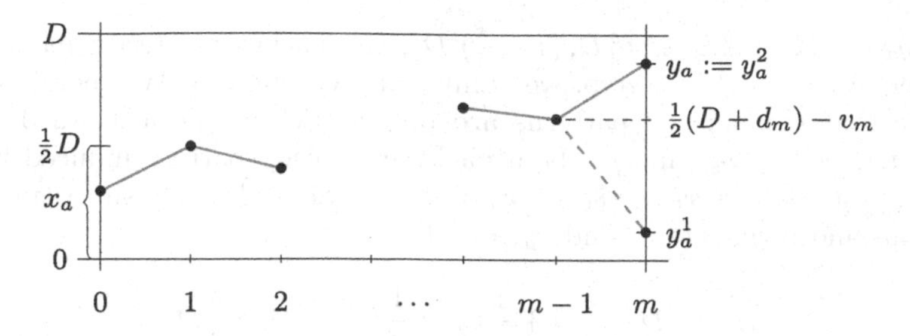

Fig. 6. An example of the construction step in Lemma 5. If $x_a \le \frac{1}{2}D$, we extend the pattern with $y_a = \frac{1}{2}(D + d_m)$.

with respect to y_a^1 and y_a^2, in fact, all arguments remain valid if we change start and end points of subsequent patterns by reflecting their value around $\frac{1}{2}D$. Let p_{z_a} denote the resulting backward greedy pattern with start point x_a and end point y_a.

We consider two cases, first that $d_m = (1-\delta)\,D$ and second that $d_m = \delta D$. For the sake of brevity, we define $\varepsilon := \frac{1}{6}D + \frac{\delta}{3}D$. In fact, we want to find a pattern with additive performance at most $D + \varepsilon$.

Case (a). If $d_m = (1-\delta)\,D$, we can rewrite $y_a = \frac{1}{2}\,(D + d_m) = D - \frac{\delta}{2}D$. The lemma follows immediately from Observation 3 if $x_a \in [\bar{y}_a - \varepsilon, \bar{y}_a + \varepsilon]$. We can therefore assume that x_a falls either into the interval $\left[0, \frac{\delta}{6}D - \frac{1}{6}D\right]$ or into the interval $\left[\frac{1}{6}D + \frac{5}{6}\delta D, \frac{1}{2}D\right]$. Recall the assumption that x_a is at most $\frac{1}{2}D$. It is easy to see that $\frac{\delta}{6}D - \frac{1}{6}D$ is negative for all $\delta \in \left[0, \frac{1}{2}\right]$. It follows that $x_a \in \left[\frac{1}{6}D + \frac{5}{6}\delta D, \frac{1}{2}D\right]$. Note that this interval is also empty for all $\delta > \frac{2}{5}$, and the lemma is trivially correct. In fact, this is exactly the argumentation used by Skutella [25] in his proof of Lemma 6.

As $y_a = D - \frac{\delta}{2}D \in \left[\frac{\delta}{4}D, (1 - \frac{\delta}{4})\,D\right]$, the backward greedy pattern p_{z_a} is proper. Because $\frac{1}{2}D \ge x_a \ge \frac{1}{6}D + \frac{5}{6}\delta D \ge \frac{\delta}{4}D$, it furthermore holds that x_a is far enough away from the boundary. We can thus apply Lemma 4 together with Lemma 3 and the fact that $x_a + y_a \in \left[\frac{7}{6}D + \frac{\delta}{3}D, \frac{3}{2}D - \frac{\delta}{2}D\right]$ to obtain a pattern with additive performance at most

$$\max\left\{\frac{17}{18}D + \frac{7}{18}\delta D, \frac{7}{6}D + \frac{\delta}{3}D\right\} = \left(\frac{7}{6} + \frac{\delta}{3}\right) D.$$

Case (b). If $d_m = \delta D$, we can rewrite $y_a = \frac{1}{2}\,(D + d_m) = \frac{1}{2}D + \frac{\delta}{2}D$. The lemma follows immediately from Observation 3 if $x_a \in [\bar{y}_a - \varepsilon, \bar{y}_a + \varepsilon]$. We can therefore assume that x_a falls either into the interval $\left[0, \frac{1}{3}D - \frac{5}{6}\delta D\right]$ or into the interval $\left[\frac{2}{3}D - \frac{\delta}{6}D, \frac{1}{2}D\right]$. Recall the assumption that x_a is at most $\frac{1}{2}D$. It is easy to see that $\frac{2}{3}D - \frac{\delta}{6}D \ge \frac{1}{2}D$ for all $\delta \in \left[0, \frac{1}{2}\right]$. It follows that $x_a \in \left[0, \frac{1}{3}D - \frac{5}{6}\delta D\right]$. Note that this interval is also empty for all $\delta > \frac{2}{5}$, and the lemma is trivially correct. We need this assumption, when arguing that we can apply Lemma 3.

As $y_a = \frac{1}{2}D + \frac{\delta}{2}D \in \left[\frac{\delta}{4}D, \left(1 - \frac{\delta}{4}\right) D\right]$, the backward greedy pattern p_{z_a} is proper. As x_a might be zero, we cannot apply Lemma 4. We therefore have to argue that the induced patterns are proper. Let p_{z_b} be a forward greedy pattern induced by p_{z_a} and p_{z_c} be a backward greedy pattern induced by p_{z_a}. By definition, we have $x_b = \frac{2}{3}\bar{y}_a + \frac{1}{3}x_a$ and $y_c = \frac{2}{3}\bar{x}_a + \frac{1}{3}y_a$. By substituting the definitions and bounds of y_a and x_a, we obtain

$$x_b = \frac{1}{3}D - \frac{\delta}{3}D + \frac{1}{3}x_a \geq \frac{1}{3}D - \frac{\delta}{3}D \geq \frac{\delta}{4}D.$$

The start point x_b is therefore far enough away from the boundary and the pattern p_{z_b} is thus proper. We similarly obtain

$$y_c = \frac{5}{6}D + \frac{\delta}{6}D - \frac{2}{3}x_a \leq \frac{5}{6}D + \frac{\delta}{6}D \leq D - \frac{\delta}{4}D,$$

for all $\delta \in \left[0, \frac{2}{5}\right]$. As we assumed that $\delta \leq \frac{2}{5}$, the backward greedy pattern p_{z_c} induced by p_{z_a} is proper. We can thus apply Lemma 3 together with the fact that $x_a + y_a \in \left[\frac{1}{2}D + \frac{\delta}{2}D, \frac{5}{6}D - \frac{\delta}{3}D\right]$ to obtain a pattern with additive performance at most

$$\max\left\{\frac{7}{6}D + \frac{\delta}{3}D, \frac{17}{18}D + \frac{7}{18}\delta D\right\} = \left(\frac{7}{6} + \frac{\delta}{3}\right) D.$$

In either case, the lemma follows.

An easy consequence of Lemma 5 is that there exists for any split routing solution a pattern with additive performance at most $\frac{4}{3}D$, which already improves upon the best known previous result of $\frac{19}{14}D$ from Skutella [25]. However, when combined with Skutellas [25] result on instances with medium demands (see Lemma 6), we obtain our main Theorem 1.

Lemma 6 ([25]). *For any δ-instance with $\delta \in \left[0, \frac{1}{2}\right]$ there exists a pattern with additive performance at most $\left(\frac{3}{2} - \frac{\delta}{2}\right) D$.*

Proof (of Theorem 1). Let $\delta \in \left[0, \frac{1}{2}\right]$ be such that the given split routing solution is a δ-instance. If $\delta \geq \frac{2}{5}$, the theorem follows from Lemma 6, as $\left(\frac{3}{2} - \frac{\delta}{2}\right) D \leq \frac{13}{10}D$. Otherwise, the theorem follows from Lemma 5, as $\left(\frac{7}{6} + \frac{\delta}{3}\right) D \leq \frac{13}{10}D$.

4 Bounds for Small Instances

In this section, we show lower and upper bounds for small instances of the *Ring Loading Problem.* The main result in this section implies in particular that in search of a lower bound example, stronger than Skutellas [25] instance with $L = L^* + \frac{11}{10}D$, one can concentrate on split routing solutions with at least $m \geq 8$ pairwise crossing demands. For full proofs, we refer to the preprint version [4].

Theorem 2. *Let $m \geq 2$ be an integer. Any split routing solution to the* Ring Loading Problem *with m split demands can be turned into an unsplittable solution without increasing the load on any edge by more than $(1 + \varepsilon) D$, if $m \leq 6$, and $\left(\frac{19}{18} + \varepsilon\right) D$, if $m = 7$, for $\varepsilon \leq 5 \times 10^{-6}$. Furthermore, there are instances of the* Ring Loading Problem *with m pairwise crossing demands with $L = L^* + D$, for $m \leq 6$, and $L = L^* + \frac{19}{18} D$, for $m = 7$.*

Proof (Proof sketch). The upper bounds follow from a mixed integer linear program (MILP) that outputs for a given integer m a split routing solution with m demands that cannot be turned into an unsplittable routing without increasing the load on some edge by at least αD, for $\alpha \geq 0$ as large as possible. Note that the dependency on ε is unavoidable, as our technique depends on solutions to large mixed integer linear programs that rely on floating point arithmetic. As the MILP is growing rapidly for increasing values of m, we are restricted to $m \leq 7$.

The (almost) matching lower bounds can also be deduced from MILP formulation.

On the way, we show that any split routing solution can be turned into an instance of the *Ring Loading Problem* while retaining the load increase. Note that this can be used in conjunction with the split routing solution of Schijver et al. [22] to provide a counterexample to their own conjecture, namely an instance of the *Ring Loading Problem* with $L = L^* + \frac{101}{100} D$.

Lemma 7. *Let $\alpha \geq 0$. Any split routing solution that cannot be turned into an unsplittable routing without increasing the load on some edge by at least αD can be turned into an instance of the* Ring Loading Problem *with $L - L^* \geq \alpha D$.*

5 Conclusions

We showed that any split routing solution to the *Ring Loading Problem* can be turned into an unsplittable solution while increasing the load on any edge by at most $\frac{13}{10} D$. We furthermore showed that split routing solutions with at most 7 pairwise crossing demands cannot yield lower bounds with additive performance worse than $\left(\frac{19}{18} + \varepsilon\right) D$, for a small ε.

The obvious open problem is the correct value of additive load increase. Skutella [25] conjectured that $L \leq L^* + \frac{11}{10} D$, which is matched by the currently best lower bound instance. After spending numerous hours on finding a stronger lower bound, unfortunately without any success, we are tempted to believe that this conjecture might be true. In any case, we highly doubt that the current best upper bound is the definitive answer.

Acknowledgements. We thank Martin Skutella for introducing us to the *Ring Loading Problem* and for many fruitful discussions and comments. We also thank Torsten Mütze for reading an early draft of the paper.

References

1. Ballart, R., Ching, Y.C.: SONET: now it's the standard optical network. IEEE Commun. Mag. **40**, 84–92 (2002)
2. Buchsbaum, A.L., Karloff, H., Kenyon, C., Reingold, N., Thorup, M.: OPT versus LOAD in dynamic storage allocation. SIAM J. Comput. **33**(3), 632–646 (2004). https://doi.org/10.1137/S0097539703423941
3. Cosares, S., Saniee, I.: An optimization problem related to balancing loads on SONET rings. Telecommun. Syst. **3**(2), 165–181 (1994). https://doi.org/10.1007/BF02110141
4. Däubel, K.: An improved upper bound for the ring loading problem (2019). https://arxiv.org/abs/1904.02119
5. Dell'Amico, M., Labbé, M., Maffioli, F.: Exact solution of the sonet ring loading problem. Oper. Res. Lett. **25**(3), 119–129 (1999). https://doi.org/10.1016/S0167-6377(99)00031-0
6. Dinitz, Y., Garg, N., Goemans, M.X.: On the single-source unsplittable flow problem. Combinatorica **19**(1), 17–41 (1999). https://doi.org/10.1007/s004930050043
7. Frank, A.: Edge-disjoint paths in planar graphs. J. Combin. Theory Ser. B **39**(2), 164–178 (1985). https://doi.org/10.1016/0095-8956(85)90046-2
8. Costain, G., Kennedy, S., Meagher, C. (eds.): Bellairs 2007 - Combinatorial Optimization Workshop Open Problems (2007). http://www.math.mcgill.ca/bshepherd/Bellairs/bellairs2007.pdf
9. Gergov, J.: Approximation algorithms for dynamic storage allocation. In: Diaz, J., Serna, M. (eds.) ESA 1996. LNCS, vol. 1136, pp. 52–61. Springer, Heidelberg (1996). https://doi.org/10.1007/3-540-61680-2_46
10. Gergov, J.: Algorithms for compile-time memory optimization. In: Proceedings of the Tenth Annual ACM-SIAM Symposium on Discrete Algorithms, pp. 907–908. Society for Industrial and Applied Mathematics (1999)
11. Goralski, W.: SONET/SDH, 3rd edn. McGraw-Hill Education, New York (2002)
12. Khanna, S.: A polynomial time approximation scheme for the SONET ring loading problem. Bell Labs Tech. J. **2**(2), 36–41 (1997). https://doi.org/10.1002/bltj.2047
13. Kierstead, H.A.: The linearity of first-fit coloring of interval graphs. SIAM J. Discrete Math. **1**(4), 526–530 (1988). https://doi.org/10.1137/0401048
14. Kierstead, H.A.: A polynomial time approximation algorithm for dynamic storage allocation. Discrete Math. **88**(2–3), 231–237 (1991). https://doi.org/10.1016/0012-365X(91)90011-P
15. Lee, C.Y., Chang, S.G.: Balancing loads on SONET rings with integer demand splitting. Comput. Oper. Res. **24**(3), 221–229 (1997). https://doi.org/10.1016/S0305-0548(96)00028-7
16. Martens, M., Salazar, F., Skutella, M.: Convex combinations of single source unsplittable flows. In: Arge, L., Hoffmann, M., Welzl, E. (eds.) ESA 2007. LNCS, vol. 4698, pp. 395–406. Springer, Heidelberg (2007). https://doi.org/10.1007/978-3-540-75520-3_36
17. Myung, Y.: An efficient algorithm for the ring loading problem with integer demand splitting. SIAM J. Discrete Math. **14**(3), 291–298 (2001). https://doi.org/10.1137/S0895480199358709
18. Myung, Y.S., Kim, H.G.: On the ring loading problem with demand splitting. Oper. Res. Lett. **32**(2), 167–173 (2004)
19. Myung, Y.S., Kim, H.G., Tcha, D.W.: Optimal load balancing on SONET bidirectional rings. Oper. Res. **45**(1), 148–152 (1997)

20. Nong, Q., Yuan, J., Lin, Y.: The weighted link ring loading problem. J. Comb. Optim. **18**(1), 38–50 (2009). https://doi.org/10.1007/s10878-007-9136-7
21. Nong, Q., Cheng, T., Ng, C.: A polynomial-time algorithm for the weighted link ring loading problem with integer demand splitting. Theor. Comput. Sci. **411**(31), 2978–2986 (2010). https://doi.org/10.1016/j.tcs.2010.04.035
22. Schrijver, A., Seymour, P., Winkler, P.: The ring loading problem. SIAM J. Discrete Math. **11**(1), 1–14 (1998). https://doi.org/10.1137/S0895480195294994
23. Shepherd, F.B.: Single-sink multicommodity flow with side constraints. In: Cook, W., Lovász, L., Vygen, J. (eds.) Research Trends in Combinatorial Optimization, pp. 429–450. Springer, Berlin (2009). https://doi.org/10.1007/978-3-540-76796-1_20
24. Skutella, M.: Approximating the single source unsplittable min-cost flow problem. Math. Program. **91**(3, Ser. B), 493–514 (2002). https://doi.org/10.1007/s101070100260
25. Skutella, M.: A note on the ring loading problem. SIAM J. Discrete Math. **30**(1), 327–342 (2016). https://doi.org/10.1137/14099588X
26. Vachani, R., Shulman, A., Kubat, P., Ward, J.: Multicommodity flows in ring networks. INFORMS J. Comput. **8**(3), 235–242 (1996). https://doi.org/10.1287/ijoc.8.3.235
27. Vasseur, J.P., Pickavet, M., Demeester, P.: Network Recovery: Protection and Restoration of Optical, SONET-SDH, IP, and MPLS. Morgan Kaufmann Publishers Inc., Burlington (2004)
28. Wang, B.F.: Linear time algorithms for the ring loading problem with demand splitting. J. Algorithms **54**(1), 45–57 (2005). https://doi.org/10.1016/j.jalgor.2004.03.003

Parallel Online Algorithms for the Bin Packing Problem

Sándor P. Fekete, Jonas Grosse-Holz, Phillip Keldenich, and Arne Schmidt(✉)

Department of Computer Science, TU Braunschweig, Braunschweig, Germany
{s.fekete,j.grosse-holz,p.keldenich,arne.schmidt}@tu-bs.de

Abstract. We study *parallel* online algorithms: For some fixed integer k, a collective of k parallel processes that perform online decisions on the same sequence of events forms a k-*copy algorithm*. For any given time and input sequence, the overall performance is determined by the best of the k individual total results. Problems of this type have been considered for online makespan minimization; they are also related to optimization with *advice* on future events, i.e., a number of bits available in advance.

We develop PREDICTIVE HARMONIC$_3$ (PH3), a relatively simple family of k-copy algorithms for the online Bin Packing Problem, whose joint competitive factor converges to 1.5 for increasing k. In particular, we show that $k = 6$ suffices to guarantee a factor of 1.5714 for PH3, which is better than 1.57829, the performance of the best known 1-copy algorithm ADVANCED HARMONIC, while $k = 11$ suffices to achieve a factor of 1.5406, beating the known lower bound of 1.54278 for a single online algorithm. In the context of online optimization with advice, our approach implies that 4 bits suffice to achieve a factor better than this bound of 1.54278, which is considerably less than the previous bound of 15 bits.

Keywords: Online algorithms · Bin packing · Competitive analysis

1 Introduction

When dealing with unknown future events, optimization with incomplete information typically considers the competitive factor of an online algorithm as its performance measure; the objective becomes to develop a single strategy that performs reasonably well against the worst case. This focus on just *one* option is more restrictive than hedging strategies in a wide variety of other scientific and application fields; these typically make use of *several* parallel choices, thereby increasing the chance that one of them will yield satisfactory results. Examples include scenarios from biology, where a large and diverse progeny increases the odds of surviving offspring; finance and insurance, where a suitable combination of investment strategies is employed to balance a portfolio against extreme losses;

A full version is available on arxiv.org [10].

Phillip Keldenich was partially supported by DFG grant FE407/17-2 as part of the Research Group FOR 1800, “Controlling Concurrent Change”.

E. Bampis and N. Megow (Eds.): WAOA 2019, LNCS 11926, pp. 106–119, 2020.
https://doi.org/10.1007/978-3-030-39479-0_8

and engineering, where redundancy is used to protect against catastrophic failure, either on individual components (such as parts in a machine) or on whole systems (such as automata in a robot swarm or spacecraft in a group of satellites), where it suffices that just one machine delivers a good outcome.

In this paper, we consider such *parallel* online strategies: Instead of making a single sequence of decisions, we consider k parallel processes for some fixed integer k, which we call a *k-copy algorithm*; the objective is to make the best of these k outcomes as good as possible, even in the worst case. We demonstrate the potential of this approach for the well-studied Bin Packing Problem, for which it is known that no single deterministic online algorithm can achieve a competitive factor below 1.5401.

1.1 Our Results

We define a family of k-copy algorithms for the online Bin Packing Problem, called PREDICTIVE HARMONIC$_3$ (PH3), whose asymptotic competitive ratio converges to 1.5 for large k. We show that $k = 6$ suffices to guarantee a factor of 1.5714, which is better than 1.57829, the performance of the best known 1-copy algorithm ADVANCED HARMONIC [3]. Moreover, $k = 11$ suffices to achieve a competitive ratio of 1.5406 beating the known lower bound of 1.54278 for a 1-copy algorithm [4]. In the context of online optimization with advice, our approach implies that 4 bits suffice to achieve less than 1.5401, which is considerably less than the previous bound of 16 bits of REDBLUE by Angelopoulos et al. [2]; in fact, for $k = 16$ (corresponding to four bits of advice) PH3 achieves a ratio of 1.5305, compared to 3.3750 for REDBLUE, while $k = 65,536$ (corresponding to 16 bits of advice) yields a factor of 1.5001 for PH3, but 1.5293 for REDBLUE.

1.2 Related Work on Online Bin Packing

There is a wide range of online algorithms for bin packing. The Next Fit algorithm [9] achieves a competitive ratio of 2, whereas "Almost Any Fit" algorithms [13] like First Fit or Best Fit achieve competitive ratios of 1.7.

An important online bin packing algorithm is HARMONIC$_M$, introduced by Lee and Lee [15], which achieves a competitive ratio of less than 1.692 for $M \to \infty$. Based on HARMONIC$_M$, SON OF HARMONIC by Heydrich and van Stee [12] achieves a competitive ratio of 1.5816. The currently best known algorithm is ADVANCED HARMONIC, which achieves a competitive ratio of 1.57829 [3].

For lower bounds, Yao [20] established a value of 3/2 that was later improved to 1.536, independently by Brown [8] and by Liang [16]. Using a generalization of their methods, van Vliet [19] proved a lower bound of 1.5401. Balogh et al. [4] improved the lower bound to 1.54278.

1.3 Related Work on Online Bin Packing with Advice

In the context of online algorithms with advice, Boyar et al. [7] showed that an online algorithm with $n\lceil\log(OPT(I))\rceil$ bits of advice is sufficient and that at least $(n - 2OPT(I)) \cdot \log(OPT(I))$ bits of advice are necessary to achieve optimality. In the same paper, they presented an online bin packing algorithm, namely ReserveCritical, with $O(\log(n)) + o(\log(n))$ bits of advice that is 1.5-competitive and an algorithm with $2n + o(n)$ bits of advice that is $\frac{4}{3}$-competitive. Zhao and Shen [21] developed an algorithm using $3n + o(n)$ bits of advice achieving a competitive ratio of $\frac{5}{4}\text{OPT} + 2$. Renault et al. [18] developed an $(1+\varepsilon)$-competitive algorithm using $O(\frac{1}{\varepsilon}\log\frac{1}{\varepsilon})$ bits of advice per request.

Based on ReserveCritical, Angelopoulos et al. [2] developed the algorithm RedBlue with constant advice that is 1.5-competitive. Their second algorithm achieves a competitive ratio of $1.47012 + \varepsilon$ with finite advice that is exponentially dependent of ε. However, to beat the competitive ratio of 1.5 already an enormous amount of advice is needed, which makes the algorithm impractical.

In terms of lower bounds, Boyar et al. [7] proved that no competitive ratio better than 9/8 can be reached by any algorithm that uses sub-linear advice. Angelopoulos et al. [2] improved this bound to 7/6.

1.4 Related Work on Parallel Online Algorithms

Parallel algorithms have already been considered in the field of online algorithms with advice. Boyar et al. [6] presented an algorithm for the online list update problem, making use of 2 bits of advice to choose one out of three algorithms. This algorithm achieves a competitive ratio of 5/3, beating the lower bound for conventional online algorithms of 2. A practical application of this algorithm was shown by Kamali and Ortiz [14], who applied it in the Burrows-Wheeler transform compression. More work on parallel online algorithms include parallel scheduling [1], finding independent sets [11] and the "multiple-cow" version of the linear search problem [17].

While online algorithms with advice mostly focus on the amount of advice to allow classification of online algorithms and problems, k-copy online algorithms focus on small finite values for k and thus small finite amounts of advice, with more emphasis on practical application. The perspective on different algorithms running in parallel instead of abstract arbitrary information facilitates finer optimization in some cases.

Also, when considering online algorithms with advice, the number of algorithms can only be doubled by increasing the amount of advice by one bit. The perspective of k-copy algorithms allows arbitrary $k \in \mathbb{N}$ for the number of algorithms.

2 Preliminaries

2.1 k-Copy Online Algorithms

In this paper, we consider k online algorithms $A_1, \ldots, A_k$, each of them processing the same input list I in parallel. We call the set $\mathcal{A} := \{A_1, \ldots, A_k\}$ a *k-copy online algorithm.*

For an input list I and an online algorithm A, let $A(I)$ denote the number of bins used by A and $\text{OPT}(I)$ denote the number of bins used in an optimal offline solution. The absolute competitive ratio $R_{\mathcal{A}}$ for a k-copy online algorithm $\mathcal{A}$ is defined as

$$R_{\mathcal{A}} = \sup_I \left\{ \frac{\min_{A \in \mathcal{A}} A(I)}{\text{OPT}(I)} \right\}.$$

The asymptotic competitive ratio $R^{\infty}_{\mathcal{A}}$ for algorithm $\mathcal{A}$ is defined as

$$R^{\infty}_{\mathcal{A}} = \lim_{n \to \infty} \sup_I \left\{ \frac{\min_{A \in \mathcal{A}} A(I)}{\text{OPT}(I)} \,\middle|\, \text{OPT}(I) = n \right\}$$

As already stated by Boyar et al. [5], any k-copy online algorithm can be converted into an online algorithm with advice, and vice versa.

Lemma 1. *Any k-copy online algorithm can be converted into an online algorithm with $l = \lceil \log_2(k) \rceil$ bits of advice that achieves the same competitive ratio. Conversely, any online algorithm with $l \in \mathbb{N}$ bits of advice can be converted into a k-copy online algorithm without advice with $k = 2^l$ that achieves the same competitive ratio.*

Proof. Let $\mathcal{A} = \{A_1, A_2, \ldots, A_k\}$ be a k-copy algorithm. Construct the online algorithm A' that gets a value $i \in \{1, 2, \ldots, k\}$ as advice, specifying the index i of the algorithm $A_i \in \mathcal{A}$ that performs best on the given input sequence. The value i can be encoded using $\lceil \log_2(k) \rceil$ bits. A' then behaves like A_i and thus achieves the same competitive ratio as $\mathcal{A}$.

Let A be an online algorithm that gets $l \in \mathbb{N}$ bits of advice. Construct the online k-copy algorithm $\mathcal{A}'$ with $k = 2^l$ algorithms $A_i, i \in \{1, 2, \ldots, k\}$. For each $i \in \{1, 2, \ldots, k\}$, the algorithm A_i behaves like A given i encoded in binary as advice. As the values $i \in \{1, 2, \ldots, k\}$ cover every possible configuration of the advice bits, for any advice given to A, there is an algorithm $A_i \in \mathcal{A}'$, that assumes this advice. Accordingly, there is an algorithm $A_i \in \mathcal{A}'$, that performs as well as A, i.e., the best algorithm $A_j \in \mathcal{A}$ that performs at least as well as A. Thus, $\mathcal{A}'$ performs at least as well as A.

2.2 Bin Packing

In the online version of bin packing, we are given a list of items $I := \langle a_1, \ldots, a_n \rangle$ with $a_i \in (0, 1]$ for $i \in \{1, \ldots, n\}$. These items must be packed by an algorithm, one at a time, without any information on subsequent items and without the possibility to change previous decisions. The goal is to pack all items into a minimum number of bins with unit capacity.

Definition 1 (Item size).
Let $S = \left[0, \frac{1}{3}\right]$, $M = \left(\frac{1}{3}, \frac{1}{2}\right]$, $L = \left(\frac{1}{2}, \frac{2}{3}\right)$ and $XL = \left[\frac{2}{3}, 1\right]$. We call items in S small, *items in M* medium, *items in L* large *and items in XL* extra large. *For a list $I = \langle a_1, a_2, \ldots a_n \rangle$, the set of items $Set(I) \cap XL$ is noted as I_{XL} for improved readability. The subsets I_L, I_M and I_S are used analogously.*

Definition 2 (Size function). *Let S be a set (or list) of items. Then,* $\text{size}(S) := \sum_{i \in S} i$*. For a bin b, we refer to* $\text{size}(b)$ *as the size of the bin, i.e., the sum of items already packed in b.*

Definition 3 (Sub-bins).
Given a bin b, it can be split into two parts b_1 and b_2, such that the sum of their capacities is equal to the capacity of b. We refer to b_1 and b_2 as sub-bins. We call a sub-bin with capacity C a C-sub-bin.

As sub-bins are not packed with an amount larger than their capacity, each sub-bin can be packed independently from the other.

3 Predictive Harmonic$_3$

Now we introduce the algorithm Predictive Harmonic$_3$ (PH3). Although developed independently, it bears many similarities to ReserveCritical and RedBlue. PH3 uses the same classifications as the other two algorithms and tries to pack all large items with small items, such that the corresponding bins are packed to a level of at least 2/3. However, in contrast to RedBlue, the information needed by PH3 does not depend on the result of ReserveCritical, but only on the number and size of certain item types, and can be calculated in linear time.

The main idea of PH3 is to guess the ratio of how many small items must be packed with large items to obtain a packing density of 2/3. Having multiple instances of PH3, every instance can guess a different ratio to get close to a competitive ratio of 1.5.

Algorithm 1 Predictive Harmonic$_3$. Given a list $I = \langle a_1, a_2, \ldots, a_n \rangle$ of items $a_i \in (0, 1], i \in 1, \ldots, n$, and a ratio $r_L \in [0, 1]$, the algorithm packs the items as follows:

- Extra large items are packed into individual bins. These bins are called XL-bins, the set of all XL-bins is called B_{XL}.
- Large items are packed into individual bins. These bins are called L-bins, the set of all L-bins is called B_L. Furthermore, we split each L-bin into a $\frac{2}{3}$-sub-bin (for large items) and a $\frac{1}{3}$-sub-bin (for small items).
- Medium items are packed into separate bins together with other medium items (note that at most two of them fit into one bin). These bins are called M-bins, the set of all M-bins is called B_M.
- Small items are packed into a $\frac{1}{3}$-sub-bin of L-bins in a next fit manner, if the size of small items packed into L-bins is smaller than r_L times the total size of small items packed so far; otherwise we pack the small item into S-bins.

3.1 Competitive Ratio

Using simple bounds for an optimal solution and performing a case analysis, we can prove the following theorem. Due to space constraints, the proof can be found in the full version [10].

Theorem 2. *Let* $r_L^* = min\left\{\frac{|I_L|}{6\ \text{size}(I_S)}, 1\right\}$[1] *and* $\delta = r_L - r_L^*$. *PH3 achieves the asymptotic competitive ratio*

$$R_{PH3}^{\infty} \leq \begin{cases} \frac{3}{2} + min\left\{\frac{1}{4r_L^*}, \frac{3}{6r_L^* + 2}\right\}(-\delta) & \textit{for } \delta \leq 0 \\ \frac{3}{2} + min\left\{\frac{3}{4r_L^*}, \frac{9}{6r_L^* + 2}\right\}\delta & \textit{for } \delta \geq 0. \end{cases}$$

3.2 Tightness

Theorem 3. *For any* $r_L, r_L^* \in [0, 1]$, *the asymptotic competitive ratio given in Theorem 2 is tight.*

Proof sketch: Let $\langle a_1, a_2, \ldots a_k \rangle \times n$ with $n \in \mathbb{N}$ denote n repetitions of the sequence $\langle a_1, a_2, \ldots a_k \rangle$. Let I be a sequence consisting of concatenated sub-sequences I_S, I_M and I_L, where I_S is a sequence consisting of two interleaved sub-sequences I_{SL} and I_{LL}. With $N \in \mathbb{N}$ and $\varepsilon = 1/(12N + 2)$, we define

$$I_L = \left(\frac{1}{2} + \frac{\varepsilon}{2}\right) \times n_L \text{ with } n_L = \lceil 4r_L^* N \rceil$$

$$I_M = \left(\frac{1}{3} + \frac{\varepsilon}{2}\right) \times n_M \text{ with } n_M = \begin{cases} 0 & \text{for } r_L^* \leq 1/3 \\ \lfloor (6r_L^* - 2)N \rfloor & \text{for } r_L^* \geq 1/3 \end{cases}$$

$$I_{SS} = \left(\frac{1}{3} - 2\varepsilon, \frac{1}{6} - \varepsilon, \frac{1}{6} - \varepsilon, 12\varepsilon\right) \times n_{SS} \text{ with } n_{SS} = \lceil n'_{SS} \rceil = \lceil (1 - r_L)N \rceil$$

$$I_{SL} = \left(\frac{1}{6} - \varepsilon, 3\varepsilon\right) \times n_{SL} \text{ with } n_{SL} = \lceil n'_{SL} \rceil = \lceil 4r_L N \rceil$$

The proof is based on a case analysis of which item appears next and in which bin this item is packed by PH3. Due to space constraints, a full proof can be found in the full version [10].

4 Parallel PREDICTIVE HARMONIC3

4.1 Competitive Ratio for PH3 as 1-Copy Online Algorithm

To optimize the performance for PH3 as a 1-copy algorithm, we determine the optimal value for r_L with respect to minimizing the asymptotic competitive ratio over all $r_L^* \in [0, 1]$.

---

[1] The intuition of this value is that at least 1/2 of each 1/3-sub-bin must be filled to guarantee a packing density of 2/3. Therefore, for $|I_L|$ bins, we have to fill up a total capacity of $\frac{|I_L|}{6}$ with small items.

Lemma 2 (Monotonicity of competitive ratio of PH3). *For any fixed $r_L \in [0,1]$, the competitive factor is monotonically decreasing for $r_L^* \in [0, r_L]$ and monotonically increasing for $r_L^* \in [r_L, 1]$.*

Proof. Assume r_L to be fixed. Let $r_{+,<}, r_{-,<} : [0, 1/3] \to \mathbb{R}$ and $r_{+,>}, r_{-,>} : [1/3, 1] \to \mathbb{R}$ with

$$
\begin{aligned}
r_{-,<}(r_L^*) &= \frac{3}{2} + \frac{3}{6r_L^* + 2}(-\delta) && = R_{PH3}^{\infty} \text{ for } \delta \leq 0, r_L^* \leq \frac{1}{3} \\
r_{-,>}(r_L^*) &= \frac{3}{2} + \frac{1}{4r_L^*}(-\delta) && = R_{PH3}^{\infty} \text{ for } \delta \leq 0, r_L^* \geq \frac{1}{3} \\
r_{+,<}(r_L^*) &= \frac{3}{2} + \frac{9}{6r_L^* + 2}\delta && = R_{PH3}^{\infty} \text{ for } \delta \geq 0, r_L^* \leq \frac{1}{3} \\
r_{+,>}(r_L^*) &= \frac{3}{2} + \frac{3}{4r_L^*}\delta && = R_{PH3}^{\infty} \text{ for } \delta \geq 0, r_L^* \geq \frac{1}{3}
\end{aligned}
$$

Consider the derivative of $r_{-,<}$ and $r_{-,>}$.

$$
\begin{aligned}
\frac{\partial}{\partial r_L^*} r_{-,<}(r_L^*) &= \frac{\partial}{\partial r_L^*}\left(\frac{3}{2} + \frac{3}{6r_L^* + 2}(-\delta)\right) \\
&= \frac{\partial}{\partial r_L^*}\left(\frac{3(r_L^* - r_L)}{6r_L^* + 2}\right) \\
&= \frac{18r_L + 6}{(6r_L^* + 2)^2} \geq 0 \text{ for } 0 \leq r_L \leq r_L^* \leq \frac{1}{3} \\
\frac{\partial}{\partial r_L^*} r_{-,>}(r_L^*) &= \frac{\partial}{\partial r_L^*}\left(\frac{3}{2} + \frac{1}{4r_L^*}(-\delta)\right) \\
&= \frac{\partial}{\partial r_L^*}\left(\frac{r_L^* - r_L}{4r_L^*}\right) \\
&= \frac{r_L}{4(r_L^*)^2} \geq 0 \text{ for } 0 \leq r_L \leq r_L^* \text{ and } \frac{1}{3} \leq r_L^* \leq 1
\end{aligned}
$$

As the derivatives of $r_{-,<}$ and $r_{-,>}$ are both non-negative in their respective domains, they are both monotonically increasing. Because $r_{-,<}(\frac{1}{3}) = r_{-,>}(\frac{1}{3})$, we conclude that the competitive ratio is monotonically increasing for $r_L^* \in [r_L, 1]$.

Now consider the derivative of $r_{+,<}$ and $r_{+,>}$.

$$\begin{aligned}
\frac{\partial}{\partial r_L^*} r_{+,<}(r_L^*) &= \frac{\partial}{\partial r_L^*}\left(\frac{3}{2} + \frac{9}{6r_L^* + 2}\delta\right) \\
&= \frac{\partial}{\partial r_L^*}\left(\frac{9(r_L - r_L^*)}{6r_L^* + 2}\right) \\
&= \frac{-54r_L - 18}{(6r_L^* + 2)^2} \leq 0 \text{ for } r_L^* \leq r_L \leq 1 \text{ and } 0 \leq r_L^* \leq \frac{1}{3} \\
\frac{\partial}{\partial r_L^*} r_{+,>}(r_L^*) &= \frac{\partial}{\partial r_L^*}\left(\frac{3}{2} + \frac{3}{4r_L^*}\delta\right) \\
&= \frac{\partial}{\partial r_L^*}\left(\frac{3(r_L - r_L^*)}{4r_L^*}\right) \\
&= \frac{-3r_L}{4(r_L^*)^2} \leq 0 \text{ for } \frac{1}{3} \leq r_L^* \leq r_L \leq 1
\end{aligned}$$

As the derivatives of $r_{+,<}$ and $r_{+,>}$ are both non-positive in their respective domains, they are both monotonically decreasing. Because $r_{+,<}(\frac{1}{3}) = r_{+,>}(\frac{1}{3})$, we conclude that the competitive ratio is monotonically decreasing for $r_L^* \in [0, r_L]$.

Because of Lemma 2, the competitive ratio does not decrease with r_L^* increasing for $\delta \leq 0$. Thus, as an upper bound on the competitive ratio for $\delta \leq 0$, only the competitive ratio for $r_L^* = 1$ has to be considered.

$$\begin{aligned}
R_{PH3}^{\infty} &\leq \frac{3}{2} + \frac{1}{4}(-\delta) \text{ for } \delta \leq 0 \\
&= \frac{3}{2} + \frac{1}{4}(1 - r_L) \\
&= \frac{7}{4} - \frac{r_L}{4}
\end{aligned}$$

For $\delta \geq 0$, the competitive ratio does not decrease with r_L^* decreasing. In this case, the competitive ratio for $r_L^* = 0$ is an upper bound on the competitive ratio.

$$\begin{aligned}
R_{PH3}^{\infty} &\leq \frac{3}{2} + \frac{9}{2}\delta \text{ for } \delta \geq 0 \\
&= \frac{3}{2} + \frac{9}{2}(r_L - 0) \\
&= \frac{3}{2} + \frac{9}{2}r_L
\end{aligned}$$

At the same time, these values are lower bounds on the overall competitive ratio. Given these bounds, this linear program can be formulated to minimize the competitive ratio:

$$\begin{aligned} \text{Minimize } & R^{\infty}_{PH3} \\ \text{Subject to } & R^{\infty}_{PH3} \geq \frac{7}{4} - \frac{r_L}{4} \\ & R^{\infty}_{PH3} \geq \frac{3}{2} + \frac{9}{2} r_L \\ & r_L \geq 0 \\ & r_L \leq 1 \end{aligned}$$

The optimal solution for this linear program is $r_L = 1/19$ and $R^{\infty}_{PH3} = 33/19 < 1.7369$. Figure 1 shows the asymptotic competitive ratio of PH3 over r^*_L for $r_L = 1/19$.

Compared to other known algorithms for online bin packing, PH3 is not a good choice for worst-case behavior. Among the classical algorithms, only NF and WF, both of which are 2-competitive, are worse than PH3. Any AAF algorithm achieves an asymptotic competitive ratio $R^{\infty}_{AAF} = 1.7$ [9] and thus performs slightly better than PH3. The best-performing online algorithm for bin packing currently known, Son Of Harmonic, is 1.5816-competitive and thus clearly superior to PH3 [12].

However, if we know in advance that r^*_L is restricted to some interval $I_r = [a, b] \subset [0, 1]$, the above argument can be used to prove a better competitive ratio.

4.2 Competitive Ratio for PH3 as *k*-Copy Online Algorithm

PH3's property of achieving a better competitive ratio for r^*_L being further restricted can be used to create a set of $k \in \mathbb{N}$ algorithms achieving a better competitive ratio. For this purpose, the interval $[0, 1]$ is split into k sub-intervals $I_1, \ldots, I_k \subset [0, 1]$ with $\cup_{i \in \{1,\ldots,k\}} I_i = [0, 1]$. Each interval I_i is covered by one instance of the algorithm PH3 A_i, such that A_i achieves a targeted competitive ratio $R \in (3/2, 33/19)$ for $r^*_L \in I_i$.

R is restricted to $(3/2, 33/19)$, because any competitive ratio above or equal to 33/19 can be achieved with the instance of PH3 shown above, and k-copy PH3 cannot achieve a competitive ratio of 3/2 or less with finitely many algorithms.

To calculate the number k of algorithms needed to achieve a given competitive ratio R, the following iterative approach can be used.

Let $\mathcal{A}$ be a set of algorithms. Initially, $\mathcal{A} := \emptyset$. We initialize our iterative approach with $i = 0$ and set $r^0_{max} = 0$. Then, while $r^i_{max} < 1$, we increase i by one and we compute three values r^i_{min}, r^i_L and r^i_{max}. With these three values we can define algorithm A_i for which r^i_L denotes the value of r_L, r^i_{min} denotes the minimal and r^i_{max} denotes the maximal value for r^*_L for which A_i is still R-competitive. By Lemma 2, A_i will be R-competitive for the interval $[r^i_{min}, r^i_{max}]$. All three values are computed as follows. We set $r^i_{min} = r^{i-1}_{max}$. Given r^i_{min}, r^i_L can be computed:

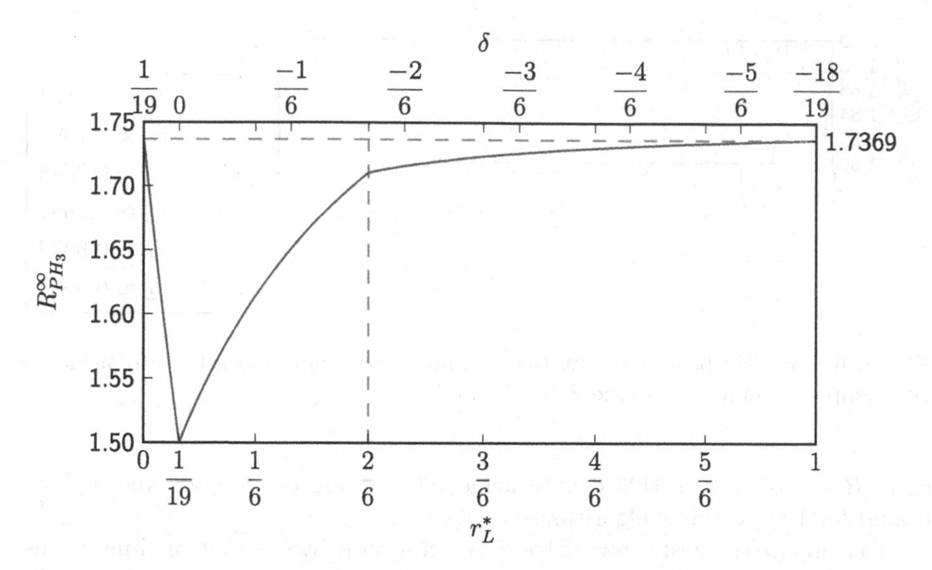

Fig. 1. Competitive ratio of the optimal 1-copy PH3 algorithm dependent on r^*_L for a fixed r_L.

If $r^i_{min} \leq 1/3$, we have $R = \frac{3}{2} + \frac{9}{2+6r^i_{min}}(r^i_L - r^i_{min})$. Solving this equation for r^i_L we get $r^i_L = r^i_{min} + \left(R - \frac{3}{2}\right)\left(\frac{2+6r^i_{min}}{9}\right)$. If $r^i_{min} \geq 1/3$, we have $R = \frac{3}{2} + \frac{3}{4r^i_{min}}(r^i_L - r^i_{min})$. Solving this equation for r^i_L yields $r^i_L = r^i_{min} + \left(R - \frac{3}{2}\right)\left(\frac{4r^i_{min}}{3}\right)$.

Having r^i_L, we can compute r^i_{max}. Because the competitive ratio is the minimum of two values, we get two candidates $r^i_{max,1}$ and $r^i_{max,2}$ for r^i_{max}. We can take the maximum of those two candidates, i.e., $r^i_{max} = \max(r^i_{max,1}, r^i_{max,2})$, because it is sufficient to be R-competitive in one case. In the first case $\left(\frac{3}{6r^*_L+2} < \frac{1}{4r^*_L}\right)$ we obtain $r^i_{max,1} = \frac{3r^i_L-3+2R}{12-6R}$ and in the second case we get $r^i_{max,2} = \frac{r^i_L}{7-4R}$.

Now consider the case when $r^i_{max} \geq 1$. Because each algorithm A_ℓ with $1 \leq \ell \leq i$ is R-competitive for the interval $[r^\ell_{min}, r^\ell_{max}] = [r^{\ell-1}_{max}, r^\ell_{max}]$ with $r^0_{min} = 0$, there is an algorithm A_m for any $r^*_L \in [0,1]$ that is R-competitive. Therefore, we have a i-copy online algorithm for bin packing achieving the competitive factor R.

Following this method, we see that $k = 6$ algorithms are sufficient to guarantee a competitive ratio $R = 1.5815$. This beats the currently best 1-copy online algorithm Son Of Harmonic with a competitive ratio of 1.5816. Figure 2 shows the competitive ratio achieved by the individual algorithms over $r^*_L \in [0,1]$ for $R = 1.5815$. Note that 1.5815 is not the best competitive ratio achievable by 6-copy PH3, as shown below in Fig. 3. Using $k = 12$ algorithms, a competitive

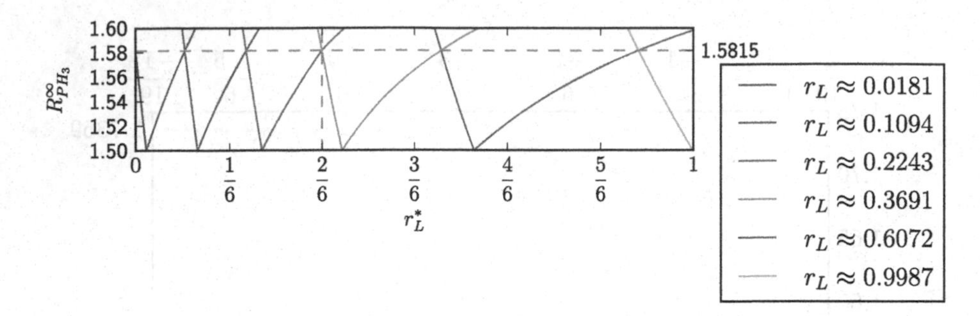

Fig. 2. 6-copy PH3 beats the best 1-copy online algorithm known to date, achieving an asymptotic competitive ratio $R^{\infty}_{PH3} < 1.5815$.

ratio $R = 1.5402 < 1.5403$ can be achieved, beating the highest known lower bound for 1-copy online algorithms.

To compute the best competitive ratio achievable by $k \in \mathbb{N}$ algorithms, we use binary search on R starting in the interval $[3/2, 33/19]$ and test in each iteration if we can guarantee R-competitiveness with at most k algorithms. Figure 3 shows the best competitive ratios achievable by k-copy PH3.

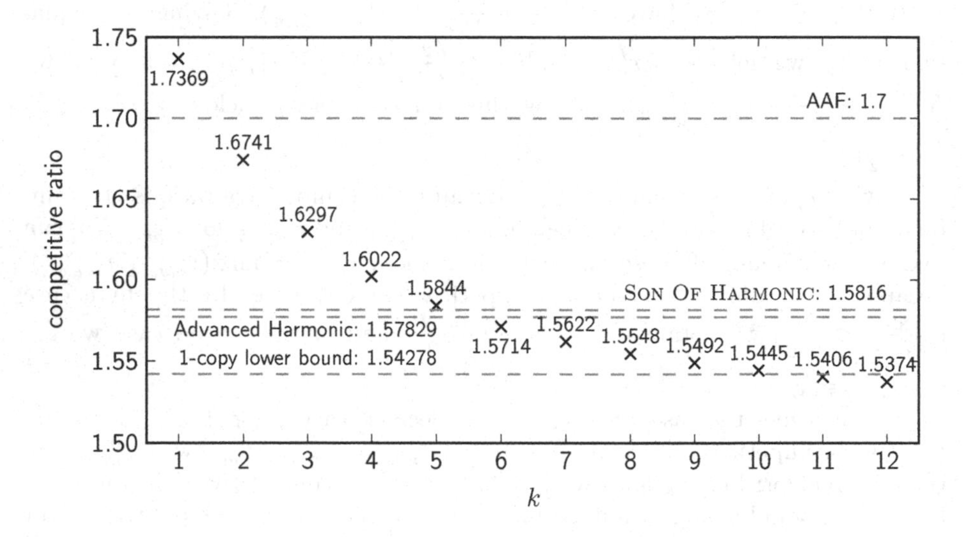

Fig. 3. k-copy PH3 performance dependent on k.

4.3 Comparison to Related Algorithms

Because k-copy online algorithms can be translated to an online algorithm with advice and vice versa (see Lemma 1), it seems natural to compare these two variants, even though k-copy allows a more precise analysis on the competitive ratio. In this subsection we compare our algorithm to the best known online algorithm with constant advice, namely RedBlue introduced by Angelopoulos et al. [2]. Their second algorithm is 1.47012-competitive (and thus beats our algorithm), but the amount of advice needed by this algorithm is too large. As the focus of k-copy algorithms is to provide good solutions for small k, it is reasonable to only compare k-copy PH3 to RedBlue.

Table 1 shows a comparison between RedBlue and k-copy PH3 for small amounts of advice. The competitive ratios given are rounded up to the fourth decimal place. The competitive ratios for RedBlue are computed using the upper bound on the competitive ratio $1.5 + 15/(2^{\ell/2+1})$. The competitive ratios for k-copy PH3 are calculated using binary search as described above.

Table 1. Comparison of the performance of k-copy PH3 and RedBlue.

Advice in bits	k	$R^{\infty}_{\text{RedBlue}}$	R^{∞}_{PH3}
4	16	3.3750	1.5305
5	32	2.8258	1.5155
6	64	2.4375	1.5078
7	128	2.1629	1.5040
8	256	1.9688	1.5020
9	512	1.8315	1.5010
10	1024	1.7344	1.5005
11	2048	1.6657	1.5003
12	4096	1.6172	1.5002
13	8192	1.5829	1.5001
14	16384	1.5586	1.5001
15	32768	1.5414	1.5001
16	65536	1.5293	1.5001

Table 1 clearly shows the advantage of k-copy PH3 over RedBlue for few bits of advice. With as few as 5 bits of advice, or $k = 32$, k-copy PH3 achieves a better competitive ratio than RedBlue with 16 bits of advice, which corresponds to $k = 65536$ algorithms when used as k-copy algorithm.

Although RedBlue and k-copy PH3 work in a similar way, k-copy PH3 achieves a better competitive ratio due to the more precise analysis of the intervals for r^*_L, in which each algorithm achieves the competitive ratio. By avoiding overlaps in these intervals, fewer algorithms are needed.

On the other hand, RedBlue simply splits an interval for its parameter β evenly into $2^{\ell/2}$ intervals; translated into a k-copy setting, this leads to overlaps in the intervals covered by each algorithm.

5 Conclusion

We studied the concept of parallel online algorithms for the Bin Packing Problem. We developed a k-copy online algorithm named PH3 and showed that PH3 has an asymptotic competitive ratio of 1.5 for large k; in particular, $k = 11$ suffices to break through the lower bound of a single online algorithm. We also considered the relationship to online algorithms with advice and achieved a considerable improvement compared to a previous algorithm.

There are various directions for future work. We saw that PH3 is $(1.5 + \varepsilon)$-competitive if $\frac{|I_L|}{6\,\mathrm{size}(I_S)} \leq 1$, i.e., when there is a surplus of small items. If there are too few small items, PH3 is asymptotically $(1.5 + \varepsilon)$-competitive. Can we make better use of the second case for an improvement? Can we guarantee an absolute competitive ratio of $1.5(+\varepsilon)$?

How does the asymptotic competitive ratio of PH3 depend on k? It seems to be something like $\frac{3}{2} + O\left(\frac{1}{k+\log_2(k+1)}\right)$. Translated to an online algorithm with ℓ bits of advice, this would yield an asymptotic competitive ratio of $\frac{3}{2} + O\left(\frac{1}{2^\ell+\ell}\right)$.

We also believe that the concept of k-copy algorithms is useful for a wide range of other problems.

References

1. Albers, S., Hellwig, M.: Online makespan minimization with parallel schedules. Algorithmica **78**(2), 492–520 (2017)
2. Angelopoulos, S., Dürr, C., Kamali, S., Renault, M., Rosén, A.: Online bin packing with advice of small size. In: Dehne, F., Sack, J.-R., Stege, U. (eds.) WADS 2015. LNCS, vol. 9214, pp. 40–53. Springer, Cham (2015). https://doi.org/10.1007/978-3-319-21840-3_4
3. Balogh, J., Békési, J., Dósa, G., Epstein, L., Levin, A.: A new and improved algorithm for online bin packing. In: 26th Annual European Symposium on Algorithms (ESA 2018). Schloss Dagstuhl-Leibniz-Zentrum fuer Informatik (2018)
4. Balogh, J., Békési, J., Dósa, G., Epstein, L., Levin, A.: A new lower bound for classic online bin packing. arXiv preprint arXiv:1807.05554 (2018). (To appear at 17th Workshop on Approximation and Online Algorithms (WAOA))
5. Boyar, J., Favrholdt, L.M., Kudahl, C., Larsen, K.S., Mikkelsen, J.W.: Online algorithms with advice: a survey. ACM Comput. Surv. (CSUR) **50**(2), 19 (2017)
6. Boyar, J., Kamali, S., Larsen, K.S., López-Ortiz, A.: On the list update problem with advice. In: 8th Conference on Language and Automata Theory and Applications (LATA), pp. 210–221 (2014)
7. Boyar, J., Kamali, S., Larsen, K.S., López-Ortiz, A.: Online bin packing with advice. Algorithmica **74**(1), 507–527 (2016)
8. Brown, D.M.: A lower bound for on-line one-dimensional bin packing algorithms. Technical report (1979)
9. Csirik, J., Woeginger, G.J.: On-line packing and covering problems. In: Fiat, A., Woeginger, G.J. (eds.) Online Algorithms: The State of the Art. LNCS, vol. 1442, pp. 147–177. Springer, Heidelberg (1998). https://doi.org/10.1007/BFb0029568
10. Fekete, S.P., Grosse-Holz, J., Keldenich, P., Schmidt, A.: Parallel online algorithms for the bin packing problem (2019). arXiv preprint (1910.03249)

11. Halldórsson, M.M., Iwama, K., Miyazaki, S., Taketomi, S.: Online independent sets. Theor. Comput. Sci. **289**(2), 953–962 (2002)
12. Heydrich, S., van Stee, R.: Beating the harmonic lower bound for online bin packing. In: The 43rd International Colloquium on Automata, Languages, and Programming (ICALP), pp. 41:1–41:14 (2016)
13. Johnson, D.S.: Fast algorithms for bin packing. J. Comput. Syst. Sci. **8**(3), 272–314 (1974)
14. Kamali, S., Ortiz, A.L.: Better compression through better list update algorithms. In: 2014 Data Compression Conference, pp. 372–381 (2014)
15. Lee, C.C., Lee, D.T.: A simple on-line bin-packing algorithm. J. ACM **32**, 562–572 (1985)
16. Liang, F.M.: A lower bound for on-line bin packing. Inf. Process. Lett. **10**(2), 76–79 (1980)
17. López-Ortiz, A., Schuierer, S.: On-line parallel heuristics, processor scheduling and robot searching under the competitive framework. Theor. Comput. Sci. **310**(1–3), 527–537 (2004)
18. Renault, M.P., Rosén, A., van Stee, R.: Online algorithms with advice for bin packing and scheduling problems. Theor. Comput. Sci. **600**, 155–170 (2015)
19. van Vliet, A.: An improved lower bound for on-line bin packing algorithms. Inf. Proc. Lett. **43**(5), 277–284 (1992)
20. Yao, A.C.-C.: New algorithms for bin packing. J. ACM **27**(2), 207–227 (1980)
21. Zhao, X., Shen, H.: On the advice complexity of one-dimensional online bin packing. In: Chen, J., Hopcroft, J.E., Wang, J. (eds.) FAW 2014. LNCS, vol. 8497, pp. 320–329. Springer, Cham (2014). https://doi.org/10.1007/978-3-319-08016-1_29

Managing Multiple Mobile Resources

Björn Feldkord(✉), Till Knollmann, Manuel Malatyali, and Friedhelm Meyer auf der Heide

Heinz Nixdorf Institut & Departement of Computer Science, Paderborn University, Fürstenallee 11, 33102 Paderborn, Germany
{bjoern.feldkord,till.knollmann,manuel.malatyli,fmadh}@upb.de

Abstract. We extend the Mobile Server Problem introduced in [8] to a model where k identical mobile resources, here named servers, answer requests appearing at points in the Euclidean space. In order to reduce communication costs, the positions of the servers can be adapted by a limited distance m_s per round for each server. The costs are measured similar to the classical Page Migration Problem, i.e., answering a request induces costs proportional to the distance to the nearest server, and moving a server induces costs proportional to the distance multiplied with a weight D.

We show that, in our model, no online algorithm can have a constant competitive ratio, i.e., one which is independent of the input length n, even if an augmented moving distance of $(1+\delta)m_s$ is allowed for the online algorithm. Therefore we investigate a restriction of the power of the adversary dictating the sequence of requests: We demand *locality of requests*, i.e., that consecutive requests come from points in the Euclidean space with distance bounded by some constant m_c. We show constant lower bounds on the competitiveness in this setting (independent of n, but dependent on k, m_s and m_c).

On the positive side, we present a deterministic online algorithm with bounded competitiveness when augmented moving distance and locality of requests is assumed. Our algorithm simulates any given algorithm for the classical k-Page Migration problem as guidance for its servers and extends it by a greedy move of one server in every round. The resulting competitive ratio is polynomial in the number of servers k, the ratio between m_c and m_s, the inverse of the augmentation factor $1/\delta$ and the competitive ratio of the simulated k-Page Migration algorithm.

Keywords: Online Algorithms · K-Server Problem · Page Migration Problem · Resource augmentation

---

This work was partially supported by the German Research Foundation (DFG) within the Collaborative Research Centre "On-The-Fly Computing" under the project number 160364472 — SFB 901/3. A full version of this paper is available at https://arxiv.org/abs/1907.09834.

E. Bampis and N. Megow (Eds.): WAOA 2019, LNCS 11926, pp. 120–137, 2020.
https://doi.org/10.1007/978-3-030-39479-0_9

1 Introduction

We consider a scenario where several devices continuously access a common set of k identical resources. The devices pose *requests* for a resource which must be answered by communicating with the resources, incurring cost for communication. The placement of resources is managed by an algorithm whose goal it is to reduce the costs for communication and for moving the resources as much as possible. Typically requests do not concern the complete resource (which may be rather large) and it is cheaper to answer a request for a resource by communicating with the resource instead of moving it. We assume requests appear in an online fashion, i.e., it is unknown to the algorithm where the next requests will arrive while newly arriving requests must be answered instantly to provide latency guarantees.

The scenario described above can be modeled based on the classical Page Migration problem [5]: A single resource can be moved between two points a and b for costs $D \cdot d(a, b)$, where $d(a, b)$ is the distance between a and b and $D \geq 1$ is a constant. In every round a request appears at some point r, and if the current position of the resource is p, it is served for costs $d(p, r)$. This problem was extended to the Mobile Server Problem [8], which puts a limit on how much the resource (called server) can move in each time step and therefore introduces the idea that the positions of resources can not be arbitrarily changed in each time step.

In our work, we extend this idea to multiple resources: We consider k identical servers located in the Euclidean space (of arbitrary dimension). Each of them may move a distance of at most m_s per time step. In each time step, a request appears which has to be served by one of the servers by the end of the time step. The cost function is the same as in the Page Migration Problem.

1.1 Related Work

Besides being a direct extension of the Mobile Server Problem [8], our work builds on and is related to results surrounding the k-Server and Page Migration problems. These problems have been examined in many variants and especially for the k-Server Problem there are many algorithms for special metrics. In this overview we only focus on most relevant results for our problem, which are mostly algorithms with an (asymptotically) optimal competitive ratio.

In the classical k-Server Problem as introduced by Manasse et al. [13], k identical servers are located in a metric space and requests are answered by moving at least one of the servers to the point of the request. The associated costs are equal to the total distance moved. Manasse et al. showed that no online algorithm could be better than k-competitive on every metric with at least $k+1$ points. They stated as the k-Server Conjecture that there is a k-competitive online algorithm for every metric space. Further, the conjecture is shown to hold for $k = 2$ and $k = n - 1$ where n is the number of points in the metric space.

Since its introduction, many algorithms have been designed for special cases of the problem. Most notable is the Double-Coverage Algorithm [7], which is

k-competitive on trees. For general metrics, the best known result is the Work-Function Algorithm, which is shown to be $2k-1$-competitive [11]. Although this algorithm seems generally inefficient in case of runtime and memory, there have been studies showing that an efficient implementation of this algorithm is indeed possible [14,15]. It was also shown that the algorithm has an optimal competitive ratio of k on line and star metrics, as well as metrics with $k+2$ points [3].

The study of randomized online algorithms was initiated by Fiat et al. [10] who gave a $\log(k)$-competitive algorithm for the complete graph. It is speculated that this factor can be obtained for all metrics, however the question is still open. For general metrics, the first algorithm with polylogarithmic competitive ratio was an $\mathcal{O}(\log^3 n \cdot \log^2 k)$-competitive algorithm by Bansal et al. [1]. This was recently improved by Bubeck et al. [6] who gave an $\mathcal{O}(\log^2 k)$-competitive algorithm for HSTs which can be turned into an $\mathcal{O}(\log^9(k) \cdot \log\log(k))$-competitive one for general metrics by a dynamic embedding of general metrics into HSTs [12].

Regarding the Page Migration Problem [5] (also known as File Migration Problem), most results focus on online algorithms which handle only a single page. Contrary to the k-Server Problem, the design of such algorithms is not trivial for the Page Migration Problem. To the best of our knowledge, the current best results are a 4-competitive deterministic algorithm by Bienkowski et al. [4] and a collection of randomized algorithms with competitive ratio of at most 3 by Westbrook [16]. The most relevant results for our problem are two constructions by Bartal et al. [2] who give both a deterministic and a randomized algorithm which transform a given algorithm for the k-Server problem into a deterministic/randomized algorithm for the k-Page Migration Problem. If the given k-Server algorithm is c-competitive, the deterministic algorithm is $\mathcal{O}(c^2)$-competitive, the randomized algorithm is $\mathcal{O}(c)$-competitive. Conversely, we use the resulting algorithms as a black box in our constructions.

1.2 Our Results and Outline of the Paper

In [8] it was already shown that no online algorithm for our problem can be competitive even on the real line and with just $k=1$ server. As a consequence, we employ the following methods to derive reasonable results for the problem: On the one hand we restrict the adversary to the case with *locality of requests*, i.e., we introduce a parameter m_c by which we can define families of instances classified by the maximum distance between two consecutive requests. On the other hand we apply *resource augmentation* as in [8], i.e., we allow the online algorithm to use a maximum movement distance of $(1+\delta)m_s$. We show that, for $k \geq 2$, both methods are needed to yield competitive bounds independent of the length of the instance. For $k=1$, it was shown in [9] that a locality of requests can improve the competitiveness, but is not necessary to achieve a constant upper bound.

The parameters m_c and m_s have a crucial impact on the resulting competitiveness and thus separate simple from hard instances. We are able to show that these parameters seem to naturally describe the problem, since we can prove a lower bound of $\Omega(\frac{m_c}{m_s})$. For fast moving resources ($m_c < (1+\delta)m_s$), our algorithm has an almost optimal competitive ratio when given an optimal k-Page Migration algorithm. For the case of slow moving resources ($m_c \geq (1+\delta)m_s$), we can achieve bounds independent of the length of the input stream. In detail, we obtain a bound of $\mathcal{O}(\frac{1}{\delta^4} \cdot k^2 \cdot \frac{m_c}{m_s} + \frac{1}{\delta^3} \cdot k^2 \cdot \frac{m_c}{m_s} \cdot c(\mathcal{K}))$, where $c(\mathcal{K})$ is the competitiveness of a given k-Page Migration algorithm. For the case $D = 1$, which we call the *unweighted problem*, the k-Page Migration algorithm can be replaced by a k-Server algorithm. Note that the parameter ε in Table 1 is indirectly given as the relative difference between m_c and m_s. If $m_c < m_s$, then in the first row we have $\varepsilon > \delta$. Alternatively, if $\delta = 0$, this case still yields an almost optimal upper bound up to a factor of $1/\varepsilon$.

Table 1. An overview of the results, using the best known deterministic algorithms for k-Server/k-Page Migration. The results in the first row also hold without resource augmentation when $m_c \leq (1-\varepsilon)m_s$.

	Lower bound	Unweighted ($D = 1$), Weighted ($D > 1$) Case
$m_c \leq (1+\delta-\varepsilon)m_s$	$\Omega(k)$	$\mathcal{O}(1/\varepsilon \cdot k)$, $\mathcal{O}(1/\varepsilon \cdot k^2)$
$m_c \geq (1+\delta)m_s$	$\Omega(k + \frac{m_c}{m_s})$	$\mathcal{O}(1/\delta^4 \cdot k^3 \cdot \frac{m_c}{m_s})$, $\mathcal{O}(1/\delta^4 \cdot k^4 \cdot \frac{m_c}{m_s})$

The paper is structured as follows: A formal definition of our model can be found in Sect. 2. All relevant lower bounds are established in Sect. 3. In terms of upper bounds, we first give an algorithm for the *unweighted problem* in Sect. 4. The analysis for instances with $m_c < (1+\delta)m_s$ consists of a simple potential function argument found in Sect. 4.1. The analysis of the other case is much more challenging and is conducted in Sect. 4.2. The weighted case ($D > 1$) is discussed in Sect. 5. While the basic approach stays the same, we need to modify the movement of the online algorithm due to the higher movement costs. We show how the algorithm can be adapted and present the resulting competitive ratio following a similar structure as in the unweighted case.

2 Model and Notation

In this section we formally describe the model and some common notation used throughout the paper.

Time is considered discrete and divided into time steps $1, \ldots, n$. An input to the k-Mobile Server Problem is given by a sequence of requests $r_1, \ldots, r_n$ where each r_t occurs in time step t and is represented by a point in the Euclidean space of arbitrary dimension. We are given k servers $a_1, \ldots, a_k$ controlled by our online algorithm. At each point in time, one server occupies exactly one point in the Euclidean space. We denote by $a_i^{(t)}$ the position of server a_i at end of time step t, and by $d(a, b)$ the Euclidean distance between two points a and b. For the distance between two servers $a_i^{(t)}$ and $a_j^{(t)}$ in the same time step t, we also use the notation $d_t(a_i, a_j)$. We may also leave out the time t entirely if it is clear from the context.

In each time step t, the current request r_t is revealed to the online algorithm. The algorithm may then move each server, such that $d(a_i^{(t-1)}, a_i^{(t)}) \leq m_s$ for all servers a_i. The movement incurs cost of $D \cdot \sum_{i=1}^{k} d(a_i^{(t-1)}, a_i^{(t)})$ for a constant $D \geq 1$. The request r_t is then served by the closest server $a_i^{(t)}$, which incurs cost of $d(a_i^{(t)}, r_t)$. Note that the variables indexed with the time t represent the configuration at the end of the time step t.

In our model, we consider the locality of requests dictated by a parameter m_c limiting the distance between consecutive requests, i.e., $d(r_t, r_{t+1}) \leq m_c$. We also consider a resource augmentation setting, where the maximum distance an online algorithm may move is in fact $(1 + \delta)m_s$ for some $\delta \in (0, 1)$. The cost of our online algorithm is denoted by C_{Alg}. We compare the costs of an online algorithm to an offline optimum, whose servers are denoted by $o_1, \ldots, o_k$ and whose cost is C_{Opt}.

3 Lower Bounds

In this section, we will prove lower bounds for the competitive ratio of our problem. They show the importance both of the resource augmentation and the locality of requests introduced above. All our lower bounds already hold on the line (and therefore in arbitrary dimensions, too). Since our model is an extension of the k-Page Migration Problem, $\Omega(k)$ is a lower bound for deterministic algorithms inherited from that problem (which itself inherits the bound from the k-Server Problem, see [2,13]). Even when m_c is restricted, the lower bound instance can simply be scaled down such that the distance limits are not relevant for the instance.

We start by discussing the model without any restriction on the distance between the requests in two consecutive time steps, i.e., the parameter m_c is unbounded. We also consider the case, that there is no resource augmentation, i.e., the maximum movement distance of the online algorithm and of the offline solution are the same. The following lower bound, originally formulated for $k = 1$, carries over from [8]:

Theorem 1. *Every randomized online algorithm for the Mobile Server Problem (with $k = 1$) has a competitive ratio of $\Omega(\frac{\sqrt{n}}{D})$ against an oblivious adversary, where n is the length of the input sequence.*

For more than one server, we obtain an additional bound which can not be resolved with the help of resource augmentation. The proofs of the following theorems can be found in the full version of the paper.

Theorem 2. *For $k \geq 2$, every randomized online algorithm for the k-Mobile Server Problem has a competitive ratio of at least $\Omega(\frac{n}{Dk^2})$, where n is the length of the input sequence.*

Since we often consider input sequences for problems such as ours to be potentially infinite, we deem competitive ratios dependent on the input length undesirable. Hence, as a consequence of the bounds shown so far, we apply two modifications to our model which help us to achieve a competitive ratio independent of the length of the input sequence. We use the concept of resource augmentation just as in [8] to allow the online algorithm to utilize a maximum movement distance of $(1+\delta)m_s$ for some $\delta \in (0,1)$ as opposed to the distance m_s used by the optimal offline solution. This measure alone does not address the bound from Theorem 2 (the ratio shrinks, but still depends on n). Hence, we introduce the locality of requests, i.e., restrict the distance between two consecutive requests to a maximum distance of m_c. Note, that only restricting the distance between consecutive requests does also not remove the dependence on n, as was shown in [9]. The following theorem can be obtained in a similar way as Theorem 2:

Theorem 3. *For $k \geq 2$, every randomized online algorithm for the k-Mobile Server Problem, where the distance between consecutive requests is bounded by m_c, has a competitive ratio of at least $\Omega(\frac{m_c}{m_s})$.*

4 An Algorithm for the Unweighted Problem

In this section we consider the unweighted problem ($D = 1$). Our algorithm does the following: We mainly follow around a simulated k-Server algorithm, but always move the closest server greedily towards the request.

We use the following notation in this section: Denote by $a_1, \ldots, a_k$ the servers of the online algorithm, $c_1, \ldots, c_k$ the servers of the simulated k-server algorithm and $o_1, \ldots, o_k$ the servers of the optimal solution. For an offline server o_i, we denote by o_i^a the closest server of the online algorithm to o_i (this might be the same server for multiple offline servers). Furthermore, we denote by a^*, c^* and o^* the closest server to the request of the algorithm, the k-server algorithm, and the optimal solution respectively. For a fixed time step t, we add a "$'$" to any variable to denote the state at the end of the current time step, e.g., $a_1 = a_1^{t-1}$ is the position of the server at the beginning of the time step and $a_1' = a_1^t$ is the position at the end of the current step.

Our algorithm *Unweighted-Mobile Servers (UMS)* works as follows:
Take any k-Server algorithm $\mathcal{K}$ with bounded competitiveness in the Euclidean space. Upon receiving the next request r', simulate the next step of $\mathcal{K}$. Calculate a minimum weight matching (with the distances as weights) between the servers $a_1, \ldots, a_k$ of the online algorithm and the servers $c_1', \ldots, c_k'$ of $\mathcal{K}$. There must be a server c_i for which $c_i' = r'$. If the server matched to c_i' can reach r' in this turn, move all servers towards their counterparts in the matching with the maximum possible speed of $(1+\delta)m_s$. Otherwise, select the server $\tilde{a}$ which is closest to r' and move it to r' with speed at most $(1+\frac{\delta}{2})m_s$. All other servers move towards their counterparts in the matching with speed $(1+\delta)m_s$.

We briefly want to discuss the fact that both steps of our algorithm are necessary for a bounded competitiveness. For the classical k-Server Problem, a simple greedy algorithm, which always moves the closest server onto the request has an unbounded competitive ratio. We can show, that a simple algorithm which just tries to imitate any k-Server algorithm as best as possible is also not successful. Intuitively, the simulated algorithm can move many servers towards the request within one time step and serve the following sequence with them, while the online algorithm needs multiple time steps to get the corresponding servers in position due to the speed limitation.

Simple algorithm: Let $\mathcal{K}$ be any given k-Server algorithm. The k-Mobile Server algorithm does the following: Simulate $\mathcal{K}$. Compute a minimum weight matching (with the distances as weights) between the own servers and the servers of $\mathcal{K}$. Move every server towards the matched server at maximum speed.

Theorem 4. *For $k \geq 2$, there are competitive k-Server algorithms such that the simple algorithm for the k-Mobile Server Problem does not achieve a competitive ratio independent of n.*

The remainder of this section is devoted to the analysis of the competitive ratio of the UMS algorithm. In Sect. 4.1, we first consider the case that the distance between consecutive requests m_c is smaller than the movement speed of the algorithm's servers. This case is easier than the case of slower servers since we can always guarantee that the online algorithm has one server on the position of the request. In the other case ($m_c \geq (1+\delta)m_s$), described in Sect. 4.2, we need to extend our analysis to incorporate situations in which our online algorithm has no server near the request although the optimal offline solution might have such a server. Details left out due to space constraints can be found in the full version of the paper.

4.1 Fast Resource Movement

We first deal with the case that $m_c \leq (1-\varepsilon)\cdot m_s$ for some $\varepsilon \in (0,1)$. We show that we can achieve a result independent of the input length, even without resource augmentation. At the end of this section, we briefly discuss how to extend the result to incorporate resource augmentation, i.e., if the online algorithm has a maximum movement distance of $(1+\delta)m_s$, we handle all cases with $m_c \leq (1+\delta-\varepsilon)\cdot m_s$.

Theorem 5. *If $m_c \leq (1-\varepsilon)\cdot m_s$ for some $\varepsilon \in (0,1)$, the algorithm UMS is $2/\varepsilon \cdot c(\mathcal{K})$-competitive, where $c(\mathcal{K})$ is the competitive ratio of the simulated k-server algorithm $\mathcal{K}$.*

Proof. We assume the servers adapt their ordering $a_1, \ldots, a_k$ according to the minimum matching in each time step. Based on the matching, we define the potential $\psi := \frac{2}{\varepsilon} \cdot \sum_{i=1}^{k} d(a_i, c_i)$. Note that the algorithm reaches the point of r in each time step, and hence only pays for the movement of its servers, i.e., $C_{Alg} = \sum_{i=1}^{k} d(a_i, a_i')$. We assume, that c_1 is on the request after the current time step, i.e., $c_1' = r'$.

First, consider the case that a_1 can reach r' in this time step. Since each server moves directly towards their counterpart in the matching, we have

$$\begin{aligned}\Delta\psi &= \tfrac{2}{\varepsilon}\cdot\textstyle\sum_{i=1}^{k} d(a_i',c_i') - \tfrac{2}{\varepsilon}\cdot\textstyle\sum_{i=1}^{k} d(a_i,c_i)\\ &\leq \tfrac{2}{\varepsilon}\cdot\textstyle\sum_{i=1}^{k} d(c_i,c_i') - \tfrac{2}{\varepsilon}\cdot\textstyle\sum_{i=1}^{k} d(a_i,a_i')\\ &= \tfrac{2}{\varepsilon}\cdot C_{\mathcal{K}} - \tfrac{2}{\varepsilon}\cdot C_{Alg}.\end{aligned}$$

Now assume that a_1 cannot reach r' in this time step. The server moves at full speed and hence $d(a_1',c_1') - d(a_1,c_1') = -m_s$. Now, let a_2 be the server which is at range at most m_c to r' and does the greedy move possibly away from c_2' onto r'. It holds $d(a_2',c_2') - d(a_2,c_2') \leq m_c$. In total, we get

$$\begin{aligned}\Delta\psi &\leq \tfrac{2}{\varepsilon}(\textstyle\sum_{i=1}^{k} d(a_i',c_i') - \textstyle\sum_{i=1}^{k} d(a_i,c_i')) + \tfrac{2}{\varepsilon}\textstyle\sum_{i=1}^{k} d(c_i,c_i')\\ &\leq \tfrac{2}{\varepsilon}(d(a_1',c_1') - d(a_1,c_1') + d(a_2',c_2') - d(a_2,c_2'))\\ &\quad -\tfrac{2}{\varepsilon}\textstyle\sum_{i=3}^{k} d(a_i,a_i') + \tfrac{2}{\varepsilon}\textstyle\sum_{i=1}^{k} d(c_i,c_i')\\ &\leq -2m_s - \tfrac{2}{\varepsilon}\textstyle\sum_{i=3}^{k} d(a_i,a_i') + \tfrac{2}{\varepsilon}\cdot C_{\mathcal{K}}\\ &\leq -\textstyle\sum_{i=1}^{k} d(a_i,a_i') + \tfrac{2}{\varepsilon}\cdot C_{\mathcal{K}}.\end{aligned}$$

We can extend this bound to the resource augmentation scenario, where the online algorithm may move the servers a maximum distance of $(1+\delta)\cdot m_s$. When relaxing the condition appropriately to $m_c \leq (1+\delta-\varepsilon)\cdot m_s$, we get the following result:

Corollary 1. *If* $m_c \leq (1+\delta-\varepsilon)\cdot m_s$ *for some* $\varepsilon \in (0,1)$*, the algorithm UMS is* $\frac{2\cdot(1+\delta)}{\varepsilon}\cdot c(\mathcal{K})$*-competitive, where* $c(\mathcal{K})$ *is the competitive ratio of the simulated* k*-server algorithm* $\mathcal{K}$*.*

The proof works the same as above by replacing occurrences of m_s by $(1+\delta)m_s$ and changing the potential to $\frac{2\cdot(1+\delta)}{\varepsilon}\sum_{i=1}^{k} d(a_i,c_i)$.

At first glance, the result seems to become weaker with increasing δ if ε stays the same. The reason is that by fixing ε the relative difference $((1+\delta)m_s - m_c)/m_s$ between m_c and $(1+\delta)m_s$ actually decreases, i.e., relatively speaking, m_c gets closer to $(1+\delta)m_s$. It can be seen that if instead we fix the value of m_c and increase δ, the value of ε increases by the same amount and hence the competitive ratio tends towards $2\cdot c(\mathcal{K})$.

4.2 Slow Resource Movement

This section considers the case $m_c \geq (1+\delta)m_s$ and is structured as follows: To support our potential argument, we first introduce a transformation of the simulated k-Server algorithm which ensures that the simulated servers are always located near the request. We then introduce an abstraction of the offline solution, reducing it to the positioning of a single server $\hat{o}$ which acts as a reference point for a new potential function. The server $\hat{o}$ approximates the optimal positioning of the servers while at the same time obeys certain movement restrictions necessary in our analysis. Finally, we complete the analysis by combining the new derived potential function with the methods from the previous section.

The k-Server Projection. Our goal is to transform a k-Server algorithm $\mathcal{K}$ into a k-Server algorithm $\hat{\mathcal{K}}$ which serves the requests of a k-Mobile Server instance such that all servers keep relatively close to the current request r. For the case $m_c \geq (1+\delta)m_s$, we want our algorithm to use this projection as a simulated algorithm as opposed to a regular k-Server algorithm, hence we must ensure that this projection is computable online with the information available to our online algorithm. The servers of $\mathcal{K}$ are denoted as $c_1, \dots, c_k$ and the servers of $\hat{\mathcal{K}}$ as $\hat{c}_1, \dots, \hat{c}_k$.

We define two circles around r: The inner circle $inner(r)$ has a radius of $4k \cdot m_c$ and the outer circle $outer(r)$ has a radius of $(8k+1) \cdot m_c$. We will maintain $\hat{c}_i \in outer(r)$ for the entirety of the execution. The time is divided into phases, where the phase starting at time t with the request at point r_t ends on the smallest $t' > t$ such that $d(r_t, r'_t) \geq 4k \cdot m_c$. During a phase the simulated servers move to preserve the following: If $c_i \in inner(r)$, then $\hat{c}_i = c_i$. At the end of the phase, in addition to the previous condition, it should hold: If $c_i \notin inner(r)$, then $\hat{c}_i$ is on the boundary of $inner(r)$ such that $d(c_i, \hat{c}_i)$ is minimized. It is obvious that the definition of the algorithm guarantees $\hat{c}_i \in outer(r)$ for all i at each point in time.

Proposition 1. *For the servers $\hat{c}_1, \dots, \hat{c}_k$ of $\hat{\mathcal{K}}$ it holds $d(\hat{c}_i, r) \leq (8k+1) \cdot m_c$ during the whole execution.*

The costs of $\hat{\mathcal{K}}$ are at most $\mathcal{O}(k)$ times the costs of $\mathcal{K}$.

The Offline Helper. We define a new offline server $\hat{o}$, which approximates the optimal position o^* while managing the role change of o^* in a smooth manner. By $\hat{a}$, we denote the server of the online algorithm with minimal distance to $\hat{o}$. For a formal description of the behavior, we need the following definitions:

- The inner circle $inner_t(o_i)$ contains all points p with $d_t(o_i, p) \leq \frac{\delta^2}{48960k} \cdot d_t(o_i, o_i^a)$.
- The outer circle $outer_t(o_i)$ contains all points p with $d_t(o_i, p) \leq \frac{\delta}{48} \cdot d_t(o_i, o_i^a)$.

Abusing notation, we also refer to $inner_t(o_i)$ and $outer_t(o_i)$ as distances equal to the radius defined above. This section is devoted to proving the following:

Proposition 2. *There exists a virtual server $\hat{o}$ which moves at a speed of at most $(2+\frac{1020k}{\delta}) \cdot m_c$ per time step, for which $d(\hat{a}, \hat{o}) \leq 2 \cdot d(o^*, o^{*a}) + d(a^*, r)$ at all times, and for which the following conditions hold as long as $d_t(o^*, o^{*a}) \geq 2 \cdot 51483 \frac{km_c}{\delta^2}$:*

1. *If $r \in inner(o^*)$ at the end of the current time step, $\hat{o}$ moves at a maximum speed of $(1+\frac{\delta}{8})m_s$, i.e., $r_t \in inner_t(o^*) \Rightarrow d(\hat{o}^{(t-1)}, \hat{o}^{(t)}) \leq (1+\frac{\delta}{8})m_s$.*
2. *If $r \in inner(o^*)$ at the end of the current time step, then $\hat{o} \in outer(o^*)$ at the end of the current time step, i.e., $r_t \in inner_t(o^*) \Rightarrow \hat{o}^{(t)} \in outer_t(o^*)$.*

In the following, we show that it is possible to define a movement pattern for $\hat{o}$ in a way, such that invariants 1 and 2 of Proposition 2 hold as long as

$d(o^*, o^{*a}) \geq 51483\frac{km_c}{\delta^2}$. Otherwise, $\hat{o}$ will simply follow r and restore the properties once $d(o^*, o^{*a}) \geq 2 \cdot 51483\frac{km_c}{\delta^2}$. In order to describe the movement in detail, we introduce the concept of transitions.

In the input sequence and a given optimal solution, we define a *transition* between two steps $t_1 < t_2$, if there are o_i, o_j such that $o_i = o^*$ and $r \in inner_{t_1}(o_i)$ at time step t_1 and $o_j = o^*$ and $r \in inner_{t_2}(o_j)$ at time step t_2. In between these two time steps, $r \notin inner(o^*)$. For such a transition, we define the transition time as $t^* := t_2 - t_1$. If $t^* > inner_{t_1}(o^*)/m_c + 2$, we call this a *long transition.* Otherwise, we call it a *short transition.* We say that o_i *passes* the request after t_1 and o_j *receives* the request in t_2. The concept is illustrated in Fig. 1.

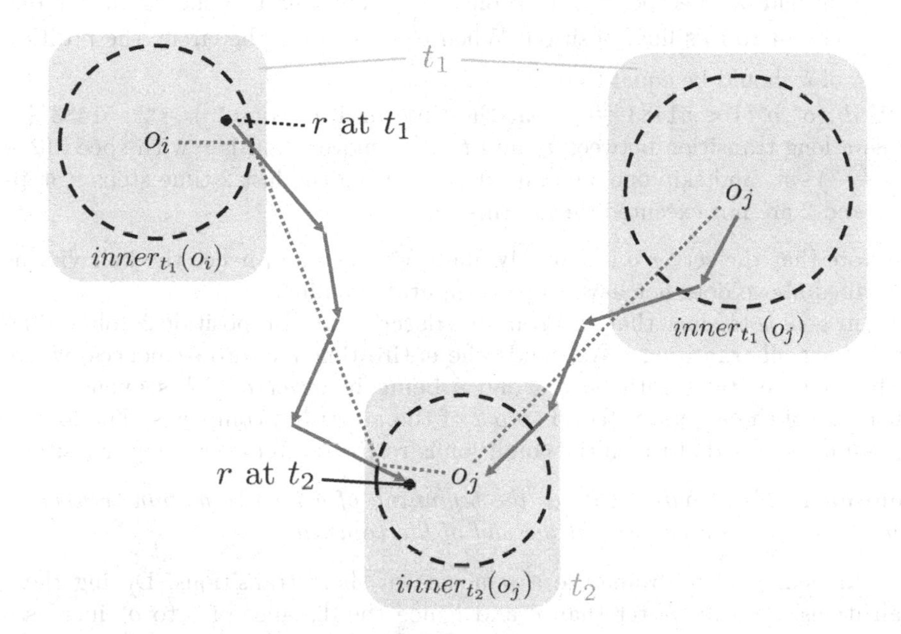

Fig. 1. Example for a transition from o_i to o_j. By definition, r crosses the border of $inner(o_i)$ after time step t_1 (o_i passes r after t_1). The transition stops at step t_2 when r has entered $inner_{t_2}(o_j)$ (o_j receives r in t_2). Note that o_j's position and the radius of its inner circle may change from t_1 to t_2. The distance moved by r is at most $(t_2 - t_1) \cdot m_c$. The dotted line represents the estimation of $d_{t_1}(o_i, o_j)$ used in Lemma 2.

The behavior of $\hat{o}$ can be computed as follows:

1. During a long transition between time steps t_1 and t_2, move with speed $d(\hat{o}^{(t-1)}, \hat{o}^{(t)}) \leq (2 + \frac{1020k}{\delta}) \cdot m_c$ towards r_t whenever $r_t \notin inner_t(o^*)$. In the last two steps $t_2 - 1$ and t_2, move such that $\hat{o}^{(t_2-1)} = r_{t_2}$ at time $t_2 - 1$ and do not move in t_2 at all. Informally, $\hat{o}$ moves one step ahead of r such that $\hat{o} = r$ after the transition, as soon as $r \in inner(o^*)$.
2. For a sequence of short transitions starting with $o^* = o_i$ in t_1, determine which of the following events terminating the current sequence occurs first:

(a) A long transition from a server o_ℓ to o_j between time t_2 and t_3 occurs. In this case, $\hat{o}$ simply moves towards $o_\ell^{(t)}$ in each step t with speed at most $(1+\frac{\delta}{8})m_s$ until t_2.

(b) A short transition from a server o_ℓ to o_j between time t_2 and t_3 occurs, where at one point prior in the sequence $d(o_j, o^*) > outer(o^*)/3$. If $\hat{o}$ can move straight towards the final position of o_j in t_3 with speed $(1+\frac{\delta}{8})m_s$ without ever leaving $outer(o^*)$, then do that. Otherwise move towards a point p with $d(p, o_\ell) = \frac{2\delta}{145} \cdot d(o_\ell, o_\ell^a)$. Among those candidates, p minimizes $d(p, o_\ell^{(t_3)})$. When this point is reached, keep the invariant $d(\hat{o}, o_\ell) = \frac{2\delta}{145} \cdot d(o_\ell, o_\ell^a)$ whenever the final position of o_j is not within $\frac{2\delta}{145} \cdot d(o_\ell, o_\ell^a)$ around o_ℓ. The position of $\hat{o}$ on the circle around o_ℓ should be the one closest to o_j's final position. When $o_j^{(t_3)}$ is inside the circle, the position of $\hat{o}$ should be equal to $o_j^{(t_3)}$.

3. If $d_{t_1}(o^*, o^{*a}) < 51483\frac{km_c}{\delta^2}$, treat the time until $d_{t_2}(o^*, o^{*a}) \geq 2 \cdot 51483\frac{km_c}{\delta^2}$ as a long transition between t_1 and t_2, i.e., move towards r with speed $(2+\frac{1020k}{\delta}) \cdot m_c$ and skip one step ahead of r during the last 2 time steps. (Steps 1 and 2 are not executed during this time.)

Note that the server $\hat{o}$ is a purely analytical tool and hence the behavior as described above does not have to be computable online.

Our goal is to show that all invariants described in Proposition 2 hold inductively over all transitions. We divide the entire timeline into sequences, where each sequence starts with both r and $\hat{o}$ being in $inner(o^*)$. A sequence ends when one of the events stated in step 2 of the algorithm completes. The following lemma states that the initial condition is restored after every long transition.

Lemma 1. *If $\hat{o} \in outer_{t_1}(o^*)$ at the beginning of a long transition between t_1 and t_2, then $\hat{o} \in inner_{t_2}(o^*)$ at the end of the transition.*

Our next goal is to analyze a sequence of short transitions. During these transitions, r moves faster than $\hat{o}$ and hence the distance of $\hat{o}$ to o^* increases due to the role change after a transition. The next lemma establishes an upper bound on that increase. Since we use the lemma in another context as well, the formulation is slightly more general.

Lemma 2. *Every short transition between o_i in step t_1 and o_j in step t_2 can increase the distance of some server s, which moves at speed at most $(1+\delta)m_s$, to o^* by at most*
$\min\{6.001 \cdot \frac{\delta^2}{48960k} \cdot d_{t_1}(o^*, o^{*a}) + 8.001m_c\ ,\ 6.002 \cdot \frac{\delta^2}{48960k} \cdot d_{t_2}(o^*, o^{*a}) + 8.002m_c\}$.
Likewise, s decreases its distance to o^* by at most
$\min\{6.001 \cdot \frac{\delta^2}{48960k} \cdot d_{t_1}(o^*, o^{*a}) + 8.001m_c\ ,\ 6.002 \cdot \frac{\delta^2}{48960k} \cdot d_{t_2}(o^*, o^{*a}) + 8.002m_c\}$.

We want to show, that $\hat{o} \in inner(o^*)$ holds after a sequence of short transitions is terminated by one of the conditions described in step 2 of the algorithm. During the sequence, we must also show that $\hat{o} \in outer(o^*)$. The main idea for the following lemma is that o_ℓ never leaves $outer(o^*)/3$ per definition and hence following it keeps $\hat{o}$ inside $outer(o^*)$.

Lemma 3. *Consider a sequence of short transitions which is terminated by a long transition. If $\hat{o} \in inner(o^*)$ at the beginning of the sequence, then $\hat{o} \in inner(o^*)$ after the long transition. During the sequence of short transitions, $\hat{o} \in outer(o^*)$.*

We show with the help of Lemma 2 that during the sequence of transitions, $\hat{o}$ does not loose too much distance to o^*, while o_j, since at one point $d(o_j, o^*) > outer(o^*)/3$, takes enough time to get into position for a short transition such that $\hat{o}$ can reach the final position of o_j in time.

Lemma 4. *Consider a sequence of short transitions which is terminated by a short transition from o_ℓ to o_j, where at one point prior in the sequence $d(o_j, o^*) > outer(o^*)/3$. If $\hat{o} \in inner(o^*)$ at the beginning of the sequence and $d(o^*, o^{*a}) \geq 51483\frac{km_c}{\delta^2}$ at all times, then $\hat{o} \in inner(o^*)$ after the transition to o_j. During the sequence, $\hat{o} \in outer(o^*)$.*

Our analysis of the movement pattern of $\hat{o}$ leads directly to the following lemma, in which we mostly need to argue that either $\hat{o} \in outer(o^*)$ or $\hat{o} = r$.

Lemma 5. *During the execution of the algorithm, $d(\hat{a}, \hat{o}) \leq 2 \cdot d(o^*, o^{*a}) + d(a^*, r)$ as long as the algorithm is in step 1 or 2.*

So far we have shown that all claims of Proposition 2 hold as long as the algorithm is not in step 3. It remains to analyze step 3 of the algorithm, using similar arguments as for analyzing the long transitions earlier.

Lemma 6. *After the execution of step 3 it holds $\hat{o} = r$. Furthermore, $d(\hat{a}, \hat{o}) \leq 2 \cdot d(o^*, o^{*a}) + d(a^*, r)$ during step 3 of the algorithm.*

Algorithm Analysis. We now turn our attention back to the analysis of the UMS algorithm. In the following, we assume $\mathcal{K}$ to be a k-Server algorithm obtained from Proposition 1. We use a potential composed of two major parts which balance the main ideas of our algorithm against each other: ϕ will measure the costs of the greedy strategy, while ψ will cover the matching to the simulated k-Server algorithm.

Let $\hat{o}$ be an offline server which fulfills the invariants stated in Proposition 2. Recall that $\hat{a}$ denotes the currently closest server of the online algorithm to $\hat{o}$. The first part of the potential is then defined as

$$\phi := \begin{cases} 4 \cdot d(\hat{a}, \hat{o}) & \text{if } d(\hat{a}, \hat{o}) \leq 107548 \cdot \frac{km_c}{\delta^2} \\ 4 \cdot \frac{1}{\delta m_s} d(\hat{a}, \hat{o})^2 - A & \text{if } 107548 \cdot \frac{km_c}{\delta^2} < d(\hat{a}, \hat{o}) \end{cases}$$

with $A := 4 \cdot (\frac{1}{\delta m_s}(107548\frac{km_c}{\delta^2})^2 - 107548\frac{km_c}{\delta^2})$.

For the second part, we set

$$\psi := Y \cdot \frac{m_c}{\delta m_s} \sum_{i=1}^{k} d(a_i, c_i)$$

where the online servers a_i are always sorted such that they represent a minimum weight matching to the simulated servers c_i. We choose $Y = \Theta(\frac{k}{\delta^2})$ to be sufficiently large.

If we understand ϕ as a function in $d(\hat{a}, \hat{o})$, then we can rewrite it as

$$\phi(d(\hat{a}, \hat{o})) = \max\{4 \cdot d(\hat{a}, \hat{o}), 4 \cdot \frac{1}{\delta m_s} d(\hat{a}, \hat{o})^2 - A\}.$$

Hence, when estimating the potential difference $\Delta\phi = \phi(d(\hat{a}', \hat{o}')) - \phi(d(\hat{a}, \hat{o}))$, we can upper bound it by replacing the term $\phi(d(\hat{a}, \hat{o}))$ with the case identical to $\phi(d(\hat{a}', \hat{o}'))$. This mostly reduces estimating $\Delta\phi$ to bounding the difference $d(\hat{a}', \hat{o}') - d(\hat{a}, \hat{o})$.

For some of our estimations we use a slightly altered result from [8].

Lemma 7. *Let s be some server with $d(s', r') \leq \frac{\sqrt{\delta}}{2} \cdot d(a_i', r')$ and a_i moves towards r' a distance of $d(a_i, a_i')$, then $d(a_i, s') - d(a_i', s') \geq \frac{1+\frac{1}{4}\delta}{1+\frac{1}{2}\delta} d(a_i, a_i')$.*

We start the analysis by bounding the second potential difference $\Delta\psi$. The bounds can be obtained by similar arguments as in the proof of Theorem 5.

Lemma 8. $\Delta\psi \leq Y \cdot \frac{m_c}{\delta m_s} \cdot C_{\mathcal{K}} - \sum_{i=1}^{k} d(a_i, a_i')$.

Lemma 9. *If $d(a^{*'}, r') > 0$, then* $\Delta\psi \leq Y \cdot \frac{m_c}{\delta m_s} C_{\mathcal{K}} - \sum_{i=1}^{k} d(a_i, a_i') - \frac{Y-4}{2} m_c$.

Now consider the case that $r' \notin inner(o^{*'})$. We have $d(a^{*'}, r') \leq d(o^{*a'}, r') \leq d(o^{*'}, o^{*a'}) + d(o^{*'}, r') \leq (\frac{48960k}{\delta^2} + 1) \cdot d(o^{*'}, r')$. The movement cost are canceled by $\Delta\psi$ as in Lemma 8. It only remains to bound the possible increase of ϕ. We use $d(\hat{a}', \hat{o}') - d(\hat{a}, \hat{o}) \leq (3 + \frac{1020k}{\delta}) \cdot m_c$.

1. $d(\hat{a}', \hat{o}') \leq 107548 \cdot \frac{km_c}{\delta^2}$: $\Delta\phi \leq 4 \cdot d(\hat{a}', \hat{o}') \leq 8 \cdot d(o^{*'}, o^{*a'}) + 4 \cdot d(a^{*'}, r') \leq (12 \cdot \frac{48960k}{\delta^2} + 4) \cdot d(o^{*'}, r')$.
2. $107548 \cdot \frac{km_c}{\delta^2} < d(\hat{a}', \hat{o}')$: $\Delta\phi \leq \frac{4}{\delta m_s}(d(\hat{a}', \hat{o}')^2 - d(\hat{a}, \hat{o})^2) \leq \frac{4}{\delta m_s}(d(\hat{a}', \hat{o}')^2 - (d(\hat{a}', \hat{o}') - (3 + \frac{1020k}{\delta}) \cdot m_c)^2) \leq \mathcal{O}(\frac{k}{\delta}) \cdot \frac{m_c}{\delta m_s} d(\hat{a}', \hat{o}') \leq \mathcal{O}(\frac{k^2}{\delta^3}) \cdot \frac{m_c}{\delta m_s} d(o^{*'}, r')$.

In all of the above, the competitive ratio is bounded by $\mathcal{O}(\frac{k^2}{\delta^3}) \cdot \frac{m_c}{\delta m_s} + Y \cdot \frac{m_c}{\delta m_s} \cdot c(\mathcal{K})$.

Finally, we consider the case $r' \in inner(o^{*'})$. When $d(a^*, r') > 102970 \frac{km_c}{\delta^2}$, we use Lemma 7 to obtain the following:

Lemma 10. *If $d(a^*, r') > 102970 \frac{km_c}{\delta^2}$ and $r' \in inner(o^{*'})$, then $d(a_i', \hat{o}') - d(a_i, \hat{o}) \leq -\frac{\delta}{8} m_s$.*

With this lemma, ϕ can be used to cancel the costs of the algorithm in case of a high distance to r.

Lemma 11. *If $r' \in inner(o^{*'})$, then* $C_{Alg} + \Delta\phi + \Delta\psi \leq Y \cdot \frac{m_c}{\delta m_s} \cdot C_{\mathcal{K}} + 2 \cdot d(o^{*'}, r')$.

The resulting competitive ratio of $Y \cdot \frac{m_c}{\delta m_s} \cdot c(\mathcal{K}) + 2$ is less than the $\mathcal{O}(\frac{k^2}{\delta^3}) \cdot \frac{m_c}{\delta m_s} + Y \cdot \frac{m_c}{\delta m_s} \cdot c(\mathcal{K})$ bound from the former set of cases. Accounting for the loss due to the transformation of the simulated k-Server algorithm, we obtain the following result:

Theorem 6. *If $m_c \geq (1+\delta)m_s$, the algorithm UMS is $\mathcal{O}(\frac{1}{\delta^4} \cdot k^2 \cdot \frac{m_c}{m_s} + \frac{1}{\delta^3} \cdot k^2 \cdot \frac{m_c}{m_s} \cdot c(\mathcal{K}))$-competitive, where $c(\mathcal{K})$ is the competitive ratio of the simulated k-server algorithm $\mathcal{K}$.*

5 Extension to the Weighted Problem

In this section we consider our general model in which the movement costs are weighted with a factor $D > 1$. We assume throughout the section that $D \geq 2$ for convenience in the analysis. In case $D < 2$, we may just apply the algorithm from the previous section, whose costs increase by at most a factor of 2 as a result.

The main difference to the unweighted case is that our algorithm uses a k-Page Migration algorithm as guidance, whose best competitive ratio in the deterministic case so far is a factor $\Theta(k)$ worse than that of a k-Server algorithm for general metrics. The analysis is slightly more involved since unlike in the k-Server Problem, a k-Page Migration algorithm does not always have to have one page at the point of the request. In case of small distances to r, the movement costs have to be balanced against the serving costs by scaling down the movement distance by a factor of D. Throughout this section, we use the same notation as for the unweighted version.

Our algorithm *Weighted-Mobile Servers (WMS)* works as follows:
Take any k-Page Migration algorithm $\mathcal{K}$. Upon receiving the next request r', simulate the next step of $\mathcal{K}$. Calculate a minimum weight matching (with the distances as weights) between the servers $a_1, \ldots, a_k$ of the online algorithm and the pages $c_1', \ldots, c_k'$ of $\mathcal{K}$. Select the closest server $\tilde{a}$ to r' and move it to r' at most a distance $\min(m_c, \frac{1}{D}(1-\varepsilon) \cdot d(\tilde{a}, r'))$ in case $m_c \leq (1+\delta-\varepsilon)m_s$ and at most $\min((1+\frac{\delta}{2})m_s, \frac{1}{D}(1-\frac{\delta}{2}) \cdot d(\tilde{a}, r'))$ in case $m_c \geq (1+\delta)m_s$. All other servers a_i move towards their counterparts in the matching c_i' with speed $\min((1+\delta)m_s, \frac{1}{D} \cdot d(\tilde{a}, r'))$. If another server than $\tilde{a}$ is closer to r' after movement, then move all servers towards their counterpart in the matching with speed m_s instead.

The remainder of this section is devoted to the analysis of the WMS algorithm and is structured similar to Sect. 4. Due to space constraints, we can only give a brief overview and refer for the details to the full version.

We start by analyzing the case that $m_c \leq (1-\varepsilon) \cdot m_s$ for some $\varepsilon \in (0, \frac{1}{2}]$. For $\varepsilon \geq \frac{1}{2}$, our algorithm simply assumes $\varepsilon = \frac{1}{2}$. It can be easily verified, that this does not hinder the analysis.

Theorem 7. *If $m_c \leq (1-\varepsilon) \cdot m_s$ for some $\varepsilon \in (0, \frac{1}{2}]$, the algorithm WMS is $\sqrt{2} \cdot 11/\varepsilon \cdot c(\mathcal{K})$-competitive, where $c(\mathcal{K})$ is the competitive ratio of the simulated k-Page Migration algorithm $\mathcal{K}$.*

We can extend this bound to the resource augmentation scenario, where the online algorithm may move the servers a maximum distance of $(1+\delta)\cdot m_s$. When relaxing the condition appropriately to $m_c \leq (1+\delta-\varepsilon)\cdot m_s$, then we get the following result:

Corollary 2. *If* $m_c \leq (1+\delta-\varepsilon)\cdot m_s$ *for some* $\varepsilon \in (0, \frac{1}{2}]$, *the algorithm WMS is* $\frac{\sqrt{2}\cdot 11\cdot(1+\delta)}{\varepsilon}\cdot c(\mathcal{K})$*-competitive, where* $c(\mathcal{K})$ *is the competitive ratio of the simulated* k*-Page Migration algorithm* $\mathcal{K}$*.*

Similar to the k-Server Projection discussed in Sect. 4.2, we obtain the following result which gives us a new k-Page Migration algorithm needed for the case $m_c \geq (1+\delta)m_s$.

Proposition 3. *Let* $\mathcal{K}$ *be an online algorithm for the* k*-Page Migration Problem. There exists an online algorithm* $\hat{\mathcal{K}}$ *for the* k*-Page Migration Problem with pages* $\hat{c}_1, \ldots, \hat{c}_k$ *such that it holds* $d(\hat{c}_i, r) \leq (32kD+1)\cdot m_c$ *during the whole execution. The costs of* $\hat{\mathcal{K}}$ *are at most* $\mathcal{O}(k)$ *times the costs of* $\mathcal{K}$*.*

From here on we assume $\mathcal{K}$ to be a k-Page Migration algorithm obtained from the transformation in Proposition 3. The offline helper and its invariants as stated in Proposition 2 do not depend on the simulated algorithm and therefore all insights gained from the corresponding section are still valid. We use a potential composed of two major parts just as for the unweighted case.

Let $\hat{o}$ be an offline server which fulfills the invariants stated in Proposition 2. The first part of the potential is then defined as

$$\phi := \begin{cases} 4\cdot d(\hat{a},\hat{o}) & \text{if } d(\hat{a},\hat{o}) \leq 107548D\cdot\frac{km_c}{\delta^2} \\ 4\cdot\frac{1}{\delta m_s}d(\hat{a},\hat{o})^2 + A & \text{if } 107548D\cdot\frac{km_c}{\delta^2} < d(\hat{a},\hat{o}) \end{cases}$$

with $A := 4\cdot(107548D\frac{km_c}{\delta^2} - \frac{1}{\delta m_s}(107548D\frac{km_c}{\delta^2})^2)$.

For the second part, we set

$$\psi := Y\cdot D\frac{m_c}{\delta m_s}\sum_{i=1}^{k} d(a_i, c_i)$$

where the online servers a_i are always sorted such that they represent a minimum weight matching to the simulated servers c_i. We choose $Y = \Theta(\frac{k}{\delta^2})$ to be sufficiently large.

We begin by analyzing ψ, reusing ideas from the proof of Theorem 7.

Lemma 12. $\Delta\psi \leq \mathcal{O}(1)\cdot Y\cdot\frac{m_c}{\delta m_s}\cdot C_{\mathcal{K}} - D\cdot\sum_{i=1}^{k} d(a_i, a_i')$.

Lemma 13. *If* $d(a^{*'}, r') > d(c^{*'}, r')$*, then*

$$\Delta\psi \leq Y\cdot\frac{m_c}{\delta m_s}C_{\mathcal{K}} - D\cdot\sum_{i=1}^{k} d(a_i, a_i') - \frac{Y-4}{2}D\frac{m_c}{\delta m_s}\cdot\min(m_s, \frac{1}{D}\cdot d(\tilde{a}, r')).$$

Now consider the case that $r' \notin inner(o^{*'})$. We have $d(a^{*'}, r') \leq d(o^{*a'}, r') \leq d(o^{*'}, o^{*a'}) + d(o^{*'}, r') \leq (\frac{48960k}{\delta^2} + 1) \cdot d(o^{*'}, r')$. The movement costs are canceled by $\Delta\psi$ as in Lemma 12. It only remains to bound the possible increase of ϕ. We use $d(\hat{a}', \hat{o}') - d(\hat{a}, \hat{o}) \leq (3 + \frac{1020k}{\delta}) \cdot m_c$.

1. $d(\hat{a}', \hat{o}') \leq 107548D \cdot \frac{km_c}{\delta^2}$: $\Delta\phi \leq 4 \cdot d(\hat{a}', \hat{o}') \leq 8 \cdot d(o^{*'}, o^{*a'}) + 4 \cdot d(a^{*'}, r') \leq (12 \cdot \frac{48960k}{\delta^2} + 4) \cdot d(o^{*'}, r')$.
2. $107548D \cdot \frac{km_c}{\delta^2} < d(\hat{a}', \hat{o}')$: $\Delta\phi \leq \frac{4}{\delta m_s}(d(\hat{a}', \hat{o}')^2 - d(\hat{a}, \hat{o})^2) \leq \frac{4}{\delta m_s}(d(\hat{a}', \hat{o}')^2 - (d(\hat{a}', \hat{o}') - (3 + \frac{1020k}{\delta}) \cdot m_c)^2) \leq \mathcal{O}(\frac{k}{\delta}) \cdot \frac{m_c}{\delta m_s} d(\hat{a}', \hat{o}') \leq \mathcal{O}(\frac{k^2}{\delta^3}) \cdot \frac{m_c}{\delta m_s} d(o^{*'}, r')$.

In all of the above, the competitive ratio is bounded by $\mathcal{O}(\frac{k^2}{\delta^3}) \cdot \frac{m_c}{\delta m_s} + Y \cdot \frac{m_c}{\delta m_s} \cdot c(\mathcal{K})$.

Finally, we consider the case $r' \in inner(o^{*'})$. As in the previous Section, whenever $d(a^*, r') > 102970D\frac{km_c}{\delta^2}$, we make use of Lemma 7 to obtain the following result, which then helps us bound $\Delta\phi$:

Lemma 14. *If* $d(a^*, r') > 102970D\frac{km_c}{\delta^2}$ *and* $r' \in inner(o^{*'})$, *then* $d(a_i', \hat{o}') - d(a_i, \hat{o}') \leq -\frac{\delta}{8} m_s$.

Lemma 15. *If* $r' \in inner(o^{*'})$, *then* $C_{Alg} + \Delta\phi + \Delta\psi \leq Y \cdot \frac{m_c}{\delta m_s} \cdot C_{\mathcal{K}} + 2 \cdot d(o^{*'}, r')$.

The resulting competitive ratio $Y \cdot \frac{m_c}{\delta m_s} \cdot c(\mathcal{K}) + 2$ is less than the $\mathcal{O}(\frac{k^2}{\delta^3}) \cdot \frac{m_c}{\delta m_s} + Y \cdot \frac{m_c}{\delta m_s} \cdot c(\mathcal{K})$ bound from the former set of cases. Accounting for the loss due to the transformation of the simulated k-Page Migration algorithm, we obtain the following upper bound:

Theorem 8. *If* $m_c \geq (1 + \delta)m_s$, *the algorithm WMS is* $\mathcal{O}(\frac{1}{\delta^4} \cdot k^2 \cdot \frac{m_c}{m_s} + \frac{1}{\delta^3} \cdot k^2 \cdot \frac{m_c}{m_s} \cdot c(\mathcal{K}))$*-competitive, where* $c(\mathcal{K})$ *is the competitive ratio of the simulated* k*-Page Migration algorithm* $\mathcal{K}$.

6 Open Problems

The gap between the upper and lower bound is closely related to the question of the deterministic upper bound for k-Page Migration: Not only would an $\mathcal{O}(k)$-competitive algorithm for k-Page Migration directly improve the bound for $D > 1$, it could also give an idea how to improve the analysis of the greedy step in our algorithm, such that the costly transformation of the simulated algorithm would no longer be needed. This would potentially reduce the upper bound by another factor of k. On the other hand, if $\Omega(k^2)$ is a lower bound for k-Page Migration, this carries over to our model as well. We believe that the main algorithmic idea is suitable to reach an asymptotically optimal competitive ratio, but it remains an open problem to derive a proof of that. The high constants in our proofs are partially due to allowing easier argumentation in certain segments of the proof. There is however also great potential in reducing constants by trying to extend the potential analysis to operate in longer phases instead of doing a step-by-step analysis.

If we allow randomization, we can get an $\mathcal{O}(k)$-competitive k-Page Migration algorithm from [2]. As discussed in the related work section, the question of the best possible competitive ratio of randomized algorithms for the k-Server problem is still open, however we know that a result polylogarithmic in k can be achieved [12]. As our construction is entirely deterministic, apart from potentially the simulated algorithm, it would be interesting whether randomization can be used to significantly improve the competitive ratio. The desired result would be an algorithm with a competitive ratio polylogarithmic in k. More generally, the problem of finding a randomized algorithm with competitiveness $o(k)$ is still open for the classical k-Page Migration problem.

References

1. Bansal, N., Buchbinder, N., Madry, A., Naor, J.: A polylogarithmic-competitive algorithm for the k-server problem. J. ACM **62**(5), 40:1–40:49 (2015). https://doi.org/10.1145/2783434
2. Bartal, Y., Charikar, M., Indyk, P.: On page migration and other relaxed task systems. Theor. Comput. Sci. **268**(1), 43–66 (2001). https://doi.org/10.1016/S0304-3975(00)00259-0
3. Bartal, Y., Koutsoupias, E.: On the competitive ratio of the work function algorithm for the k-server problem. Theor. Comput. Sci. **324**(2–3), 337–345 (2004). https://doi.org/10.1016/j.tcs.2004.06.001
4. Bienkowski, M., Byrka, J., Mucha, M.: Dynamic beats fixed: on phase-based algorithms for file migration. In: Proceedings of the 44th International Colloquium on Automata, Languages, and Programming (ICALP), pp. 13:1–13:14 (2017). https://doi.org/10.4230/LIPIcs.ICALP.2017.13
5. Black, D.L., Sleator, D.D.: Competitive algorithms for replication and migration problems. Technical Report CMU-CS-89-201, Department of Computer Science, Carnegie-Mellon University (1989)
6. Bubeck, S., Cohen, M.B., Lee, Y.T., Lee, J.R., Madry, A.: k-server via multiscale entropic regularization. In: Proceedings of the 50th Annual ACM SIGACT Symposium on Theory of Computing (STOC), pp. 3–16 (2018). https://doi.org/10.1145/3188745.3188798
7. Chrobak, M., Karloff, H.J., Payne, T.H., Vishwanathan, S.: New results on server problems. SIAM J. Discrete Math. **4**(2), 172–181 (1991). https://doi.org/10.1137/0404017
8. Feldkord, B., der Heide, F.M.A.: The mobile server problem. In: Proceedings of the 29th ACM Symposium on Parallelism in Algorithms and Architectures (SPAA), pp. 313–319 (2017). https://doi.org/10.1145/3087556.3087575
9. Feldkord, B., der Heide, F.M.A.: The mobile server problem. CoRR abs/1904.05220 (2019). https://arxiv.org/abs/1904.05220
10. Fiat, A., Karp, R.M., Luby, M., McGeoch, L.A., Sleator, D.D., Young, N.E.: Competitive paging algorithms. J. Algorithms **12**(4), 685–699 (1991). https://doi.org/10.1016/0196-6774(91)90041-V
11. Koutsoupias, E., Papadimitriou, C.H.: On the k-server conjecture. J. ACM **42**(5), 971–983 (1995). https://doi.org/10.1145/210118.210128
12. Lee, J.R.: Fusible HSTs and the randomized k-server conjecture. In: 59th IEEE Annual Symposium on Foundations of Computer Science, FOCS 2018, Paris,

France, 7–9 October 2018, pp. 438–449 (2018). https://doi.org/10.1109/FOCS.2018.00049
13. Manasse, M.S., McGeoch, L.A., Sleator, D.D.: Competitive algorithms for server problems. J. Algorithms **11**(2), 208–230 (1990). https://doi.org/10.1016/0196-6774(90)90003-W
14. Rudec, T., Baumgartner, A., Manger, R.: A fast work function algorithm for solving the k-server problem. CEJOR **21**(1), 187–205 (2013). https://doi.org/10.1007/s10100-011-0222-7
15. Rudec, T., Manger, R.: A fast approximate implementation of the work function algorithm for solving the k-server problem. CEJOR **23**(3), 699–722 (2015). https://doi.org/10.1007/s10100-014-0349-4
16. Westbrook, J.R.: Randomized algorithms for multiprocessor page migration. SIAM J. Comput. **23**(5), 951–965 (1994). https://doi.org/10.1137/S0097539791199796

On the Cycle Augmentation Problem: Hardness and Approximation Algorithms

Waldo Gálvez[1(✉)], Fabrizio Grandoni[1], Afrouz Jabal Ameli[1], and Krzysztof Sornat[2]

[1] IDSIA, Lugano, Switzerland
{waldo,fabrizio,afrouz}@idsia.ch
[2] University of Wrocław, Wrocław, Poland
krzysztof.sornat@cs.uni.wroc.pl

Abstract. In the k-Connectivity Augmentation Problem we are given a k-edge-connected graph and a set of additional edges called *links*. Our goal is to find a set of links of minimum cardinality whose addition to the graph makes it $(k+1)$-edge-connected. There is an approximation preserving reduction from the mentioned problem to the case $k = 1$ (a.k.a. the Tree Augmentation Problem or TAP) or $k = 2$ (a.k.a. the Cactus Augmentation Problem or CacAP). While several better-than-2 approximation algorithms are known for TAP, nothing better is known for CacAP (hence for k-Connectivity Augmentation in general).

As a first step towards better approximation algorithms for CacAP, we consider the special case where the input cactus consists of a single cycle, the *Cycle Augmentation* Problem (CycAP). This apparently simple special case retains part of the hardness of the general case. In particular, we are able to show that it is APX-hard.

In this paper we present a combinatorial $\left(\frac{3}{2}+\varepsilon\right)$-approximation for CycAP, for any constant $\varepsilon > 0$. We also present an LP formulation with a matching integrality gap: this might be useful to address the general case of the problem.

Keywords: Approximation algorithms · Connectivity Augmentation · Cactus Augmentation · Cycle Augmentation

1 Introduction

The basic goal of *Survivable Network Design* is to construct low cost networks that provide connectivity guarantees between pre-specified sets of nodes even after the failure of a few edges/nodes (in the following we will focus on the egde

Partially supported by the SNSF Grant 200021_159697/1, the SNSF Excellence Grant 200020B_182865/1, the National Science Centre, Poland, grant numbers 2015/17/N/ST6/03684, 2015/18/E/ST6/00456 and 2018/28/T/ST6/00366. K. Sornat was also supported by the Foundation for Polish Science (FNP) within the START programme.

E. Bampis and N. Megow (Eds.): WAOA 2019, LNCS 11926, pp. 138–153, 2020.
https://doi.org/10.1007/978-3-030-39479-0_10

failure case). This has many applications, e.g., in transportation and telecommunication networks.

A relevant subclass of these problems is given by *Network Augmentation* problems. Here the goal is to *augment* a given graph $G = (V, E)$ by adding extra edges taken from a given set L (*links*), so as to satisfy given (edge-)connectivity requirements. Several such problems are NP-hard, and in most cases the best known approximation factor is 2 due to Jain [14].

In this paper we focus on the following k-Connectivity Augmentation Problem (k-CAP). Given a k-(edge)-connected[1] undirected graph $G = (V, E)$ and a collection L of extra edges (*links*), the goal is to find a minimum cardinality subset $A \subseteq L$ such that $G' = (V, E \cup A)$ is $(k + 1)$-connected. Dinitz et al. [8] presented an approximation preserving reduction from this problem to the case $k = 1$ for odd k, and $k = 2$ for even k. This motivates a deeper understanding of the latter two special cases.

The case $k = 1$ is also known as the Tree Augmentation Problem (TAP). The reason for this name is that any 2-connected component of the input graph G can be contracted, hence leading to a tree. For this problem several better-than-2 approximation algorithms are known [1,6,9,10,13,17,20]. The case $k = 2$ is also known as the Cactus Augmentation Problem (CacAP) since, similarly to the previous case, the input graph can be assumed to be a cactus[2] [8]. However, here the best-known approximation factor is still 2 [14] (implying the same for k-CAP in general).

For all the mentioned problems it makes sense to consider the weighted version, where links have non-negative integral weights, and the goal is to find a minimum weight (rather than minimum cardinality) subset of links A with the desired properties. In particular we will speak about Weighted TAP (WTAP) and Weighted CacAP (WCacAP). Here the best-known approximation factor is 2 in both cases [14]. Moreover, improving on that approximation factor for WTAP is considered as a major open problem in the area. We also notice that we can turn a WTAP instance into an equivalent WCacAP instance by replacing each edge with two parallel edges. Hence, approximating WCacAP is not any easier than approximating WTAP (and the same holds for the corresponding unweighted versions).

1.1 Our Results

As mentioned before, CacAP contains TAP as a special case when the cactus consists of several short cycles. Hence, in order to make progress on CacAP, it makes sense to consider the somehow complementary case where the input cactus consists of a single cycle of n nodes. We call the corresponding subproblem

---

[1] We recall that $G = (V, E)$ is k-connected if for every set of edges $F \subseteq E$, $|F| \leq k - 1$, the graph $G' = (V, E \setminus F)$ is connected.

[2] We recall that a *cactus* G is a connected undirected graph in which every edge belongs to exactly one cycle. For technical reasons it is convenient to allow cycles of length 2 consisting of parallel edges.

the Cycle Augmentation Problem (CycAP), and its weighted version Weighted CycAP (WCycAP). To the best of our knowledge, these special cases were not studied before. However, as we will see, they still retain part of the difficulties of the general cactus case. In more detail, we achieve the following main results.

Hardness of Approximation. We are able to show that WCycAP is as hard to approximate as WCacAP. Therefore, improving on a 2-approximation for WCycAP would imply a major breakthrough in the area (in particular, it would imply the same for WTAP). This also justifies a more careful investigation of CycAP. In our opinion it is a priori not so obvious that CycAP is even NP-hard. Indeed, the special case of TAP (and even of WTAP) where the input graph is a path can be solved exactly in polynomial time. The case of an input cycle might closely remind the path case. Here we show that this intuition is not correct: we prove that CycAP is NP-hard and even APX-hard via a simple but non-trivial adaptation of the proofs in [11,16]. In particular, we need one extra step in the reduction where we turn an intermediate CacAP instance into a CycAP one while maintaining certain properties of the optimal solution.

Approximation Algorithms. As discussed, the best we can hope for CycAP is some constant $c > 1$ approximation. We present better-than-2 approximation algorithms for this problem. In particular, we present a simple $\frac{5}{3}$-approximation, and a slightly more complex $(3/2 + \varepsilon)$-approximation for any constant $\varepsilon > 0$. Notice that the latter approximation factor is not far from the best known approximation factor for TAP which is equal to 1.458 [13]. Our algorithms are purely combinatorial, and they consist of two main phases. In the first phase, we *greedily* add some links to the solution under construction and *contract* them. At the end of this phase we achieve an instance of CacAP that can be solved exactly in polynomial time. In particular, for the $\frac{5}{3}$-approximation this reduces to computing a spanning tree, while for the $(3/2 + \varepsilon)$-approximation we use an FPT algorithm parameterized by a proper notion of maximum *length* of a link.

LP Gaps. The recent literature on TAP approximation [1,10,13] shows that finding strong LP relaxations for the problem can be very helpful to design improved approximation algorithms. In the same spirit, we tried to address the problem of finding LP relaxations for CycAP with small integrality gap. For both TAP and CacAP (hence CycAP) one can define a natural and simple standard cut LP (more details later). While for TAP it was recently shown that the standard cut LP has integrality gap smaller than 2 [21], interestingly for CycAP (hence for CacAP) the standard cut LP has integrality gap 2. Here we present a stronger LP that, for any $\varepsilon > 0$, has integrality gap at most $\frac{3}{2} + \varepsilon$ (hence matching the approximation ratio of our algorithm). In our opinion this could be useful for future work on CacAP approximation.

1.2 Related Work

As mentioned before, the best known result in terms of polynomial time approximation algorithms for k-CAP is a 2-approximation proposed by Jain [14]. However, if the set of links is equal to $V \times V$ it is possible to solve this problem optimally [22]. More recently, this problem has been studied in the framework of Fixed-Parameter Tractability: Végh and Marx [19] proved that this problem is in FPT when parameterized by the size of the optimal solution, and later the running time of their algorithm was further improved [2].

Tree Augmentation has been extensively studied over the past few decades. It was first shown that WTAP is NP-hard by Frederickson and Jájá [11], then that TAP is NP-hard by Cheriyan et al. [5], and later that TAP is APX-hard by Kortsarz et al. [16]. For WTAP, the best-known approximation guarantee is 2 and was first established by Frederickson and Jájá [11]. Their algorithm was later simplified by Khuller and Thurimella [15]. A 2-approximation can also be achieved by various other techniques developed later on, including a primal-dual approach [12] and iterative rounding [14]. Improvements on the factor 2 have only been obtained for restricted cases, including bounded diameter trees [7] and bounded weights [1,10,13,21].

Regarding TAP, the first algorithm beating the approximation guarantee of 2 is due to Nagamochi [20], achieving an approximation factor of $1.815 + \varepsilon$. This factor was subsequently improved to 1.8 [9] and to 1.5 [6,17]. These results are combinatorial in nature, but LP-based results have been achieved as well. As an example, recently Nutov [21] showed that the standard cut LP for TAP has an integrality gap of at most 28/15 while a lower bound of 3/2 was known [6]. An LP-based $(\frac{5}{3} + \varepsilon)$-approximation was given by Adjiashvili [1] and then refined by Fiorini et al. [10] to obtain a $(\frac{3}{2} + \varepsilon)$-approximation (see also [3,4,18]). Both results are obtained by adding a proper family of extra constraints to the standard cut LP. Recently, Grandoni et al. [13] achieved a 1.458 approximation for TAP, which is smaller than the integrality gap of the standard cut LP.

The rest of this paper is organized as follows. In Sect. 2 we give some preliminary definitions and results. The approximation algorithms and LP-gaps are discussed in Sects. 3 and 4 respectively. Due to space limitations, some of the results and proofs are deferred to the full version of the paper.

2 Preliminaries

For a set X and element y, we use the shortcut $X \setminus y$ for $X \setminus \{y\}$, and similarly for other set operations.

Given a graph $G = (V, E)$, we let $V(G) = V$ and $E(G) = E$. Recall that in WCacAP we are given a cactus $G = (V, E)$, a set of links $L \subseteq \binom{V}{2}$ and a non-negative weight function $c : L \to \mathbb{R}_{\geq 0}$. The task is to compute a subset of links $A \subseteq L$ such that the graph $(V, E \cup A)$ is 3-edge-connected while minimizing $c(A) := \sum_{\ell \in A} c(\ell)$. The special case where G is a cycle is called WCycAP, and the unweighted versions of the above problems are called CacAP and CycAP respectively. By n we will denote the number of nodes of the considered instance of the problem.

Notice that, given an instance (G, L) of CacAP, we can check in polynomial time if the graph $(V(G), E(G) \cup L)$ is 3-edge-connected by exhaustively checking if the removal of any pair of elements from $E(G) \cup L$ disconnects the graph. Hence we will assume along this work that the instance always admits a feasible solution.

Remark 1. The 2-edge cuts of a cactus G are identified by pairs $S = \{e, e'\}$ of distinct edges belonging to the same cycle, and consist of the node sets (V', V'') of the two connected components obtained by removing S from G. A necessary and sufficient condition for a subset of links A to be a feasible solution for WCacAP is that, for any such cut S, there is at least one $\ell \in A$ crossing the cut (in which case ℓ **satisfies** the $\{e, e'\}$-cut).

Note that in the case of CycAP, Remark 1 implies that any feasible solution must be an edge cover as 2-edge cuts defined by neighboring edges of the cycle must be satisfied. Given a 2-edge cut $S = \{e, e'\}$, let L_S be the subset of links satisfying S. The standard cut LP for CycAP is as follows:

$$\begin{aligned} \min \sum_{\ell \in L} x_\ell & \qquad \text{(standard cut LP)} \\ s.t. \sum_{\ell \in L_S} x_\ell \geq 1 & \qquad \forall S : S \text{ is a 2-edge cut} \\ 0 \leq x_\ell \leq 1 & \qquad \forall \ell \in L \end{aligned}$$

Now we proceed to define a standard building block for our algorithms, the *contraction* of a link.

Definition 1. *Contracting a subset of nodes W consists of the following operations: (i) remove the nodes in W and all edges/links incident to them; (ii) add a new node w and, for each original edge/link of type (y, x), $x \in W, y \notin W$, add the edge/link (y, w) (of the same weight for the case of links). Note that we do not create loops this way but may introduce parallel links. We say that (y, w) is the image of (y, x) and (y, x) is the preimage of (y, w).*

We will sometimes slightly abuse notation and use the same label to denote a link and its image: the meaning will be clear from the context.

For a link $\ell = (u, v)$, we define a sequence $w_0, \ldots, w_q$ of boundary nodes $B(\ell)$ as follows. Consider a simple path from u to v in the cactus, and let $C_1, C_2, \ldots, C_q$ be the ordered sequence of cycles visited[3] by this path (possibly $q = 1$). We define w_i, $i = 1, \ldots, q-1$ as the unique common node between C_i and C_{i+1}, and set $w_0 = u$ and $w_q = v$.

Definition 2. *Contracting a link ℓ is the operation of contracting its boundary nodes $B(\ell)$. We denote by $G|\ell$ the graph obtained by this operation. Contracting a set of links A is the operation of contracting any $\ell \in A$, and then continue recursively on $G|\ell$ and on the image of $A \setminus \ell$ until A becomes empty.*

Note that contracting a link in a cactus yields again a cactus. We will extensively use the following standard fact, whose proof is given in the full version of the paper.

[3] A path visits a cycle iff it includes an edge from the cycle.

Lemma 1. *Let (G, L) be a CacAP instance, $A \subseteq L$, and $\ell \in A$. Then A is a feasible solution for (G, L) iff the image of $A \setminus \ell$ is a feasible solution for $(G|\ell, L \setminus \ell)$.*

3 Approximation Algorithms for Cycle Augmentation

In this section we present improved approximation algorithms for CycAP. We start with a simple $\frac{5}{3}$-approximation to illustrate the main ideas, and then present a slightly more complex $(\frac{3}{2} + \varepsilon)$-approximation. The approach we will follow in both cases is as follows: in a first phase we iteratively add a properly chosen subset of a few links to the solution under construction, and then contract them. Notice that, after the first contraction, the cycle structure may be lost and we obtain a CacAP instance instead. These choices are designed so that, at the end of the first phase, the remaining CacAP instance can be solved efficiently, which is done in a second phase with an ad-hoc algorithm. We remark that the running times of the presented algorithms is not analyzed in detail, and indeed such a task may require to devise carefully crafted data structures.

3.1 A $\frac{5}{3}$-approximation

We next describe a simple greedy algorithm that provides a $\frac{5}{3}$-approximation for CycAP. We need the following definitions.

Definition 3. *A link $\ell = (u, v)$ of a CacAP instance is **internal** if both its endpoints belong to a common cycle, and* external *otherwise.*

Definition 4. *Given a CacAP instance, a pair of internal links $\{(u_1, v_1), (u_2, v_2)\}$ of a cycle C is **crossing** if they are node disjoint and deleting u_2 and v_2 disconnects u_1 from v_1 in C.*

The kind of links that we want to add in the first stage of the algorithm are external links plus crossing pairs of links. More in detail, the algorithm has two main stages. The first stage consists of a set of rounds, where in each round we first check if there exists an external link ℓ, in which case we add it to our solution, contract it and proceed to the next round. Otherwise, if there exists a pair of (internal) crossing links ℓ' and ℓ'', we add them to our solution, contract them and proceed to the next round. If none of the two cases above applies, we are left with a CacAP instance without neither external links nor crossing pairs of links which we address in the second stage of the algorithm. We refer to this algorithm as CROSSING-FIRST. As the following lemma states, in the second stage we can efficiently compute the optimal solution.

Lemma 2. *Consider an instance $(G = (V, E), L)$ of CacAP. If there are no external links and no crossing pairs of links, then every minimal solution has size exactly $|V| - 1$ and induces a spanning tree over V.*

Proof. We prove the first part of the claim by induction on $n = |V|$. The base case $n = 2$ is trivial since in this case the instance is just a cycle consisting of two parallel edges and any link must be incident to the two nodes of G (hence defining a feasible solution). For the inductive case, assume the claim is true up to instances having $n-1$ nodes, and consider an instance of the problem defined by a cactus G having n nodes with optimal solution OPT. If G is not a cycle of length n, then it is defined by a set of cycles of length at most $n-1$ where every link is internal, so we can apply the inductive hypothesis to each cycle independently. If G is a cycle of n nodes, then let $\ell = (u, v) \in \text{OPT}$. Contracting ℓ leads to a CacAP instance on two cycles C_1 and C_2 sharing a common node w, with $|V(C_1)| + |V(C_2)| = n$. Let OPT′ be the optimal solution for the new instance. By Lemma 1, $|\text{OPT}| = |\text{OPT}'|+1$. Observe that any remaining link ℓ' must have both endpoints in the same C_i (otherwise ℓ and ℓ' would be crossing). Thus by the inductive hypothesis the optimum solution for the problem induced by C_i has size $|V(C_i)|-1$. It then follows that $|\text{OPT}'| = |V(C_1)|-1+|V(C_2)|-1 = n-2$. Hence $|\text{OPT}| = n-1$ as desired.

For the second part of the claim, it is sufficient to show that a minimal solution does not induce a cycle. By contradiction, consider a minimal solution containing a simple cycle L', and consider now a solution where we remove precisely one arbitrary link $\ell = (u, v)$ from L'. Consider any pair of edges e_1, e_2 belonging to the same cycle such that ℓ satisfies the $\{e_1, e_2\}$-cut. Since $L' \setminus \ell$ induces a simple u-v path, then some $\ell' \in L' \setminus \ell$ must satisfy the cut. Thus $L' \setminus \ell$ is a feasible solution, contradicting the minimality of L'.

Now we proceed to prove the approximation guarantee of the algorithm.

Theorem 1. *The* CROSSING-FIRST *algorithm is a $\frac{5}{3}$-approximation for CycAP.*

Proof. Let OPT be the optimal solution and APX the computed solution. Let also n'' be the number of nodes remaining at the end of the first stage, and APX′ (resp. APX″) be the set of links added to the solution during the first (resp. second) stage. Since contracting an external link decreases the number of nodes by at least 2 and contracting any pair of crossing links decreases the number of nodes by at least 3, we have that $|\text{APX}'| \leq \frac{2}{3}(n - n'')$.

By Lemma 2, $|\text{APX}''| = n'' - 1$, and hence $|\text{APX}| \leq \frac{2}{3}(n - n'') + n'' - 1 = \frac{2n+n''-3}{3}$. On the other hand, since any feasible solution must be an edge cover, we have that $|\text{OPT}| \geq n/2$. Observe also that $|\text{OPT}| \geq n'' - 1$ since by Lemma 1 contracting links cannot increase the cost of the optimum solution. Thus $|\text{OPT}| \geq \max\{n/2, n'' - 1\}$. We can conclude that $\frac{|\text{APX}|}{|\text{OPT}|} \leq \frac{(2n+n''-3)/3}{\max\{n/2, n''-1\}} \leq \frac{5}{3}$, being $n'' - 1 = n/2$ the worst case.

We complement this result with an asymptotically matching lower bound.

Lemma 3. *The approximation ratio of the* CROSSING-FIRST *algorithm is not better than $\frac{5}{3}$.*

Proof. Consider the following construction: for each $k \geq 2$ consider an instance (G_k, L_k) of CycAP defined by a cycle of $n = 6k$ nodes (assume that the cycle is defined by the order of the nodes $v_1, v_2, \ldots, v_{6k}$) and the following set of links (see Fig. 1 (Left)):

- $(v_1, v_{\frac{n}{2}+1}) \in L_k$;
- For each $i = 1, \ldots, \frac{n}{2} - 1$, $(v_{i+1}, v_{n+1-i}) \in L_k$;
- For each $i = 1, \ldots, \frac{n}{6}$, $(v_{3(i-1)+1}, v_{3(i-1)+3}) \in L_k$ and $(v_{3(i-1)+2}, v_{3(i-1)+4}) \in L_k$;

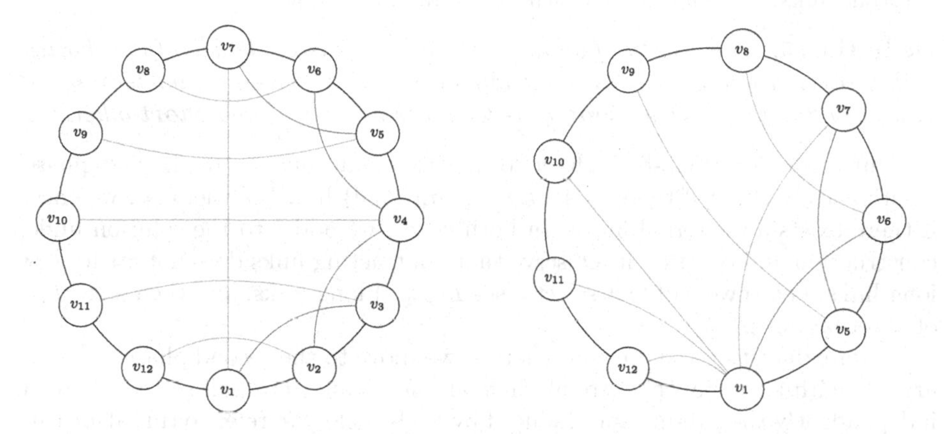

Fig. 1. Left: Instance (G_2, L_2) from the lower bound construction in Lemma 3. Red links define an optimal solution. Right: If the algorithm in the first phase picks and contracts the crossing links $\{(v_1, v_3), (v_2, v_4)\}$, this is the obtained CacAP instance. (Color figure online)

Notice that the first and second set of links define a feasible solution of size $\frac{n}{2}$, hence being optimal: if we remove any two edges of the cycle, then we are either satisfying the corresponding cut via $(v_1, v_{\frac{n}{2}+1})$, or one side of the partition is contained in either $\{v_2, \ldots, v_{\frac{n}{2}}\}$ or in $\{v_{\frac{n}{2}+2}, \ldots, v_n\}$ but the links selected form a matching between those sets.

We will now prove that there exists a sequence of choices performed by our algorithm that outputs a solution of size $\frac{5n}{6} - 1$, which implies that the approximation ratio is at least $\frac{5}{3} - \frac{2}{n}$ and this value approaches $\frac{5}{3}$ as k goes to infinity. Notice first that the pair of links $\{(v_1, v_3), (v_2, v_4)\} \subseteq L_k$ is crossing, and hence the algorithm can include them in the solution in the first round (and finish the round). Furthermore, after these links are contracted no link becomes external as the new cactus instance consists of a cycle of length $n - 3$, and also the links with endpoints v_n, v_{n-1} and v_{n-2} are not part of any pair of crossing links (see Fig. 1 (Right)). If we now iteratively pick all the pairs of crossing links $\{(v_{3(i-1)+1}, v_{3(i-1)+3}), (v_{3(i-1)+2}, v_{3(i-1)+4})\} \subseteq L_k$, $i = 2, \ldots, \frac{n}{6}$, after $\frac{n}{6}$ rounds we end up with a cycle of length $\frac{n}{2}$ without crossing links, and the algorithm must now take the remaining $\frac{n}{2} - 1$ links to complete the solution. Thus, the size of the computed solution is $2 \cdot \frac{n}{6} + \frac{n}{2} - 1 = \frac{5}{6}n - 1$, proving the claim.

3.2 A $(\frac{3}{2}+\varepsilon)$-approximation

The family of instances from Lemma 3 suggests that "short" crossing pairs of links, although being locally profitable, may enforce the algorithm to take expensive decisions in the end. In this section we present a more involved $(\frac{3}{2}+\varepsilon)$-approximation for CycAP that tries to avoid this kind of situation. Like in the previous algorithm, there is a certain kind of links that we want to iteratively add to our solution in a first phase, and in this case such links correspond to external links and *long* links, which are defined as follows.

Definition 5. *The* ***length*** *of an internal link* (u, v) *is the length of the shortest path between* u *and* v *in the corresponding cycle. For a given parameter* $0 < \varepsilon < 1$*, an internal link is called* ***long*** *if its length is at least* $\frac{1}{\varepsilon}$*, and* ***short*** *otherwise.*

Our algorithm consists of the following two main phases. In the first phase, we iteratively check if there exists a long (internal) link ℓ. Otherwise, we check if there exists an external link ℓ. In both cases, we add ℓ to the solution under construction and contract it. Observe that contracting links does not create new long links, hence we will first select a set L_{long} of long links, and then a set L_{ext} of external links.

After exhausting the previous choices, we move to the second phase. Here we are left with an instance where all links are short and internal, so we can solve independently the sub-instance induced by each cycle. We refer to this algorithm as LONG-FIRST. This second stage can be solved efficiently, due to the lack of long links, by means of the following lemma[4].

Lemma 4. *Given a CycAP instance, there exists an algorithm that returns the optimal solution in time poly(n)* $\cdot\, 2^{O(h_{\max}^2)}$*, where* $h_{\max}$ *is the maximum length among the links.*

Let L_{short} be the collection of edges obtained in the second stage. The final solution is $L_{\text{long}} \cup L_{\text{ext}} \cup L_{\text{short}}$.

Theorem 2. *The previous algorithm is a* $(\frac{3}{2}+\varepsilon)$*-approximation algorithm for CycAP.*

Proof. The running time of the algorithm is upper-bounded by $\text{poly}(n)2^{O(1/\varepsilon^2)}$. Consider next the approximation factor. Note first that $|L_{\text{long}}| \leq \varepsilon n$. Indeed, contracting a long link always increases the number of cycles in the cactus by one without decreasing the number of edges, and all these cycles always have size at least $1/\varepsilon$, so there are at most εn of them. Similarly to Theorem 1, we have that $|\text{OPT}| \geq |L_{\text{short}}|$ and $|\text{OPT}| \geq \frac{n}{2}$.

If $|L_{\text{long}}| + |L_{\text{ext}}| + |L_{\text{short}}| \leq \frac{(3+2\varepsilon)n}{4}$ then we already have a $(\frac{3}{2}+\varepsilon)$-approximation as $|\text{OPT}| \geq \frac{n}{2}$. Otherwise, since the contraction of each external link reduces the number of nodes by at least 2 and the contraction of any other link reduces the number of nodes by at least 1, we have that

[4] This lemma implies that CycAP is FPT with parameter $h_{\max}$.

$|L_{\text{long}}| + 2|L_{\text{ext}}| + |L_{\text{short}}| \leq n$. So $|L_{\text{ext}}| \leq n - \frac{(3+2\varepsilon)n}{4} = \frac{(1-2\varepsilon)n}{4}$ and hence $|L_{\text{ext}}| + |L_{\text{long}}| \leq \frac{n+2\varepsilon n}{4} \leq \left(\frac{1}{2}+\varepsilon\right)|\text{OPT}|$. Since $|\text{OPT}| \geq |L_{\text{short}}|$, we have that in this case the size of the solution is also at most $(\frac{3}{2}+\varepsilon)|\text{OPT}|$, concluding the proof.

Remark 2. By replacing ε with $1/\sqrt{\log n}$ in the above construction, we can obtain a slightly improved approximation factor of $3/2 + o(1)$ which still runs in polynomial time.

It remains to prove Lemma 4. We need some more notation. Recall that a 2-cut $\{e, e'\}$ is satisfied iff there is some link crossing the cut. Given a link $\ell = (u, v)$, we say that the edges of the shortest path between u and v in the cycle are *covered* by ℓ (in case of multiple shortest paths we choose the one going from u to v in counter-clockwise order along the cycle). Given an edge e of the cycle, we define the *cut-neighborhood* of e, namely $\mathcal{N}(e)$, as the $2h_{\max} - 1$ edges that are closest to e, e included. We also define $\mathcal{N}_L(e)$ as the set of links in L covering at least one edge from $\mathcal{N}(e)$.

Notice that in any feasible solution to a CycAP instance, at most one edge of the cycle is not covered: if it is not the case, then the cut defined by two uncovered edges is not satisfied as any link satisfying the cut would cover one of these two edges. We can use this observation to characterize the feasibility of a solution in terms of the cut-neighborhoods.

Lemma 5. *Consider a CycAP instance and let A be a set of links such that every edge of the cycle is covered by some link in A. A is feasible iff for each edge e, all the $\{e, e'\}$-cuts, where $e' \in \mathcal{N}(e)$, are satisfied.*

Proof. If A is feasible then the required properties are clearly satisfied since every cut is satisfied. On the other hand, suppose that A satisfies that every edge is covered by some link in A and the $\{e, e'\}$-cuts are satisfied for every edge e and $e' \in \mathcal{N}(e)$. Consider a pair of edges $\{e, e'\}$ such that $e' \notin \mathcal{N}(e)$. By definition of $\mathcal{N}(e)$ there is no link in A covering both edges at the same time, and as e is covered by some link, this link satisfies the $\{e, e'\}$-cut. This implies that A is feasible as every cut is satisfied.

This lemma is useful as it implies that, given an edge e and a set of links S, we can optimally complete S in order to satisfy every $\{e, e'\}$-cut in time $2^{O(h_{\max}^2)}$ just by guessing the subset of links from $\mathcal{N}_L(e)$ that must be added, which are $O(h_{\max}^2)$ only. Now we proceed to present the algorithm.

Proof (Proof of Lemma 4).

Let us assume that we deal with instances of CycAP such that there exists an optimal solution where every edge is covered by some link. If it is not the case, as there may be only one uncovered edge, we can guess this edge and contract it; this leads to an equivalent instance of the problem where we can require that the optimum solution covers all the edges. We say that an edge e is *satisfied* by a set of links A if it is covered by some link in A and furthermore every $\{e, e'\}$-cut is

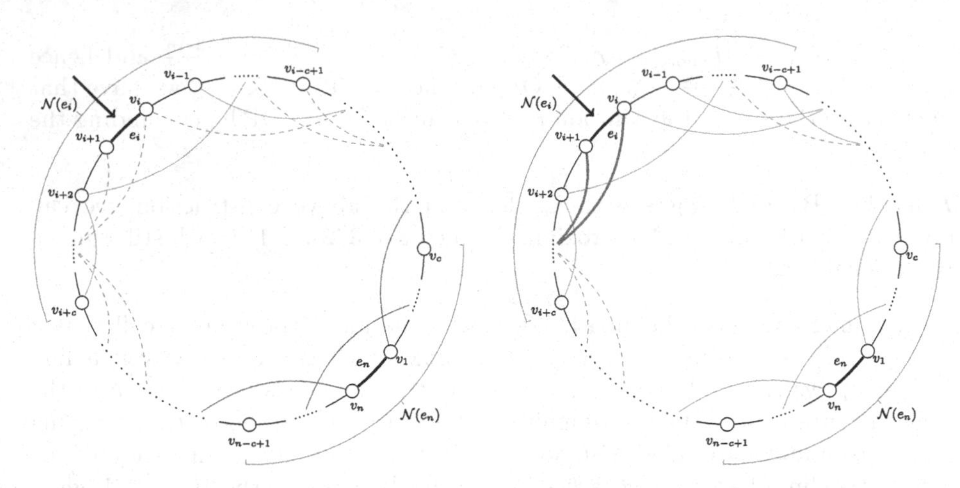

Fig. 2. Depiction of an iteration of the DP from Lemma 4, where we are currently at edge e_i. Left: Blue links correspond to L_0, green links correspond to S and at this point we must decide which extra links to add to $S \cup L_0$ in order to satisfy the edges $e_1, \ldots, e_i$. Right: This computation is done by looking at a proper previous cell in the table (orange links) which contains $S \cup L_0$ and satisfies $e_1, \ldots, e_{i-1}$, and then add the extra required links A^* (red links) in order to satisfy e_i too. (Color figure online)

satisfied by A. In particular A is a feasible solution for the problem iff it satisfies all the edges.

We next design a dynamic programming algorithm to compute a minimum cardinality feasible solution. Let us name the nodes $v_1, v_2, ..., v_n$ in counter-clockwise order starting from some arbitrary node v_1, and let the edges be $e_i = (v_i, v_{i+1})$ for each $i = 1, \ldots, n$ (assuming $v_{n+1} = v_1$). We start first by guessing the set L_0 of links from OPT that satisfy e_n. As proved in Lemma 5, L_0 is a subset of $\mathcal{N}_L(e_n)$, hence we can guess it in time $2^{O(h^2_{\max})}$.

For each edge e_i and $S \subseteq \mathcal{N}_L(e_i)$, we define a cell $T[i][S]$ which will correspond to a set S' of links of smallest cardinality such that for each $j \in \{1, \ldots, i\}$, e_j is satisfied by S', subject to $L_0 \cup S \subseteq S'$. It is then sufficient to return $T[n][\emptyset]$.

We initialize the table by computing $T[1][S]$ for each set $S \subseteq \mathcal{N}_L(e_1)$, which can be done by guessing how to complete $S \cup L_0$ in order to satisfy e_1 with links from $\mathcal{N}_L(e_1)$. Then, for each $i \geq 2$ and $S \subseteq \mathcal{N}_L(e_i)$, in order to fill the cell $T[i][S]$, we consider all the possible subsets $A \subseteq \mathcal{N}_L(e_i)$ such that $S(A) := T[i-1][(S \cup A) \cap \mathcal{N}_L(e_{i-1})] \cup (S \cup A)$ satisfies e_i. Among them we select a set A^* that minimizes $|S(A)|$, and we set $T[i][S] = S(A^*)$ (see Fig. 2 for a sketch).

The correctness of the computation follows by a simple induction on i. The table can be filled in total time $\text{poly}(n) \cdot 2^{O(h^2_{\max})}$, plus an extra factor n from the initial guessing of an uncovered edge (that is contracted).

We complement Theorem 2 with an asymptotically matching lower bound.

Lemma 6. *The approximation ratio of the* LONG-FIRST *algorithm is at least* $\frac{3}{2}$.

Proof. Consider the following construction: for each $k > \frac{1}{2\varepsilon}$ consider an instance (G_k, L_k) of CycAP defined by a cycle of $n = 4k$ nodes (assume that the cycle is defined by the order of the nodes $v_1, v_2, \ldots, v_{4k}$) and the following set of links (see Fig. 3 (Left)):

- For each $i = 1, \ldots, \frac{n}{2} - 1$, $(v_{i+1}, v_{n+1-i}) \in L_k$;
- $(v_1, v_{\frac{n}{2}+1}) \in L_k$;
- For each $i = 1, \ldots, \frac{n}{4} - 1$, $(v_{i+1}, v_{\frac{n}{2}+1-i}) \in L_k$.

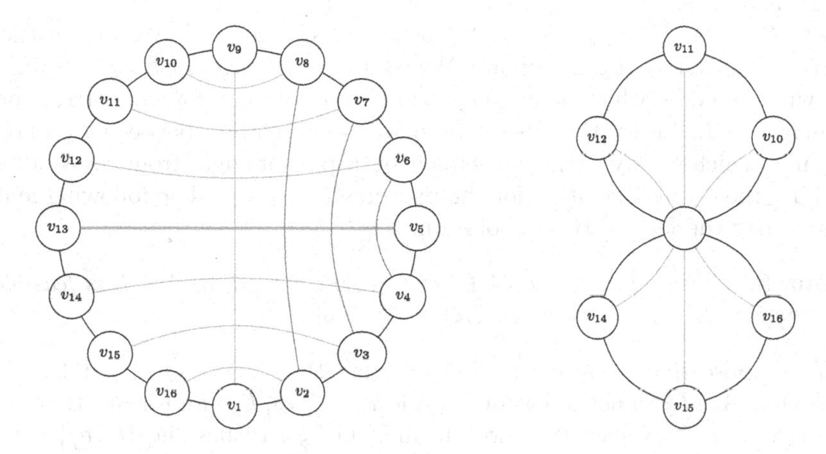

Fig. 3. Left: Instance (G_4, L_4) from the lower bound construction in Lemma 6. An optimal solution is defined by red links. Right: If the algorithm picks first the thick red link (which is long) and then the links which become external (blue links and (v_1, v_9)) we obtain this subinstance without crossing pairs of links. (Color figure online)

As argued in Lemma 3, the first and second set of links define an optimal solution of size $\frac{n}{2}$. We will now prove that there exists a sequence of choices performed by our algorithm that outputs a solution of size $\frac{3n}{4} - 1$, which implies that the approximation ratio is at least $\frac{3}{2} - \frac{2}{n}$ and this value approaches $\frac{3}{2}$ as k goes to infinity. Notice first that the link $(v_{\frac{n}{4}+1}, v_{\frac{3n}{4}+1}) \in L_k$ has length $2k > \frac{1}{\varepsilon}$ and hence it is long so the first stage of the algorithm can include it in the solution. After doing that, the second and third set of links become external and thus the algorithm will include them in the solution. Once all these links are included and contracted, we get a cactus consisting of two cycles of $\frac{n}{4}$ nodes each and without crossing links (see Fig. 3 (Right)). Hence, the algorithm must pick all the remaining links to complete the solution. The size then of this solution is $\frac{n}{4} + 1 + 2\left(\frac{n}{4} - 1\right) = \frac{3n}{4} - 1$.

4 LP Relaxations for CycAP

We start by lower-bounding the integrality gap of the standard cut LP for CycAP.

Lemma 7. *The standard cut LP for CycAP has integrality gap at least* 2.

Proof. Consider a cycle of size k and, for each edge, a parallel link. The optimum integral solution has size $k-1$, while setting each variable to $\frac{1}{2}$ gives a feasible fractional solution of cost $\frac{k}{2}$.

This shows that the standard cut LP is not strong enough even for instances without crossing nor long links, cases that we can handle optimally via combinatorial algorithms. We next present a stronger LP that exploits a more general set of constraints.

Let $(G=(V,E),L)$ be a CycAP instance and $S \subseteq E$. We define the S-*reduced* instance (G_S, L_S) as follows: We contract the edges of $E \setminus S$, obtaining a cycle with $|S|$ edges which defines G_S, and the set of links L_S will correspond to the images of L. Notice that there is a one-to-one relation between L_S and the links in L which satisfy some cut defined by a pair of edges from S. We denote by OPT_S the optimal solution for the instance (G_S, L_S)[5]. The following lemma characterizes the feasibility of a solution.

Lemma 8. *Given an instance (G,L) of CycAP, a solution $A \subseteq L$ is feasible iff for every $S \subseteq E$ it holds that $|A \cap L_S| \geq |\mathrm{OPT}_S|$.*

Proof. Suppose that there exists $S \subseteq E$ such that $|A \cap L_S| < |\mathrm{OPT}_S|$. This means that $A \cap L_S$ is not a feasible solution for (G_S, L_S) and hence there exist two edges $e_i, e_j \in S$ such that no link in $A \cap L_S$ satisfies the $\{e_i, e_j\}$-cut. As the remaining links in $A \setminus L_S$ also do not satisfy the cut by definition, this cut remains unsatisfied in the original instance, implying that A is not feasible.

On the other hand, suppose that A satisfies the claimed property for every set S. If we consider just sets S consisting of two edges this is exactly the characterization of feasibility shown in Remark 1, implying that A is feasible.

This implies that we can add the constraint $\sum_{\ell \in L_S} x_\ell \geq |\mathrm{OPT}_S|$ for $S \subseteq E$. Unfortunately there is an exponential number of such constraints and most of them require to compute $|\mathrm{OPT}_S|$ for large instances. However, if we restrict ourselves to sets of edges having constant size, we get an LP formulation with polynomially many constraints that can be written in polynomial time. We call this LP the k-edge-cut LP for a given constant $k \in \mathbb{N}$, which is similar in spirit to the *bundle-LP* for TAP introduced by Adjiashvili [1].

$$
\begin{aligned}
\min \sum_{\ell \in L} x_\ell & & & (k\text{-edge-cut LP}) \\
s.t. \sum_{\ell \in L_S} x_\ell &\geq |\mathrm{OPT}_S| & & \forall S \subseteq E, |S| \leq k \\
0 \leq x_\ell &\leq 1 & & \forall \ell \in L
\end{aligned}
$$

Notice that for $k=2$ this is exactly the standard cut LP. Now we will prove some properties of this relaxation and bound its integrality gap.

[5] For $|S| \leq 1$, we simply set $\mathrm{OPT}_S = \emptyset$.

Lemma 9. *Given* $\varepsilon > 0$, *for* $k = \frac{1}{\varepsilon^2}$ *the* k*-edge-cut LP restricted to instances with links of length at most* $\frac{1}{\varepsilon}$ *has integrality gap at most* $(1 + 2\varepsilon)$.

Proof. We will assume w.l.o.g. that the set of links L contains every possible link of length 1. If it is not the case, let us include them obtaining a new set of links $L' \supseteq L$. The optimal LP value can only decrease while the size of the optimal solution cannot decrease, implying that the integrality gap can only increase due to this operation. To see this last fact, assume by contradiction that there exists a solution OPT′ for the new instance having strictly smaller size than OPT. Consider now a solution S consisting of $\text{OPT}' \cap L$ plus a minimal set of links from L that makes S feasible (this is possible since the instance admits a feasible solution). If we in parallel iteratively contract the common links in S and OPT′ we arrive to the same CacAP instance, but now the remaining links from OPT′ have length 1 and the contraction of each of them reduces the number of nodes in the instance by exactly one node while the contraction of the remaining links in S reduces the number of nodes by at least 1. Thus $|S| \leq |\text{OPT}'|$ which is not possible since $S \subseteq L$.

Let $X = (x_\ell)_{\ell \in L}$ be an optimal solution for the k-edge-cut LP. We will construct an integral feasible solution of size at most $(1 + \varepsilon)\sum_{\ell \in L} x_\ell$. To do so, we will partition the cycle into disjoint intervals as follows: We will first define an interval of size k (which we will call a *long* interval) and then an interval of size $\frac{1}{\varepsilon}$ (which we will call a *short* interval), and then continue with this procedure until it is not possible to continue. If in the end there are at most $\frac{1}{\varepsilon}$ edges we define a last short interval consisting of these remaining edges, otherwise we define a short interval consisting of the last $\frac{1}{\varepsilon}$ edges and a long interval consisting of the remaining edges (which will have size at most k). The number of short intervals is upper bounded by $1 + \left\lfloor \frac{n}{1/\varepsilon^2 + 1/\varepsilon} \right\rfloor \leq 1 + \frac{\varepsilon^2 n}{1+\varepsilon} \leq \varepsilon^2 n$ assuming w.l.o.g. that n is lower bounded by a large enough constant.

Notice that $\sum_{\ell \in L} x_\ell \geq n/2$ by a simple averaging argument over the n constraints corresponding to all the pairs of consecutive edges: every link appears in exactly two such constraints and the right-hand side of each constraint is 1. Since the total number of links of length 1 having both endpoints in a short interval is at most $\varepsilon^2 n \cdot \frac{1}{\varepsilon} = \varepsilon n \leq 2\varepsilon \sum_{\ell \in L} x_\ell$, we can add them to our solution at a negligible cost.

Consider now the set of long intervals $S_1, S_2, \ldots, S_T$. Notice that no link has endpoints in different long intervals, and hence the LP constraints associated to such intervals do not share common variables. This implies that $\sum_{\ell \in L} x_\ell \geq \sum_{i=1}^{T} |\text{OPT}_{S_i}|$. Our feasible solution will consist of all the links of length 1 with both endpoints in a short interval plus the optimal solutions OPT_{S_i} for each long interval S_i. As argued before, the size of this solution is at most $(1+2\varepsilon)\sum_{\ell \in L} x_\ell$ and the feasibility of the solution follows since every $\{e, e'\}$-cut where e is in a short interval is satisfied by a link of length 1, while the remaining cuts are satisfied by the links computed optimally.

Lemma 10. *Given* $\varepsilon > 0$, *for* $k = \frac{1}{\varepsilon^2}$ *the* k*-edge-cut formulation has integrality gap at most* $(1 + 4\varepsilon)$ *restricted to instances without crossing pairs of links.*

Proof. Let $X = (x_\ell)_{\ell \in L}$ be an optimal solution for the k-edge-cut LP. Suppose that the instance does not contain links of length at least $\frac{1}{\varepsilon}$, then we can conclude the claim thanks to Lemma 9. Otherwise, we will pick any link of length at least $\frac{1}{\varepsilon}$ and contract it, obtaining a CacAP instance consisting of two cycles without external links (as there are no crossing links), both of size at least $\frac{1}{\varepsilon}$. If any cycle still contains some long link, we iterate this procedure. Let L_{long} be the set of long links we picked during this procedure and $C_1, C_2, \ldots, C_T$ be the set of cycles at the end. By the same argument as in Theorem 2, we have that $|L_{\text{long}}| \leq \varepsilon n \leq 2\varepsilon \sum_{\ell \in L} x_\ell$.

If we apply now Lemma 9 to each cycle we obtain a feasible solution of size at most $(1+2\varepsilon)\sum_{i=1}^{T} \text{OPT}_{\text{LP}_i} + |L_{\text{long}}|$, where LP_i is the k-edge-cut LP defined by each cycle C_i and its internal links. As there are no external links, the sum of the previous LP solutions is the optimal solution for the following LP:

$$\begin{aligned} \min \quad & \sum_{\ell \in L \setminus L_{\text{long}}} x_\ell \\ s.t. \quad & \sum_{\ell \in L_S} x_\ell \geq |\text{OPT}_S| \; \forall i \in \{1, \ldots, T\}, \forall S \subseteq E(C_i), |S| \leq \tfrac{1}{\varepsilon^2} \\ & 0 \leq x_\ell \leq 1 \qquad \forall \ell \in L \setminus L_{\text{long}} \end{aligned}$$

The set of constraints of this LP is a subset of the constraints of the original LP as links in L_{long} do not appear in these constraints and the set of variables is a subset of the original one. Thus we have $\sum_{i=1}^{T} \text{OPT}_{\text{LP}_i} \leq \sum_{\ell \in L} x_\ell$, and then we can conclude that the constructed solution has size at most $(1+4\varepsilon)\sum_{\ell \in L} x_\ell$.

Following the proof of Theorem 2 plus the previous results we can get the following bound on the integrality gap for general instances of CycAP.

Corollary 1. *For any $\varepsilon > 0$, the integrality gap of the k-edge-cut LP for $k = \frac{1}{\varepsilon^2}$ is at most $\frac{3}{2} + O(\varepsilon)$.*

Proof. Let $X = (x_\ell)_{\ell \in L}$ be an optimal solution for the k-edge cut LP and consider the output of the $(\frac{3}{2} + \varepsilon)$-approximation from Sect. 3.2 decomposed into L_{long}, L_{ext} and L_{short} as in the proof of Theorem 2. As argued before, we know that $\sum_{\ell \in L} x_\ell \geq \frac{n}{2}$ and analogously to the proof of Lemma 10 we have that $|L_{\text{short}}| \leq (1+2\varepsilon)\sum_{\ell \in L} x_\ell$. Hence essentially the same analysis as in Theorem 2 provides the same bound of $3/2 + O(\varepsilon)$ up to an extra $(1+\varepsilon)$ factor.

Acknowledgments. We would like to thank the anonymous reviewers for their helpful comments.

References

1. Adjiashvili, D.: Beating approximation factor two for weighted tree augmentation with bounded costs. SODA **2017**, 2384–2399 (2017)

2. Basavaraju, M., Fomin, F.V., Golovach, P., Misra, P., Ramanujan, M.S., Saurabh, S.: Parameterized algorithms to preserve connectivity. In: Esparza, J., Fraigniaud, P., Husfeldt, T., Koutsoupias, E. (eds.) ICALP 2014. LNCS, vol. 8572, pp. 800–811. Springer, Heidelberg (2014). https://doi.org/10.1007/978-3-662-43948-7_66
3. Cheriyan, J., Gao, Z.: Approximating (unweighted) tree augmentation via lift-and-project, part I: stemless TAP. Algorithmica **80**, 530–559 (2018)
4. Cheriyan, J., Gao, Z.: Approximating (unweighted) tree augmentation via lift-and-project, part II. Algorithmica **80**(2), 608–651 (2018)
5. Cheriyan, J., Jordán, T., Ravi, R.: On 2-coverings and 2-packings of laminar families. In: Nešetřil, J. (ed.) ESA 1999. LNCS, vol. 1643, pp. 510–520. Springer, Heidelberg (1999). https://doi.org/10.1007/3-540-48481-7_44
6. Cheriyan, J., Karloff, H., Khandekar, R., Könemann, J.: On the integrality ratio for tree augmentation. Oper. Res. Lett. **36**, 399–401 (2008)
7. Cohen, N., Nutov, Z.: A (1 + Ln 2)-approximation algorithm for minimum-cost 2-edge-connectivity augmentation of trees with constant radius. APPROX/RANDOM **2011**, 147–157 (2011)
8. Dinitz, E., Karzanov, A., Lomonosov, M.: On the structure of the system of minimum edge cuts of a graph. Stud. Discrete Optim. 290–306 (1976)
9. Even, G., Feldman, J., Kortsarz, G., Nutov, Z.: A 1.8 approximation algorithm for augmenting edge-connectivity of a graph from 1 to 2. ACM Trans. Algorithms **5**, 21:1–21:17 (2009)
10. Fiorini, S., Groß, M., Könemann, J., Sanità, L.: Approximating weighted tree augmentation via Chvátal-Gomory cuts. SODA **2018**, 817–831 (2018)
11. Frederickson, G.N., JáJá, J.: Approximation algorithms for several graph augmentation problems. SIAM J. Comput. **10**(2), 270–283 (1981)
12. Goemans, M.X., Goldberg, A.V., Plotkin, S., Shmoys, D.B., Tardos, E., Williamson, D.P.: Improved approximation algorithms for network design problems. SODA **1994**, 223–232 (1994)
13. Grandoni, F., Kalaitzis, C., Zenklusen, R.: Improved approximation for tree augmentation: saving by rewiring. STOC **2018**, 632–645 (2018)
14. Jain, K.: A factor 2 approximation algorithm for the generalized steiner network problem. Combinatorica **21**(1), 39–60 (2001)
15. Khuller, S., Thurimella, R.: Approximation algorithms for graph augmentation. J. Algorithms **14**, 214–225 (1993)
16. Kortsarz, G., Krauthgamer, R., Lee, J.R.: Hardness of approximation for vertex-connectivity network design problems. SIAM J. Comput. **33**(3), 704–720 (2004)
17. Kortsarz, G., Nutov, Z.: A simplified 1.5-approximation algorithm for augmenting edge-connectivity of a graph from 1 to 2. ACM Trans. Algorithms **12**, 23:1–23:20 (2015)
18. Kortsarz, G., Nutov, Z.: LP-relaxations for tree augmentation. In: APPROX/RANDOM 2016, pp. 13:1–13:16 (2016)
19. Marx, D., Végh, L.A.: Fixed-parameter algorithms for minimum-cost edge-connectivity augmentation. ACM Trans. Algorithms **11**, 27:1–27:24 (2015)
20. Nagamochi, H.: An approximation for finding a smallest 2-edge-connected subgraph containing a specified spanning tree. Discrete Appl. Math. **126**, 83–113 (2003)
21. Nutov, Z.: On the tree augmentation problem. In: ESA 2017, pp. 61:1–61:14 (2017)
22. Watanabe, T., Nakamura, A.: Edge-connectivity augmentation problems. J. Comput. Syst. Sci. **35**, 96–144 (1987)

Approximate Strong Edge-Colouring of Unit Disk Graphs

Nicolas Grelier[1(✉)], Rémi de Joannis de Verclos[2], Ross J. Kang[2], and François Pirot[2,3]

[1] Department of Computer Science, ETH Zürich, Zürich, Switzerland
nicolas.grelier@inf.ethz.ch
[2] Department of Mathematics, Radboud University Nijmegen, Nijmegen, The Netherlands
r.deverclos@math.ru.nl, ross.kang@gmail.com
[3] LORIA, Université de Lorraine, Nancy, France
francois.pirot@loria.fr

Abstract. We show that the strong chromatic index of unit disk graphs is efficiently 6-approximable. This improves on 8-approximability as shown by Barrett, Istrate, Kumar, Marathe, Thite, and Thulasidasan [1]. We also show that strong edge-6-colourability is NP-complete for the class of unit disk graphs. Thus there is no polynomial-time $(7/6 - \varepsilon)$-approximation unless P = NP.

1 Introduction

A *strong edge-k-colouring* is a partition of the edges of a graph G into k parts so that each part induces a matching (meaning that there exists no edge in G between two edges of the same matching). The *strong chromatic index* is the least k for which the graph admits a strong edge-k-colouring. If the vertices of the graph represent communicating nodes, say, in a wireless network, then an optimal strong edge-colouring may represent an optimal discrete assignment of frequencies to transmissions in the network so as to avoid both primary and secondary interference [1,18,20]. It is then relevant to model the network geometrically, i.e. as a *unit disk graph* [9]. Our interest is in approximative algorithmic aspects of strong edge-colouring in this model. This was considered by Barrett, Istrate, Kumar, Marathe, Thite, and Thulasidasan [1] who showed that the strong chromatic index of unit disk graphs is efficiently 8-approximable. We revisit the problem and make some further advances.

- We prove efficient 6-approximability.
- We prove an efficient online 8-competitive algorithm.
- We show impossibility of efficient $(7/6 - \varepsilon)$-approximability unless P=NP.

It is $\exists\mathbb{R}$-complete to decide if a given graph has an embedding as a unit disk graph [11], but both of the approximation algorithms we use are *robust*, in the sense that they efficiently output a valid strong edge-colouring upon the input of any abstract graph. Our contribution is to prove that they are guaranteed to

E. Bampis and N. Megow (Eds.): WAOA 2019, LNCS 11926, pp. 154–169, 2020.
https://doi.org/10.1007/978-3-030-39479-0_11

output a colouring with good approximation ratio upon the input of a unit disk graph (regardless of any embedding).

Our work parallels and contrasts with work on the chromatic number of unit disk graphs, for which the best approximation ratio known has stubbornly remained 3 since 1991 [19]. Finding an *optimal* approximation for strong chromatic index may be similarly difficult.

1.1 Graph Colouring Preliminaries

In this subsection, we highlight some graph theoretic notation, concepts and observations that are relevant to our study. For other standard background, consult e.g. [6]. Given a graph $G = (V, E)$, the *minimum degree*, *clique number*, *chromatic number* and *maximum degree* of G are denoted by $\delta(G)$, $\omega(G)$, $\chi(G)$ and $\Delta(G)$, respectively. The *degeneracy* of G is defined as $\delta^*(G) = \max\{\delta(H) \mid H \subseteq G\}$ and G is called *k-degenerate* if $\delta^*(G) \le k$. A simple but useful set of inequalities for graph colouring is as follows. For any graph G,

$$\omega(G) \le \chi(G) \le \delta^*(G) + 1 \le \Delta(G) + 1. \tag{1}$$

Note that the second inequality in (1) is algorithmic, in the sense that it follows from the use of an efficient greedy algorithm that always assigns the least available colour, provided we consider the vertices one by one in a suitable order, namely, according to degeneracy. Moreover, a greedy algorithm taking any ordering uses at most $\Delta(G) + 1$ colours.

The *line graph* $L(G)$ of G is the graph where a vertex in $L(G)$ corresponds to an edge in G and there is an edge between two vertices in $L(G)$ if the corresponding edges in G share a vertex. The *square* G^2 of G is the graph formed from G by adding all edges between pairs of vertices that are connected by a 2-edge path in G. The *strong chromatic index* of G (as defined above) is denoted $\chi'_2(G)$. Note that $\chi'_2(G) = \chi(L(G)^2)$. The *strong clique number* $\omega'_2(G)$ of G is $\omega(L(G)^2)$. Obviously, (1) implies that

$$\omega'_2(G) \le \chi'_2(G) \le \delta^*(L(G)^2) + 1 \le \Delta(L(G)^2) + 1. \tag{2}$$

It is worth reiterating that the following greedy algorithm efficiently generates a strong edge-$(\delta^*(L(G)^2)+1)$-colouring: order the edges of G by repeatedly removing from G an edge e for which $\deg_{L(G)^2}(e)$ is lowest, and then colour the edges sequentially according to the *reverse* of this ordering, at each step assigning as a colour the least positive integer that does not conflict with previously coloured edges. Again similarly, with an arbitrary ordering of the edges the greedy algorithm produces a strong edge-$(\Delta(L(G)^2) + 1)$-colouring. Our main results then follow from (2) by suitable bounds on $\delta^*(L(G)^2)$ and $\Delta(L(G)^2)$.

The strong chromatic index is a well-studied parameter in graph theory. Most notably, Erdős and Nešetřil conjectured in the 1980s that $\chi'_2(G) \le 1.25\Delta(G)^2$ for all graphs G [7]. About a decade later, Molloy and Reed [17] proved the existence of some minuscule but fixed $\varepsilon > 0$ such that $\chi'_2(G) \le (2-\varepsilon)\Delta(G)^2$ for all graphs G. Recently there have been improvements [2,3] and extensions [10,21], but all rely on Molloy and Reed's original approach, a reduction to a Ramsey-type colouring result. The conjecture remains wide open.

1.2 Unit Disk Graph Preliminaries

A graph $G = (V, E)$ is said to be a *unit disk graph* if there exists a mapping $p : V \to \mathbb{R}^2$ from its vertices to the plane such that $uv \in E$ if and only if the Euclidean distance between $p(u)$ and $p(v)$ is at most 1. Any explicit mapping p that certifies that G is a unit disk graph is called an *embedding.* When we have an embedding p, we often make no distinction between a vertex u and its corresponding point $p(u)$ in the plane.

The class of unit disk graphs is popular due to its elegance and its versatility in capturing real-world optimisation problems [5]. For example, an embedded unit disk graph may represent placement of transceivers so that circles of radius 1/2 centred at the points represent transmission areas. Indeed, the class was originally introduced in 1980 to model frequency assignment [9], with chromatic number being one of the first studied parameters. Clark, Colbourn and Johnson [5] published a proof that it is NP-hard to compute the chromatic number of unit disk graphs. They also showed the clique number of unit disk graphs is polynomial-time computable. Therefore, any upper bound C on the extremal ratio $r := \sup\{\chi(G)/\omega(G) \mid G \text{ is a unit disk graph}\}$ (algorithmic or not) implies an efficient C-approximation of the chromatic number: simply output $C \cdot \omega(G)$. In 1991, Peeters [19] noted a simple 3-approximation which also shows $r \leq 3$: after lexicographically ordering the vertices of G according to any fixed embedding, a basic geometric argument proves that G is $3(\omega(G) - 1)$-degenerate (and then apply (1)). Since 3-colourability of unit disk graphs is NP-complete, there is no efficient $(4/3 - \varepsilon)$-approximation unless P=NP. It is known that $r \geq 3/2$ [15]. The best approximation ratio known is 3.

1.3 Approximate Strong Edge-Colouring Preliminaries

Mahdian [13,14] showed in 2000 that it is NP-hard to compute the strong chromatic index, even restricted to bipartite graphs of large fixed girth. More recently, Chalermsook, Laekhanukit and Nanongkai [4] showed that in general there is no polynomial-time $(n^{1/3-\varepsilon})$-approximation algorithm (where n is the number of vertices in the input) unless NP = ZPP.

To the best of our knowledge, no previous work has shown NP-hardness upon restriction to the class of unit disk graphs. Nevertheless, Barrett *et al.* [1] have initiated the study of approximate strong edge-colouring for unit disk graphs. With an argument similar to in [19], they showed that $\delta^*(L(G)^2) \leq 8\omega_2'(G)$ for any unit disk graph G, which by (2) certifies an 8-approximation for the strong chromatic index. Kanj, Wiese and Zhang [12] noted an efficient online 10-competitive algorithm with essentially the same analysis as in [1].

1.4 Main Results and Outline

Our work improves significantly on [1] in several ways. In Sect. 2, we describe the following.

Theorem 1. *For any unit disk graph G, $\delta^*(L(G)^2) \leq 6(\omega_2'(G) - 1)$.*

Corollary 1. *The greedy algorithm under a reverse degeneracy ordering of the edges is an efficient 6-approximation for the strong chromatic index of unit disk graphs.*

The proof of Theorem 1 is rather involved. It shows that, for any embedded unit disk graph, some well-chosen edge-ordering certifies the required degeneracy bound. It would be very interesting to improve on the approximation ratio of 6. We prove the following in Sect. 3.

Theorem 2. *For any unit disk graph G, $\Delta(L(G)^2) \leq 8(\chi_2'(G) - 1)$.*

Corollary 2. *The greedy algorithm is an efficient online 8-competitive algorithm*[1] *for the strong chromatic index of unit disk graphs.*

The proof of Theorem 2 differs fairly from previous work [1,12] and from the proof of Theorem 1. Indeed the bound we give make use of the strong chromatic index $\chi_2'(G)$ instead of the strong clique number $\omega_2'(G)$. To prove Theorem 2, it suffices to solve the following kissing number-type problem. Given two intersecting unit disks C_1 and C_2 in $\mathbb{R}^2$, what is the size of a largest collection of pairwise non-intersecting unit disks such that each one intersects $C_1 \cup C_2$? The corresponding problem in $\mathbb{R}^3$ seems quite natural.

In Appendix, we prove the following.

Theorem 3. *Strong edge-k-colourability of unit disk graphs is NP-complete, where $k = 6$ or $k = \binom{\ell}{2} + 4\ell + 6$ for some fixed $\ell \geq 5$.*

Corollary 3. *It is NP-hard to compute the strong chromatic index of unit disk graphs. Moreover, it cannot be efficiently $(7/6 - \varepsilon)$-approximated unless P=NP.*

For $k \leq 3$, strong edge-k-colourability is polynomially-time solvable. The complexity for $k \in \{4, 5\}$ remains open. The proof of Theorem 3 borrows from ideas in the work of Gräf, Stumpf and Weißenfels [8], but with extra non-trivial difficulties for strong edge-colouring.

---

[1] To avoid any ambiguity, in the online setting *vertices* are revealed one at a time and all edges between a newly revealed vertex and previous vertices must be immediately and irrevocably assigned a colour.

1.5 Further Discussion

We can state more general versions of our approximation results that not only lend a more geometric flavour but also highlight a potential conceptual obstacle to further improvements on our approximation results. We call a graph $G = (V, E)$ a *twin unit disk graph* if there exists a mapping $p : V \to \mathbb{R}^2 \times \mathbb{R}^2, u \mapsto (p(u)_1, p(u)_2)$ from its vertices to pairs of points in the plane such that

- the Euclidean distance between $p(u)_1$ and $p(u)_2$ is at most 1 for every $u \in V$; and
- $uv \in E$ if and only if the Euclidean distance between $p(u)_1$ and $p(v)_1$, between $p(u)_1$ and $p(v)_2$, between $p(u)_2$ and $p(v)_1$, or between $p(u)_2$ and $p(v)_2$ is at most 1.

Equivalently, this is the intersection class over unions of pairs of intersecting unit disks in $\mathbb{R}^2$.

Note that, for any unit disk graph G, both G and $L(G)^2$ are twin unit disk graphs. (Indeed we represent $L(G)^2$ by setting $p(e)_1 = p_1$ and $p(e)_2 = p_2$ for any edge $e = p_1p_2$ in G.) So it is NP-hard to determine the chromatic number of twin unit disk graphs.

We have the following stronger versions of Theorems 1 and 2, which imply efficient 6-approximation and online 8-competitive algorithms for the chromatic number of twin unit disk graphs (by (1)).

Theorem 4. *For any twin unit disk graph G, $\delta^*(G) \leq 6(\omega(G) - 1)$.*

Theorem 5. *For any twin unit disk graph G, $\Delta(G) \leq 8(\chi(G) - 1)$.*

Malesińska *et al.* [15] showed that there are unit disk graphs G for which $\delta(G) = 3(\omega(G) - 1)$. In Appendix, we also show that there are twin unit disk graphs G for which $\delta(G) \geq 4(\omega(G) - 2) + 1$, and so the factor 6 in Theorem 4 cannot be improved below 4.

If we were able to efficiently compute or well approximate the clique number of twin unit disk graphs or, in particular, the strong clique number of unit disk graphs, then we would have a strong incentive to bound $r_2' := \sup\{\chi_2'(G)/\omega_2'(G) \mid G \text{ is a unit disk graph}\}$. This is a natural optimisation problem regardless. We only know $r_2' \leq 6$ by Theorem 1, and $r_2' \geq 4/3$ by considering the cycle C_7 on seven vertices (since $\chi_2'(C_7) = 4$ while $\omega_2'(C_7) = 3$). Relatedly, we believe that the following problem is worth investigating.

Conjecture 1. It is NP-hard to compute the clique number of twin unit disk graphs.

2 A 6-approximation

In this section we discuss the proof of Theorem 4, which has Theorem 1 as a special case.

To make the reader more familiar with the problem and the notations, we first present a much shorter argument for a weaker approximation. The proof is nearly the same as what Barrett *et al.* [1] used for an upper bound on the approximation ratio of 8, but with a small twist.

Proposition 1. *For any twin unit disk graph* G, $\delta^*(G) \leq 7(\omega(G) - 1)$.

Proof. Let $G = (V, E)$ be a twin unit disk graph. Fix any embedding $p : V \to \mathbb{R}^2 \times \mathbb{R}^2$ of G in the plane. Equipped with such an embedding, we first define an ordering of V and then use it to certify the promised degeneracy property.

The ordering we use for this result, a lexicographic ordering, is the same used in [1]. This lexicographic order considers first the y-coordinate and then the x-coordinate, (i.e. (a, b) is before (c, d) if and only if $b < d$ or ($b = d$ and $a \leq c$)). Throughout this paper, we simply refer to it as the lexicographic order on $\mathbb{R}^2$. Let $(x_1, y_1), (x_2, y_2), \ldots$ be a sequence of points in $\mathbb{R}^2$ defined by listing the elements of $\cup_{u \in V}\{p(u)_1, p(u)_2\}$ according to the lexicographic order on $\mathbb{R}^2$. We consider the points of this sequence in order and add vertices at the end of our current ordering of V as follows. When considering point (x_j, y_j) for some $j \geq 1$, we add all vertices $u \in V$ for which there is some $i \leq j$ such that $\{p(u)_1, p(u)_2\} = \{(x_i, y_i), (x_j, y_j)\}$, and we do so according to the lexicographic order on $\mathbb{R}^2$.

It suffices to show that each vertex $u \in V$ has at most $7(\omega(G) - 1)$ neighbours that precede it in the lexicographic ordering. To do so, we show that every such neighbour v of u satisfies that either $p(v)_1$ or $p(v)_2$ is contained in one of seven unit $(\pi/3)$-sectors (each of which is centred around either $p(u)_1$ or $p(u)_2$). This is enough, since the set of vertices that map one of their twin points into one such sector induces a clique in G that includes u. The proof differs from what Barrett *et al.* did in [1] by the fact that we use seven unit $(\pi/3)$-sectors instead of eight.

Let $u \in V$ and suppose without loss of generality that $p(u)_1$ is before $p(u)_2$ in lexicographic order. First observe that, if $v \in V$ is before u in the lexicographic order, then both $p(v)_1$ and $p(v)_2$ must be in the region of $\mathbb{R}^2$ that has smaller or equal y-coordinate compared to $p(u)_2$. If, moreover $uv \in E$, then $p(v)_1$ or $p(v)_2$ must lie in either a unit half-disk centred at $p(u)_2$ or in the unit disk centred at $p(u)_1$. We partition the unit disk centred at $p(u)_1$ into six unit $(\pi/3)$-sectors such that the line segment $[p(u)_1, p(u)_2]$ lies along the boundary between two of the sectors. Note that any of the points in the two sectors incident to $[p(u)_1, p(u)_2]$ also lies in the unit disk centred at $p(u)_2$. Figure 1 depicts the construction, with sectors separated by solid lines. Therefore, the four other sectors together with the three sectors that partition the unit half-disk centred at $p(u)_2$ are the seven unit $(\pi/3)$-sectors that we desire. □

It turns out that for Theorem 4 we can take the same approach as in Proposition 1, except with an ordering that is more subtle and an analysis that is substantially longer and more difficult. Since our arguments are "only" geometric, we feel that the ratio 6 can be improved, especially in Theorem 1. It should be possible to exploit the structural graph properties of $L(G)^2$, but our efforts have so far failed. This might be difficult.

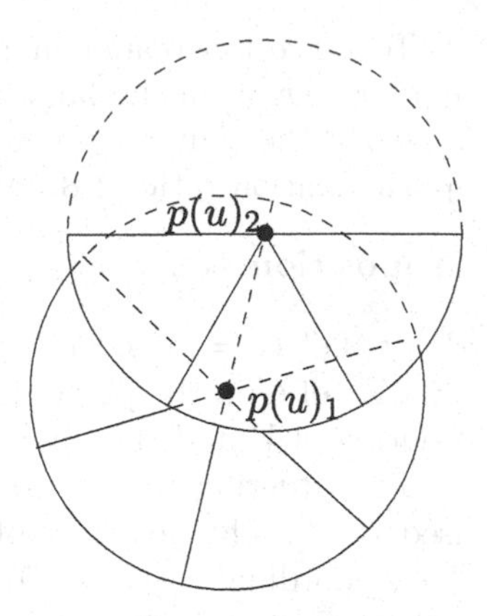

Fig. 1. The seven sectors one of which must contain $p(v)_1$ or $p(v)_2$.

2.1 Proof Outline of Theorem 4

Let $G = (V, E)$ be a twin unit disk graph. Fix any embedding $p : V \to \mathbb{R}^2 \times \mathbb{R}^2$ of G in the plane. Without loss of generality, we may assume that this embedding satisfies for all $u \in V$ that $p(u)_1$ is before $p(u)_2$ according to the lexicographic order on $\mathbb{R}^2$. We define a preorder $\preceq$ on V as follows. For any $u, v \in V$,

$$u \preceq v \text{ if and only if } p(u)_1 \text{ is not after } p(v)_1 \text{ according to the lexicographic order on } \mathbb{R}^2.$$

Note that if $p(u)_1 = p(v)_1$, then both $u \preceq v$ and $v \preceq u$.

For any $u \in V$, we define $N^-(u)$ as the set of $v \in V$, $v \neq u$ such that $v \in N(u)$ and $v \preceq u$. It suffices to show that $|N^-(u)| \leq 6(\omega(G) - 1)$ for each $u \in V$.

Fix such a vertex $u \in V$. Let h^+ be the open half-plane of points above $p(u)_1$ and h^- the closed half-plane of points not above $p(u)_1$. For a point w, let $\mathcal{C}_w$ (respectively $\mathcal{D}_w$) denote the circle (respectively the closed disk) with radius 1 centred at w. Let $\mathbf{X}(u)$ be the union of $\mathbf{X}^-(u) := (\mathcal{D}_{p(u)_1} \cup \mathcal{D}_{p(u)_2}) \cap h^-$ and the set $\mathbf{X}^+(u)$ of elements of $(\mathcal{D}_{p(u)_1} \cup \mathcal{D}_{p(u)_2}) \cap h^+$ at distance at most 1 from a point of $h^- \setminus (\mathcal{D}_{p(u)_1} \cup \mathcal{D}_{p(u)_2})$.

Similarly to the proof of Proposition 1, we aim to cover $\mathbf{X}(u)$ with six sections that each correspond to a clique of G. Instead of requiring that these sections have diameter at most 1 (as we do for proving Proposition 1), we make use of sections having the following weaker property:

a section $S \subseteq \mathbb{R}^2$ is *small* if it can be partitioned into two parts S^+ and S^- of diameter at most 1 and such that

1. $S^+ \subseteq h^+$; and
2. for all points $q \in S^-$, $p_2 \in S^+$ and $p_1 \in h^- \setminus (\mathcal{D}_{p(u)_1} \cup \mathcal{D}_{p(u)_2})$ such that p_1 and p_2 are at distance at most 1, the point q is at distance at most 1 from p_1 or p_2.

This property is always satisfied if S has diameter at most 1, as it then suffices to take $S^- := S$ and $S^+ := \varnothing$.

Let $S = S^+ \cup S^-$ be a section, with $S^+ \subseteq h^+$. Let C be the set of vertices v such that one of $p(v)_1$ and $p(v)_2$ is in S^-, or ($p(v)_2 \in S^+$ and $p(v)_1 \in h^- \setminus (\mathcal{D}_{p(u)_1} \cup \mathcal{D}_{p(u)_2})$). We say that S *induces* a clique if ($v, w \in C \implies vw$ is an edge).

Let S be a small section, with S^+ and S^- as in the definition. Then S induces a clique. Indeed, take $v, w \in C$ as defined above, and let us prove that vw is an edge. If $p(v)$ and $p(w)$ both contain a point in S^+ or both contain a point in S^-, then these points are at distance at most 1 because S^+ and S^- have diameter at most 1, which implies that vw is an edge. Using the definition of C, we may therefore assume without loss of generality that $p(v)_1 \in S^-$, $p(w)_2 \in S^+$ and $p(w)_1 \in h^- \setminus (\mathcal{D}_{p(u)_1} \cup \mathcal{D}_{p(u)_2})$, which implies that $p(v)_1$ is at distance at most 1 from $p(w)_1$ or $p(w)_2$ because S is small.

The theorem then follows from the following statement:

Claim 1. For every vertex u, $\mathbf{X}(u)$ can be covered by six sections $S_1, \dots, S_6$ that induces a clique.

Let us first see why Claim 1 implies $|N^-(u)| \leq 6(\omega(G) - 1)$ (and therefore Theorem 4). For each section S_i, let S_i^+ and S_i^- be as in the definition of inducing a clique. For every $i \in \{1, \dots, 6\}$, let C_i be the set of vertices v such that one of $p(v)_1$ and $p(v)_2$ is in S_i^-, or ($p(v)_2 \in S_i^+$ and $p(v)_1 \in h^- \setminus (\mathcal{D}_{p(u)_1} \cup \mathcal{D}_{p(u)_2})$). By definition, all C_i are cliques.

It remains to show that $\bigcup_{i=1}^6 C_i$ covers $N^-(u)$. A vertex v is in $N^-(u)$ if $p(v)_1 \in h^-$ and one of $p(v)_1$ and $p(v)_2$ is in $\mathcal{D}_{p(u)_1} \cup \mathcal{D}_{p(u)_2}$. In the case where one of $p(v)_1$ and $p(v)_2$ is in S_i^- for some $i \in \{1, \dots, 6\}$, then $v \in C_i$. We can now assume that $p(v)$ does not intersect $\bigcup_{i=1}^6 S_i^-$. We know that $S_i^+ \subseteq h^+$ for every $i \in \{1, \dots, 6\}$ and that $\bigcup_{i=1}^6 S_i$ covers $\mathbf{X}^-(u)$, so $p(v)$ does not intersect $\mathbf{X}^-(u) = (\mathcal{D}_{p(u)_1} \cup \mathcal{D}_{p(u)_2}) \cap h^-$. This enforces that $p(v)_1 \in h^- \setminus (\mathcal{D}_{p(u)_1} \cup \mathcal{D}_{p(u)_2})$ and $p(v)_2 \in (\mathcal{D}_{p(u)_1} \cup \mathcal{D}_{p(u)_2}) \cap h^+$. Since $p(v)_1$ and $p(v)_2$ are at distance at most 1, the point $p(v)_2$ belongs to $\mathbf{X}^+(u)$, so there is $i \in \{1, \dots, 6\}$ with $p(v)_i \in S_i^+$. As a consequence, the vertex v is in the clique C_i.

It remains to prove Claim 1. Due to space limitations, parts of this proof are postponed to Appendix. In the following, we mainly describe the construction of the sections.

Construction of the Sections. Let ρ be the length and θ the argument of the vector $p(u)_2 - p(u)_1$. Without loss of generality, we may assume that the position of $p(u)_1$ is $(0, 0)$ and that both coordinates of $p(u)_2$ are not negative, so that $0 \leq \theta \leq \pi/2$. The position of $p(u)_2$ is therefore $\rho(\cos(\theta), \sin(\theta))$.

We have three different constructions of the sections $S_1, \dots, S_6$ depending on θ and ρ. We distinguish a first case when $\theta \leq \pi/6$, a second case when $\pi/6 < \theta$ and $\rho \leq 2\cos\theta$, and a last one when $2\cos\theta < \rho$.

If w_1, w_2 and w_3 are three points pairwise at distance 1, the *thickened triangle* with vertices w_1, w_2 and w_3 is the area $\mathcal{D}_{w_1} \cap \mathcal{D}_{w_2} \cap \mathcal{D}_{w_3}$. A thickened triangle has diameter 1.

First Case: $0 \leq \theta \leq \pi/6$

Claim 2. Sections S_4 and S_6 have diameter at most 1 and Sections S_1 and S_2 are small.

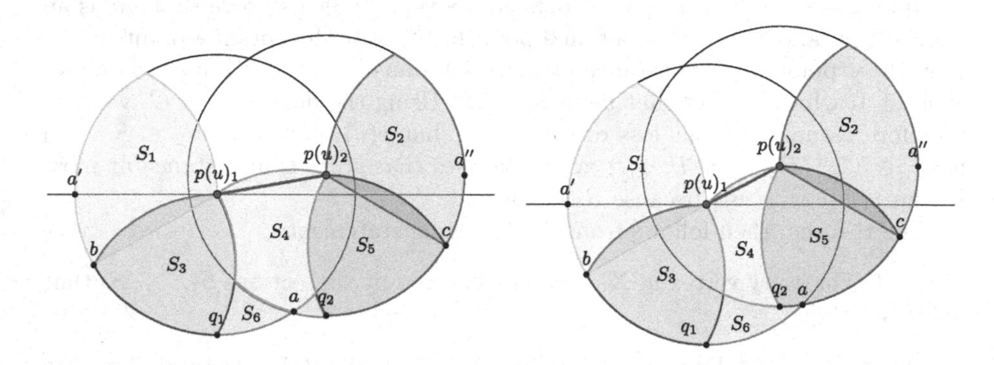

Fig. 2. The six sections when $0 \leq \theta < \pi/6$. (Color figure online)

Let a be the lower intersection of $\mathcal{C}_{p(u)_1}$ and $\mathcal{C}_{p(u)_2}$. Let a' be the point $(-1, 0)$ (so a' is the leftmost intersection of $\mathcal{C}_{p(u)_1}$ and the abscissa line). Likewise, let a'' be the point $p(u)_2 + (1, 0)$. Figure 2 shows the six sections we are defining now. Section S_1 (in yellow in Fig. 2) is defined as the intersection between $\mathcal{D}_{p(u)_1}$, $\mathcal{D}_{a'}$, and the half-plane above the line through $p(u)_1$ and the point $b := (-\sqrt{3}/2, -1/2)$. Let Section S_2 (in red in Fig. 2) be the intersection between $\mathcal{D}_{p(u)_2}$, $\mathcal{D}_{a''}$ and the half-plane above the line through $p(u)_2$ and the point $c := p(u)_2 + (\sqrt{3}/2, -1/2)$. Let Section S_3 (in purple) be the thickened triangle with vertices $p(u)_1$, b and $q_1 := (0, -1)$. Section S_4 (in green) is defined as $\mathcal{D}_{p(u)_1} \cap \mathcal{D}_{p(u)_2} \cap \mathcal{D}_a \setminus (\mathcal{D}_b \cup \mathcal{D}_c)$. Section S_5 in cyan is the thickened triangle with vertices $p(u)_2$, c and $q_2 := p(u)_2 + (0, -1)$. Lastly, let $S_6 = \mathbf{X}(u) \setminus (\bigcup_{i=1}^{5} S_i)$ be the remaining section.

These six sections cover $\mathbf{X}(u)$ by the definition of S_6. Sections S_3 and S_5 are thickened triangles so they have diameter 1. To prove the proposition in this case, it is enough to show the following property.

For some values of ρ (for instance $\rho = 1$), when θ is greater than $\pi/6$, the Euclidean distance between q_1 and a is bigger than 1, hence so is the diameter of S_6. Therefore we have different constructions when $\pi/6 < \theta$.

Second Case: $\pi/6 < \theta$ and $\rho \leq 2\cos(\theta)$. Figure 3 depicts the six sections. A simple calculus shows that the position of a is $\frac{1}{2}(\rho\cos(\theta) + \sqrt{4-\rho^2}\sin(\theta), \rho\sin(\theta) - \sqrt{4-\rho^2}\cos(\theta)$. The fact that ρ is at most $2\cos(\theta)$ implies that the point a is not above the abscissa line. We denote by r the point $(-1, 0)$. The conditions $\pi/6 < \theta$ and $\rho \leq 2\cos(\theta)$ do not imply anything on whether $(\mathcal{D}_{p(u)_2} \cap \mathcal{D}_r) \setminus \mathcal{D}_{p(u)_1}$ is empty or not. Let s_1 be the highest point in

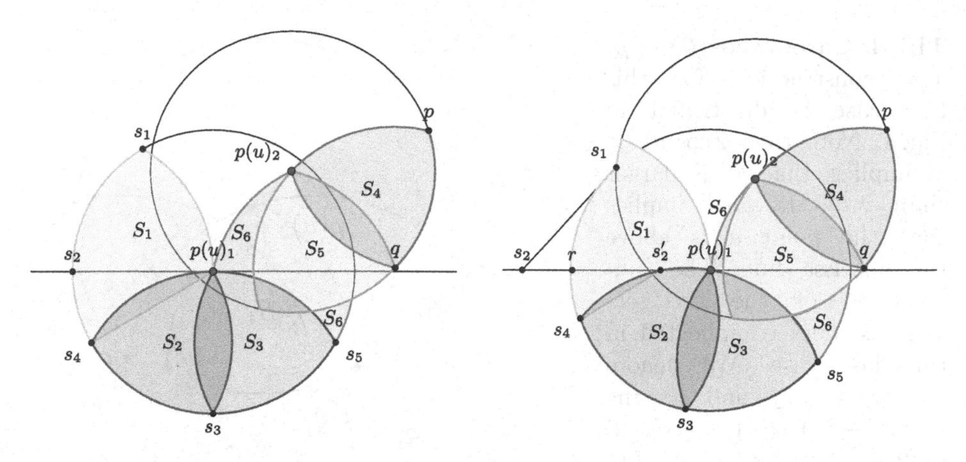

Fig. 3. The six sections when $\pi/6 < \theta$ and $\rho \leq 2\cos\theta$. (Color figure online)

$(\mathcal{D}_{p(u)_1} \cap \mathcal{D}_r) \setminus \mathcal{D}_{p(u)_2}$. Note that if $(\mathcal{D}_{p(u)_2} \cap \mathcal{D}_r) \setminus \mathcal{D}_{p(u)_1}$ is empty then the position of s_1 is $(-1/2, \sqrt{3}/2)$. Let s_2 be at the left intersection of $\mathcal{C}_{s_1}$ and the abscissa line. Let s_2' be the translated of s_2 by the vector $(1, 0)$. Let s_3 be the intersection of $\mathcal{C}_{p(u)_1}$ and $\mathcal{C}_{s_2'}$ that is below the abscissa line (if $p(u)_1$ and s_2' have the same position, the position of s_3 is set to $(0, -1)$). Let s_4 (respectively s_5) be the point at the intersection of $\mathcal{C}_{p(u)_1}$ and $\mathcal{C}_{s_3}$ that is on left side (respectively right side) of the line $(p(u)_1, s_3)$. Observe that if $(\mathcal{D}_{p(u)_2} \cap \mathcal{D}_r) \setminus \mathcal{D}_{p(u)_1}$ is empty then the positions of s_2, s_2', s_3, s_4 and s_5 are respectively $(-1, 0)$, $(0, 0)$, $(0, -1)$, $(-1/2, -\sqrt{3}/2)$ and $(1/2, -\sqrt{3}/2)$. Let S_1 (in yellow) be the section defined as the union of $\mathcal{D}_{p(u)_1} \cap \mathcal{D}_r \cap h^+$ and the intersection between $\mathcal{D}_{p(u)_1}$, h^- and the half-plane above the line $(p(u)_1, s_4)$. If $(\mathcal{D}_{p(u)_2} \cap \mathcal{D}_r) \setminus \mathcal{D}_{p(u)_1}$ is empty, then S_1 is exactly as in the precedent case. Let S_2 (in blue) be the thickened triangle with vertices $p(u)_1$, s_3 and s_4. Let S_3 (in purple) be the thickened triangle with vertices $p(u)_1$, s_3 and s_5. Let p be the rightmost point of $\mathcal{C}_{p(u)_2}$ with height 1. Let q be the point of $\mathcal{C}_{p(u)_2}$ such that $pqp(u)_2$ is a clockwise equilateral triangle (therefore with sides of length 1). Let S_4 (in red) be the thickened triangle with these three vertices. Let S_5 (in green) be the thickened triangle with vertices $p(u)_2$, q, and a third vertex inside the purple section S_3. Let $S_6 = \mathbf{X}(u) \setminus (\bigcup_{i=1}^{5} S_i)$ (in grey) be the remaining section.

It is clear from the definition of S_6 that $\bigcup_{i=1}^{6} S_i$ covers $\mathbf{X}(u)$. Sections S_2, S_3, S_4 and S_5 have diameter 1 as thickened triangles. To prove the proposition in this case, it suffices to check the following.

Claim 3. Section S_1 induces a clique and S_6 is small.

When $\theta < \pi/6$, for some values of ρ, the section S_6 is too big, and does not induce a clique. Likewise, for some values of ρ and θ with $2\cos\theta < \rho$, S_6 is too big. This is why we use different constructions for the two other cases.

Third Case: $2\cos(\theta) < \rho$. The construction for this last case is illustrated in Fig. 4. Note that $2\cos(\theta) < \rho$ implies that θ is larger than $\pi/3$. It also implies that the point a is above the abscissa line. The construction of the six sections is more complicated in this last case. We denote by r_0, r_1, r_2 and r_3 the points $(-1,0)$, $(-1/2,-\sqrt{3}/2)$, $(1/2,-\sqrt{3}/2)$ and $(1,0)$. Let S_1 (in blue) be the thickened triangle with vertices $p(u)_1$, r_0 and r_1. Let S_2 (in green) be the thickened triangle with vertices $p(u)_1$, r_1 and r_2 and let S_3 (in red) be the thickened triangle with vertices $p(u)_1$, r_2 and r_3. Let b be the upper intersection between the circles $\mathcal{C}_{p(u)_2}$ and $\mathcal{C}_{r_0}$. Note that the circles $\mathcal{C}_b$ and $\mathcal{C}_{r_1}$ intersect in r_0, and let c be their second intersection. Let c' be the translated of c by the vector $(1,0)$. Section S_4 (in purple) is defined as the thickened triangle with vertices c, c' and a third point uniquely defined with positive y-coordinate. Let S_5 (in yellow) be the thickened triangle with vertices r_0, b and a point inside the section S_4 (in purple). Set $S_6 = \mathbf{X}(u) \setminus (\bigcup_{i=1}^{5} S_i)$.

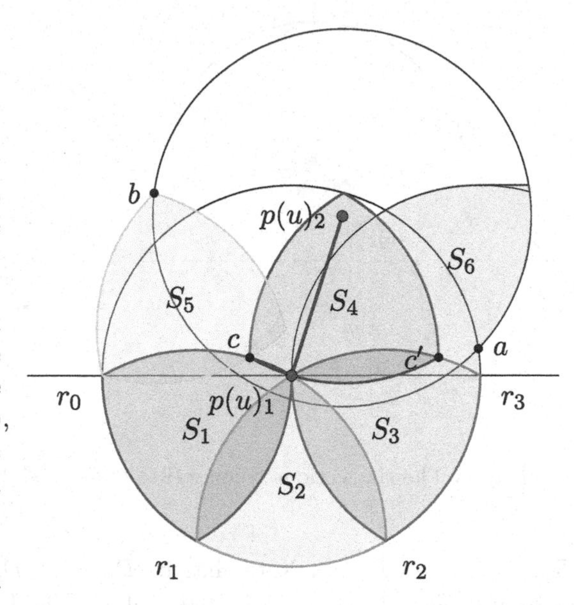

Fig. 4. The six sections when $2\cos\theta < \rho$. (Color figure online)

Sections S_1, S_2, S_3, S_4 and S_5 have diameter 1 because they are thickened triangles. To conclude this case and finish the proof, it suffices to show the following property.

Claim 4. Section S_6 is small.

For this construction, b must not be inside $\mathcal{D}_{p(u)_1}$, otherwise the distance between r_0 and c would be greater than 1. This is always true when $2\cos\theta < \rho$, but not guaranteed for other values. For instance if $\rho = 1$, then b is not inside $\mathcal{D}_{p(u)_1}$ if and only if $2\cos(\theta) \leq \rho$, i.e. $\theta \geq \pi/3$, which is why we use this construction only for this case.

The proofs of Claims 2, 3 and 4 can be found in Appendix.

3 An Online 8-competitive algorithm

Our focus in this section is to prove Theorem 5, which directly implies Theorem 2. As alluded to earlier, we make use of the following kissing number-type result, which may be of independent interest. The corresponding problem in $\mathbb{R}^3$ is interesting and may be difficult.

Theorem 6. *Let x_1 and x_2 be two points in $\mathbb{R}^2$ within Euclidean distance 1. Let Y be a collection of points in $\mathbb{R}^2$ pairwise of Euclidean distance greater than 1 such that either y and x_1 or y and x_2 are within Euclidean distance 1 for any $y \in Y$. Then $|Y| \leq 8$.*

Note that this result is sharp as illustrated in Fig. 5. Take x_1 and x_2 to be at Euclidean distance 1, and choose the 8 points in Y as in a partial optimal circle packing configuration. Now it is possible to shift one of the vertices that are at Euclidean distance 1 from both x_1 and x_2, and to perturb slightly the position of the others, so that all points in Y are pairwise of Euclidean distance greater than 1. Before giving the proof of Theorem 6, we first show how it readily implies Theorem 5.

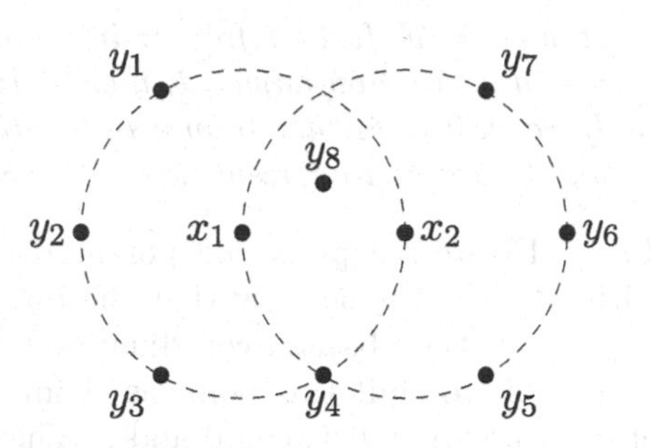

Fig. 5. Theorem 6 is tight.

Proof (Proof of Theorem 5). Let $G = (V, E)$ be a twin unit disk graph and fix an embedding $p : V \to \mathbb{R}^2 \times \mathbb{R}^2$. Let $u \in V$ be a vertex of degree $\Delta(G)$ and consider the set $N(u)$ of neighbours of u in G. Without loss of generality, we may assume that for any $v \in N(u)$ either $p(v)_1$ and $p(u)_1$ or $p(v)_1$ and $p(u)_2$ are within distance 1. It follows from Theorem 6 that the subgraph of G induced by $N(u)$ has no independent set with more than 8 vertices. We know that this induced subgraph can be properly coloured with at most $\chi(G) - 1$ colours. We therefore conclude that $\Delta(G)/8 = |N(u)|/8 \leq \chi(G) - 1$, which completes the proof. □

We prove Theorem 6 through a succession of geometric lemmas. In these lemmas, we treat points in $\mathbb{R}^2$ as hypothetical vertices of an embedded unit disk graph, so we speak of pairs of them as *adjacent*, i.e. within Euclidean distance 1, or not.

Lemma 1. *Let u, v, v' be points in $\mathbb{R}^2$ such that u and v are adjacent, u and v' are adjacent, and v and v' are non-adjacent. If we shift v further from u in the direction of the line segment $[u, v]$ until u and v are non-adjacent, then v and v' remain non-adjacent.*

Proof. Assume without loss of generality that the position of u is $(0, 0)$ and that the first position of v is $(z, 0)$ with $0 < z \leq 1$ before going to $(1, 0)$. We denote the position of v' as (x, y). As can be deduced from the proof of the Lemma 3.1 in [16], the angle between the line segments $[u, v]$ and $[u, v']$ is at least $\pi/3$. Thus $x \leq 1/2$, and so $2x \leq 1 + z$. Then we obtain $2x(1 - z) \leq 1 - z^2$ and so $z^2 - 2xz \leq 1 - 2x$. We also know $(x - z)^2 + y^2 > 1$ since v and v' are non-adjacent. Now $(x - 1)^2 + y^2 = x^2 + 1 - 2x + y^2 \geq x^2 + z^2 - 2zx + y^2 = (x - z)^2 + y^2 > 1$. Thus the distance between v and v' is still larger than 1. □

Lemma 2. *Let u, v, u', v' be points in $\mathbb{R}^2$ such that u and v are adjacent, u and u' are adjacent, v and v' are adjacent, u and v' are non-adjacent, v and u' are non-adjacent, and u' and v' are non-adjacent. Then one of the following is true.*

- *If we shift u' further from u in the direction of the line segment $[u, u']$ until u and u' are non-adjacent, then u' remains non-adjacent with v and v'.*
- *If we shift v' further from v in the direction of the line segment $[v, v']$ until v and v' are non-adjacent, then v' remains non-adjacent with u and u'.*

Proof. Figure 6 depicts one potential situation covered by Lemma 2. Assume without loss of generality that the position of u is $(0, 0)$ and the position of v is $(1, 0)$. If the abscissa coordinate of u' or v' is not between 0 and 1, then it is possible to shift this point as claimed. Assume that the abscissa coordinate of both points is between 0 and 1. Then one must be above the other. Assume that u' is above v', for consistency with Fig. 6. We may also assume that they are both above the abscissa line, otherwise the claim is immediately true. By Lemma 1, we know that shifting u' until its distance to u is 1 will not make it adjacent to v. Moreover, since the abscissa of u' and v' is between 0 and 1, the line segment $[u, u']$ goes through the disk with radius 1 centred at v'. As u' is not in this disk, and because a disk is convex, shifting u' as claimed will not return it to the disk. If instead v' is above u', then shifting will instead be possible for v' because of the same arguments. □

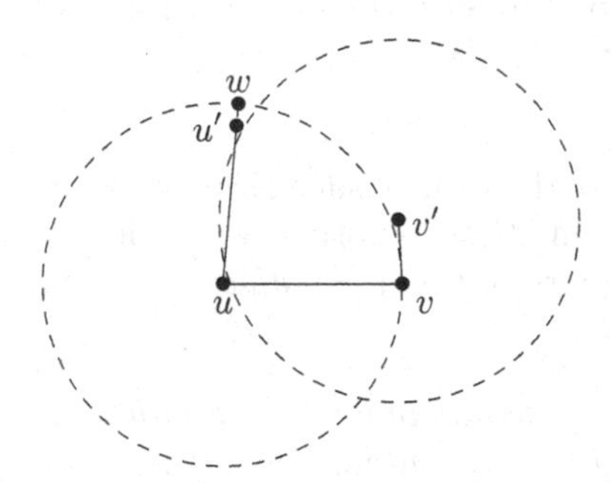

Fig. 6. By shifting u' to the position of w, it remains non-adjacent to v and to v'.

Lemma 3. *Let u and u' be points in $\mathbb{R}^2$ such that u and u' are adjacent. Let $v_1, \ldots, v_8$ be eight pairwise non-adjacent points in $\mathbb{R}^2$ each of which is adjacent to u or to u'. Then some v_i is adjacent to both u and u'.*

Proof. We assume that no v_i is adjacent to both u and u', and show that this leads to a contradiction. Without loss of generality, assume that the position of u is $(0, 0)$ and the position of u' is $(d, 0)$ with $0 < d \leq 1$. Denote by V_u (respectively $V_{u'}$) the set of those v_i adjacent to u (respectively u'). We add to the abscissa of u' and all the points in $V_{u'}$ the value $1-d$. Thus u and u' are still adjacent and the points in $V_{u'}$ are still adjacent to u'. Note that the bisector of the line segment $[u,u']$ separates the points in V_u from the points in $V_{u'}$. Indeed if a point is in V_u, it is closer to u than u', and vice versa. Thus the points $\{v_i\}_i$ are still pairwise non-adjacent. Now the position of u' is $(1,0)$.

The star graph $K_{1,6}$ is not a unit disk graph. Thus since the points in $V_u \cup \{u'\}$ are pairwise non-adjacent and adjacent to u, we have $|V_u| < 5$. By the same reasoning we obtain $|V_{u'}| < 5$. Thus we have $|V_u| = |V_{u'}| = 4$. Without loss of generality, assume that we have $V_u = \{v_1, v_2, v_3, v_4\}$ and $V_{u'} = \{v_5, v_6, v_7, v_8\}$. We order the points in V_u. For a point v we will consider the oriented angle between the line segments $[u,v]$ and $[u,u']$ taken in $[0,2\pi)$. Note that two points cannot have the same angle value. Thus we can order the points from the smallest angle value to the largest. We apply the same process to the points in $V_{u'}$. For a point v', we consider the angle between $[u',v']$ and $[u',u]$. Without loss of generality we can assume that the points in V_u appear in the right order. Figure 7a depicts the ordering. (Of course, since we are going to show a contradiction, the graph in the figure cannot satisfy the assumptions we have taken.)

Now we move the point v_2 further from u in the direction of the line segment $[u,v_2]$ until the distance between v_2 and u is 1. By Lemma 1, we know that when v_2 is shifted in this way, it will not become adjacent to any other point in $\{v_i\}_i$. We apply the same process to v_3, v_6 and v_7. By Lemma 2, it is possible to shift either v_1 or v_8, and either v_4 or v_5. Without loss of generality, assume that it is possible to shift v_1. Denote by $\theta \in [0,2\pi)$ the angle between the line segments $[u,v_1]$ and $[u,u']$. By the same arguments as above, we have $\theta > \pi/3$. Thus the angle between $[u,v_4]$ and $[u,u']$ taken in $[0,2\pi)$ is larger than $\theta+\pi$. Let us consider the angle between $[u',v_7]$ and the abscissa line. It is less than $\theta-\pi/3$, because otherwise v_8 must be adjacent to u or v_1, as we prove in the next paragraph. Thus the worst case is when the angle is exactly equal to $\theta-\pi/3$. We have u at position $(0,0)$, u' at $(1,0)$, v_1 at $(\cos(\theta),\sin(\theta))$ and v_7 at $(1+\cos(\theta-\pi/3), \sin(\theta-\pi/3))$.

We move v_8 so that it is at distance 1 from both u and v_7. This is theoretically not possible because v_8 is not adjacent to those points, but we are going to show that even in this case we have v_8 at distance 1 from v_1. Since this position is the furthest v_8 could be from v_1, this shows a contradiction, and thus that the angle between $[u',v_7]$ and the abscissa line must be less than $\theta-\pi/3$. Let us compute the position (x,y) of v_8. We have $x^2+y^2=1$ and $(x-1-\cos(\theta-\pi/3))^2+(y-\sin(\theta-\pi/3))^2=1$. There are two possible solutions. One is $(1,0)$, which is the position of u', and the other is the position of v_8: $(\cos(\theta-\pi/3),\sin(\theta-\pi/3))$. Figure 7b depicts the position of the points. But then the distance between v_1 and v_8 is equal to 1. Thus, even if we take v_8 to be the furthest possible from v_1, they are still too close. Thus we know that the angle between $[u',v_7]$ and the abscissa line taken in $[0,2\pi)$ must be larger than $\theta-\pi/3$. Hence the angle between $[u',v_6]$ and the same line taken into $[0,2\pi)$ must be less than $\theta+4\pi/3$, and the angle between $[u',v_5]$ and the same line must be less than $\theta+\pi$. We have seen that either v_4 or v_5 can be pushed until their distance to u or u' is 1.

If it is possible for v_4 we can apply what we did before to v_4 and v_6 to obtain a contradiction. If it is possible for v_5, we then apply the reasoning to v_3 and v_5. In any case we have a contradiction, which concludes the proof of the lemma. □

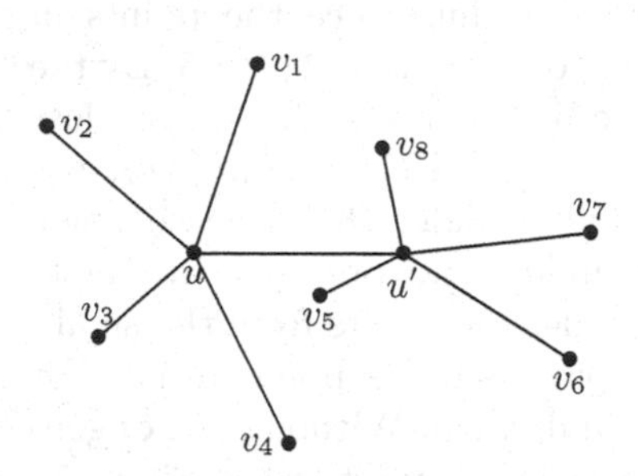

(a) Ordering of the vertices.

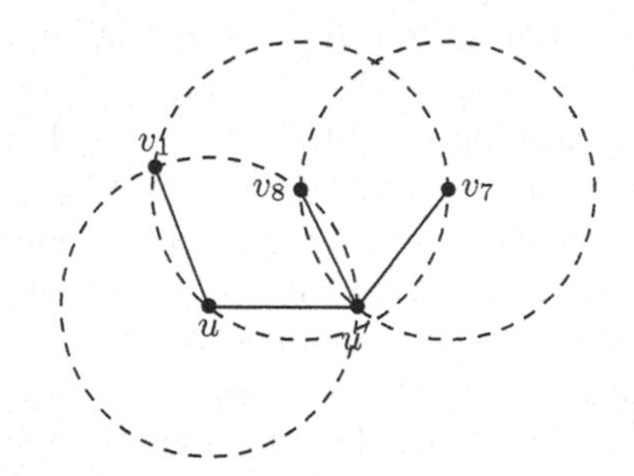

(b) Position of the considered vertices at the end of the proof.

Fig. 7. Illustration of Lemma 3

Proof (Proof of Theorem 6). Let Y in the statement of the theorem be $\{y_1, \ldots, y_9\}$ for a contradiction. For any eight points in Y, Lemma 3 guarantees that at least one of them is within Euclidean distance 1 of both x_1 and x_2. Without loss of generality, we may thus assume that both y_1 and y_2 are within Euclidean distance 1 to both x_1 and x_2. Of the seven remaining points of Y, at least four of them must be within Euclidean distance 1 of, say, x_1, by the pigeonhole principle. But then x_1 is within Euclidean distance 1 of six points that are pairwise of Euclidean distance greater than 1, which is impossible. □

Acknowledgments. The first author was supported by the Swiss National Science Foundation within the collaborative DACH project Arrangements and Drawings as SNSF Project 200021E-171681. The second and third authors were supported by a Vidi grant (639.032.614) of the Netherlands Organisation for Scientific Research (NWO).

References

1. Barrett, C.L., Istrate, G., Anil Kumar, V.S., Marathe, M.V., Thite, S., Thulasidasan, S.: Strong edge coloring for channel assignment in wireless radio networks. In: 2006 Pervasive Computing and Communications Workshops, p. 5. IEEE (2006)
2. Bonamy, M., Perrett, T., Postle, L.: Colouring graphs with sparse neighbourhoods: bounds and applications. arXiv e-prints:1810.06704 (2018)
3. Bruhn, H., Joos, F.: A stronger bound for the strong chromatic index. Combin. Probab. Comput. **27**(1), 21–43 (2018). https://doi.org/10.1017/S0963548317000244

4. Chalermsook, P., Laekhanukit, B., Nanongkai, D.: Coloring graph powers: graph product bounds and hardness of approximation. In: Pardo, A., Viola, A. (eds.) LATIN 2014. LNCS, vol. 8392, pp. 409–420. Springer, Heidelberg (2014). https://doi.org/10.1007/978-3-642-54423-1_36
5. Clark, B.N., Colbourn, C.J., Johnson, D.S.: Unit disk graphs. Discrete Math. **86**(1–3), 165–177 (1990)
6. Diestel, R.: Graph theory, Graduate Texts in Mathematics, 5th edn, vol. 173. Springer, Berlin (2017). https://doi.org/10.1007/978-3-662-53622-3
7. Erdős, P.: Problems and results in combinatorial analysis and graph theory. In: Proceedings of the First Japan Conference on Graph Theory and Applications, Hakone, 1986, vol. 72, pp. 81–92 (1988). https://doi.org/10.1016/0012-365X(88)90196-3
8. Gräf, A., Stumpf, M., Weißenfels, G.: On coloring unit disk graphs. Algorithmica **20**(3), 277–293 (1998)
9. Hale, W.K.: Frequency assignment: theory and applications. Proc. IEEE **68**(12), 1497–1514 (1980). https://doi.org/10.1109/PROC.1980.11899
10. Kaiser, T., Kang, R.J.: The distance-t chromatic index of graphs. Combin. Probab. Comput. **23**(1), 90–101 (2014). https://doi.org/10.1017/S0963548313000473
11. Kang, R.J., Müller, T.: Sphere and dot product representations of graphs. Discrete Comput. Geom. **47**(3), 548–568 (2012). https://doi.org/10.1007/s00454-012-9394-8
12. Kanj, I.A., Wiese, A., Zhang, F.: Local algorithms for edge colorings in UDGs. Theor. Comput. Sci. **412**(35), 4704–4714 (2011). https://doi.org/10.1016/j.tcs.2011.05.005
13. Mahdian, M.: The strong chromatic index of graphs. Master's thesis, University of Toronto (2000)
14. Mahdian, M.: On the computational complexity of strong edge coloring. Discrete Appl. Math. **118**(3), 239–248 (2002). https://doi.org/10.1016/S0166-218X(01)00237-2
15. Malesińska, E., Piskorz, S., Weißenfels, G.: On the chromatic number of disk graphs. Networks **32**(1), 13–22 (1998)
16. Marathe, M.V., Breu, H., Hunt III, H.B., Ravi, S.S., Rosenkrantz, D.J.: Simple heuristics for unit disk graphs. Networks **25**(2), 59–68 (1995)
17. Molloy, M., Reed, B.: A bound on the strong chromatic index of a graph. J. Combin. Theory Ser. B **69**(2), 103–109 (1997). https://doi.org/10.1006/jctb.1997.1724
18. Nandagopal, T., Kim, T.E., Gao, X., Bharghavan, V.: Achieving MAC layer fairness in wireless packet networks. In: Proceedings of the 6th Annual International Conference on Mobile Computing and Networking, pp. 87–98. ACM (2000)
19. Peeters, R.: On coloring j-unit sphere graphs. Technical report, Tilburg University (1991)
20. Ramanathan, S., Lloyd, E.L.: Scheduling algorithms for multihop radio networks. IEEE/ACM Trans. Netw. (TON) **1**(2), 166–177 (1993)
21. de Joannis de Verclos, R., Kang, R.J., Pastor, L.: Colouring squares of claw-free graphs. Can. J. Math. **71**(1), 113–129 (2019). https://doi.org/10.4153/CJM-2017-029-9

Precedence-Constrained Scheduling and Min-Sum Set Cover (Extended Abstract)

Felix Happach(✉) and Andreas S. Schulz

Operations Research, Technische Universität München, Munich, Germany
{felix.happach,andreas.s.schulz}@tum.de

Abstract. We consider a single-machine scheduling problem with bipartite AND/OR-constraints that is a natural generalization of (precedence-constrained) min-sum set cover. For min-sum set cover, Feige, Lovàsz and Tetali [15] showed that the greedy algorithm has an approximation guarantee of 4, and obtaining a better approximation ratio is NP-hard. For precedence-constrained min-sum set cover, McClintock, Mestre and Wirth [30] proposed an $O(\sqrt{m})$-approximation algorithm, where m is the number of sets. They also showed that obtaining an algorithm with performance $O(m^{1/12-\varepsilon})$ is impossible, assuming the hardness of the planted dense subgraph problem.

The more general problem examined here is itself a special case of scheduling AND/OR-networks on a single machine, which was studied by Erlebach, Kääb and Möhring [13]. Erlebach et al. proposed an approximation algorithm whose performance guarantee grows linearly with the number of jobs, which is close to best possible, unless P = NP.

For the problem considered here, we give a new LP-based approximation algorithm. Its performance ratio depends only on the maximum number of OR-predecessors of any one job. In particular, in many relevant instances, it has a better worst-case guarantee than the algorithm by McClintock et al., and it also improves upon the algorithm by Erlebach et al. (for the special case considered here).

Yet another important generalization of min-sum set cover is generalized min-sum set cover, for which a 12.4-approximation was derived by Im, Sviridenko and Zwaan [23]. Im et al. conjecture that generalized min-sum set cover admits a 4-approximation, as does min-sum set cover. In support of this conjecture, we present a 4-approximation algorithm for another interesting special case, namely when each job requires that no less than all but one of its predecessors are completed before it can be processed.

Keywords: Scheduling · Precedence constraints · Min-sum set cover

---

This work has been supported by the Alexander von Humboldt Foundation with funds from the German Federal Ministry of Education and Research (BMBF).

E. Bampis and N. Megow (Eds.): WAOA 2019, LNCS 11926, pp. 170–187, 2020.
https://doi.org/10.1007/978-3-030-39479-0_12

1 Introduction

In this paper, we consider the problem of scheduling jobs subject to AND/OR-precedence constraints on a single machine. These scheduling problems are closely related to (precedence-constrained) min-sum set cover [14,15,30] and generalized min-sum set cover [3,4,23,37]. Let $N = A \dot{\cup} B$ be the set of n jobs with processing times $p_j \geq 0$ and weights $w_j \geq 0$ for all $j \in N$. The precedence constraints are given by a directed graph $G = (N, E_\wedge \dot{\cup} E_\vee)$, where $(i,j) \in E_\wedge \cup E_\vee$ means that job i is a predecessor of job j. The arcs in $E_\wedge \subseteq (A \times A) \cup (B \times B)$ and $E_\vee \subseteq A \times B$ represent AND- and OR-precedence constraints, respectively. That is, a job in N requires that *all* its predecessors w.r.t. $E_\wedge$ are completed before it can start. A job in B, however, requires that *at least one* of its predecessors w.r.t. $E_\vee$ is completed beforehand. The set of *OR-predecessors* of job $b \in B$ is $\mathcal{P}(b) := \{a \in A \,|\, (a,b) \in E_\vee\}$. Note that $\mathcal{P}(b)$ might be empty for some $b \in B$.

A *schedule* C is an ordering of the jobs on a single machine such that each job j is processed non-preemptively for p_j units of time, and no jobs overlap. The *completion time* of $j \in N$ in the schedule C is denoted by C_j. A schedule C is *feasible* if (i) $C_j \geq \max\{C_i \mid (i,j) \in E_\wedge\} + p_j$ for all $j \in N$ (AND-constraints), and (ii) $C_b \geq \min\{C_a \mid a \in \mathcal{P}(b)\} + p_b$ for all $b \in B$ with $\mathcal{P}(b) \neq \emptyset$ (OR-constraints). The goal is to determine a feasible schedule C that minimizes the sum of weighted completion times, $\sum_{j \in N} w_j C_j$. We denote this problem by $1 \,|\, \textit{ao-prec} = A \dot{\vee} B \,|\, \sum w_j C_j$, in an extension of the notation of Erlebach, Kääb and Möhring [13] and the three-field notation of Graham et al. [18]. This scheduling problem is NP-hard. In fact, it generalizes a number of NP-hard problems, as discussed below. Therefore, we focus on approximation algorithms. Let Π be a minimization problem, and $\rho \geq 1$. Recall that a ρ-approximation algorithm for Π is a polynomial-time algorithm that returns, for every instance of Π, a feasible solution with objective value at most ρ times the optimal objective value. If ρ does not depend on the input parameters, we call the algorithm a *constant-factor approximation*.

Due to space limitations, we defer all proofs to the journal version of this paper. As already indicated, the scheduling problem we consider is motivated by its close connection to min-sum covering problems. Figure 1 gives an overview of the related problems, which we describe briefly in the following paragraphs.

(Min-Sum) Set Cover. The most basic problem is *min-sum set cover* (MSSC), where the input consists of a hypergraph with vertices V and hyperedges $\mathcal{E}$. Given a linear ordering of the vertices $f : V \to |V|$, the covering time of hyperedge $e \in \mathcal{E}$ is defined as $f(e) := \min_{v \in e} f(v)$. The goal is to find a linear ordering of the vertices that minimizes the sum of covering times, $\sum_{e \in \mathcal{E}} f(e)$. MSSC is indeed a special case of $1 \,|\, \textit{ao-prec} = A \dot{\vee} B \,|\, \sum w_j C_j$: we introduce a job in A for every vertex of V and a job in B for every hyperedge in $\mathcal{E}$, and we set $p_a = w_b = 1$ and $p_b = w_a = 0$ for all jobs $a \in A$ and $b \in B$. Further, we let $E_\wedge = \emptyset$ and introduce an arc $(a,b) \in E_\vee$ in the precedence graph, if the vertex corresponding to a is contained in the hyperedge corresponding to b.

MSSC was first introduced by Feige, Lovàsz and Tetali [14], who observed that a simple greedy heuristic due to Bar-Noy et al. [6] yields an approximation factor of 4. Feige et al. [14] simplified the analysis via a primal/dual approach based on a time-indexed linear program. In the journal version of their paper, Feige et al. [15] also proved that it is NP-hard to obtain an approximation factor strictly better than 4. The special case of MSSC where the hypergraph is an ordinary graph is called min-sum vertex cover (MSVC), and is APX-hard [15]. It admits a 2-approximation that is also based on a time-indexed linear program, and uses randomized rounding [14,15].

Munagala et al. [31] generalized MSSC by introducing non-negative costs c_v for each vertex $v \in V$ and non-negative weights w_e for each hyperedge $e \in \mathcal{E}$. Here, the goal is to minimize the sum of weighted covering costs, $\sum_{e\in\mathcal{E}} w_e f(e)$, where the covering cost of $e \in \mathcal{E}$ is defined as $f(e) := \min_{v\in e} \sum_{w:f(w)\leq f(v)} c_w$. The authors called this problem *pipelined set cover*, and proved, among other things, that the natural extension of the greedy algorithm of Feige et al. for MSSC still yields a 4-approximation. Similar to MSSC, we can model pipelined set cover as an instance of $1 \,|\, \textit{ao-prec} = A \dot{\vee} B \,|\, \sum w_j C_j$.

Munagala et al. [31] asked whether there is still a constant-factor approximation for pipelined set cover if there are AND-precedence constraints in form of a partial order $\prec$ on the vertices of the hypergraph. That is, any feasible linear ordering $f : V \to |V|$ must satisfy $f(v) < f(w)$, if $v \prec w$. This question was partly settled by McClintock, Mestre and Wirth [30]. They presented a $4\sqrt{|V|}$-approximation algorithm for *precedence-constrained MSSC*, which is the extension of MSSC where $E_\wedge = \{(a', a) \in A \times A \,|\, a' \prec a\}$. The algorithm uses a $\sqrt{|V|}$-approximative greedy algorithm on a problem called max-density precedence-closed subfamily. The authors also propose a reduction from the so-called planted dense subgraph conjecture [7] to precedence-constrained MSSC. Roughly speaking, the conjecture says that for all $\varepsilon > 0$ there is no polynomial-time algorithm that can decide with advantage $> \varepsilon$ whether a random graph on m vertices is drawn from $(m, m^{\alpha-1})$ or contains a subgraph drawn from $(\sqrt{m}, \sqrt{m}^{\beta-1})$ for certain $0 < \alpha, \beta < 1$.[1] If the conjecture holds true, then this implies that there is no $\mathcal{O}(|V|^{1/12-\varepsilon})$-approximation for precedence-constrained MSSC [30].

The ordinary set cover problem is also a special case of $1 \,|\, \textit{ao-prec} = A \dot{\vee} B \,|\, \sum w_j C_j$: we introduce a job in A with $p_a = 1$ and $w_a = 0$ for every set, a job in B with $p_b = w_b = 0$ for every element, and an arc $(a, b) \in E_\vee$ in the precedence graph, if the set corresponding to a contains the element corresponding to b. Further, we include an additional job x in B with $p_x = 0$ and $w_x = 1$, and introduce an arc $(b, x) \in E_\wedge$ for every job $b \in B \setminus \{x\}$. If the set cover instance admits a cover of cardinality k, we first schedule the corresponding set-jobs in A, so all element-jobs are available for processing at time k. Then job x can complete at time k, which gives an overall objective value of k. Similarly, any schedule with objective value equal to k implies that all element-jobs are completed before

[1] A random graph drawn from (m, p) contains m vertices and the probability of the existence of an edge between any two vertices is equal to p.

time k, so there exists a cover of size at most k. Recall that set cover admits an $\ln(m)$-approximation [25,28], where m is the number of elements, and this is best possible, unless P = NP [12].

A New Approximation Algorithm. W.l.o.g., suppose that $E_\wedge$ is transitively closed, i.e. $(i,j) \in E_\wedge$ and $(j,k) \in E_\wedge$ implies $(i,k) \in E_\wedge$. We may further assume that there are no redundant OR-precedence constraints, i.e. if $(a,b) \in E_\vee$ and $(a',a) \in E_\wedge$, then $(a',b) \notin E_\vee$. Otherwise we could remove the arc (a,b) from $E_\vee$, since any feasible schedule has to schedule a' before a. Let $\Delta := \max_{b\in B} |\mathcal{P}(b)|$ be the maximum number of OR-predecessors of a job in B. One can see that Δ is bounded from above by the cardinality of a maximum independent set in the induced subgraph on $E_\wedge \cap (A \times A)$. Note that Δ is often relatively small compared to the total number of jobs. For instance, if the precedence constraints are derived from an underlying graph, where the predecessors of each edge are its incident vertices (as in MSVC), then $\Delta = 2$. Our first result is the following.

Theorem 1. *There is a 2Δ-approximation algorithm for $1 \mid$ ao-prec $= A \dot{\vee} B$, $p_j \in \{0,1\} \mid \sum w_j C_j$. Moreover, for any $\varepsilon > 0$, there is a $(2\Delta + \varepsilon)$-approximation algorithm for $1 \mid$ ao-prec $= A \dot{\vee} B \mid \sum w_j C_j$.*

In Sect. 2, we first exhibit a randomized approximation algorithm for $1 \mid$ *ao-prec* $= A \dot{\vee} B$, $p_j \in \{0,1\} \mid \sum w_j C_j$, i.e. if all processing times are 0/1, and then we show how to derandomize it. This proves the first part of Theorem 1. A natural question that arises in the context of real-world scheduling problems is whether approximation guarantees for 0/1-problems still hold for arbitrary processing times. As observed by Munagala et al. [31], the natural extension of the greedy algorithm for MSSC still works, if the processing times of jobs in A are arbitrary, but all jobs in B have zero processing time, and there are no AND-precedence constraints. Once jobs in B have non-zero processing times, the analysis of the greedy algorithm fails. In fact, it is not clear whether there are constant-factor approximations. Our algorithm can be extended to arbitrary processing times (which proves the second part of Theorem 1) and, additionally, release dates.

Note that the result of Theorem 1 improves on the algorithm of [30] for precedence-constrained MSSC in two ways. First, the approximation factor of 2Δ does not depend on the total number of jobs, but on the maximum number of OR-predecessors of a job in B. In particular, we immediately obtain a 4-approximation for the special case of precedence-constrained MSVC. Secondly, the algorithm works for arbitrary processing times, additional AND-precedence constraints on $B \times B$, and it can be extended to non-trivial release dates of the jobs. Note that, in general, Δ and $\sqrt{|V|}$ are incomparable. In most practically relevant instances, Δ should be considerably smaller than $\sqrt{|V|}$.

It is important to highlight that the approximation factor of 2Δ in Theorem 1 does not contradict the conjectured hardness of precedence-constrained MSSC stated in [30]. The set A in the reduction of [30] from the planted dense subgraph problem contains a job for every vertex and every edge of the random graph on m vertices. Each vertex-job consists of the singleton $\{0\}$ whereas each edge-job is a (random) subset of $[q] := \{1, \ldots, q\}$, for some non-negative integer q. Every element in $[q]$ appears in expectation in mp^2 many edge-jobs, where p is a carefully chosen probability. If we interpret this as a scheduling problem, we can remove the dummy element 0 from the instance. So the maximum indegree of a job in $B = [q]$ (maximum number of appearances of the element) is $\Delta \approx mp^2 \geq m^{\frac{1}{4}}$, see [30]. Hence the gap $\Omega(m^{\frac{1}{8}})$ in the reduction translates to a gap of $\Omega(\sqrt{\Delta})$ in our setting. Therefore, if the planted dense subgraph conjecture [7] holds true, then there is no $\mathcal{O}(\Delta^{1/3-\varepsilon})$-approximation algorithm for $1 \mid \textit{ao-prec} = A \dot{\vee} B \mid \sum w_j C_j$ for any $\varepsilon > 0$.

Note that in the reduction from set cover to $1 \mid \textit{ao-prec} = A \dot{\vee} B \mid \sum w_j C_j$ the parameter Δ equals the maximum number of appearances of an element in the set cover instance. Hochbaum [22] presented an approximation algorithm for set cover with a guarantee of Δ. Hence the 2Δ-approximation of Theorem 1 does not contradict the hardness of obtaining a $(1-\varepsilon)\ln(m)$-approximation for set cover [12]. If the planted dense subgraph conjecture [7] is false, then constant-factor approximations for $1 \mid \textit{ao-prec} = A \dot{\vee} B \mid \sum w_j C_j$ with $E_\wedge \subseteq A \times A$ may be possible. However, the reduction from set cover shows that, in general, we cannot get a constant-factor approximation if $E_\wedge \cap (B \times B) \neq \emptyset$.

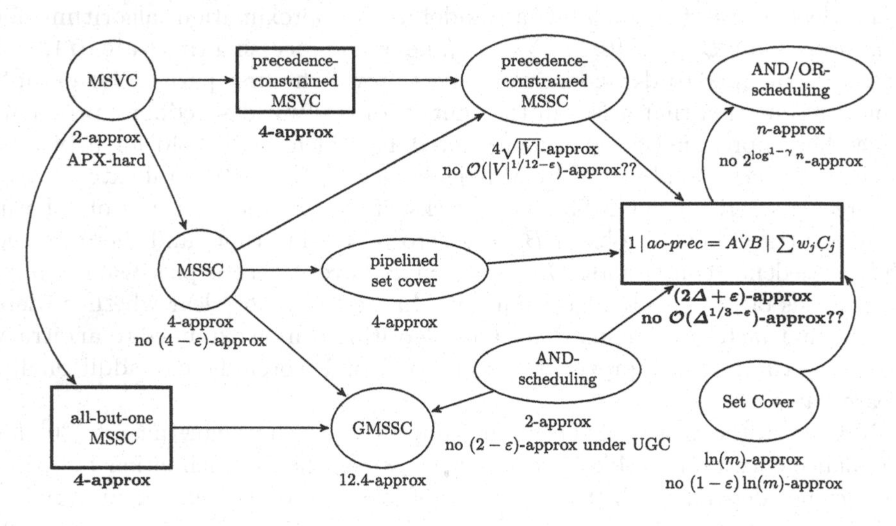

Fig. 1. Overview of related problems and results. An arrow from problem Π_1 to Π_2 indicates that Π_2 generalizes Π_1. Problems in rectangular frames are explicitly considered in this paper, and our results are depicted in bold. Lower bounds indicated with "??" are assuming hardness of the planted dense subgraph problem [7].

Generalized Min-Sum Set Cover. A different generalization of MSSC, called *generalized min-sum set cover* (GMSSC), was introduced by Azar, Gamzu and Yin [3]. The input of GMSSC is similar to MSSC, but, in addition, each hyperedge $e \in \mathcal{E}$ is associated with a covering requirement $\kappa(e) \in [|e|]$. Given a linear ordering of the vertices, the covering time of $e \in \mathcal{E}$ is now the first point in time when $\kappa(e)$ of its incident vertices appear in the linear ordering. The goal is again to minimize the sum of covering times over all hyperedges.

In our notation, this means that $E_\wedge = \emptyset$ and each job $b \in B$ requires at least $\kappa(b) \in [|\mathcal{P}(b)|]$ of its OR-predecessors to be completed before it can start. The extreme cases $\kappa(b) = 1$ and $\kappa(b) = |\mathcal{P}(b)|$ are MSSC and the minimum latency set cover problem. The latter is, in fact, equivalent to single-machine scheduling with AND-precedence constraints [40]. Over time, several constant-factor approximations for GMSSC were proposed. Bansal, Gupta and Krishnaswamy [4] presented an algorithm with an approximation guarantee of 485, which was improved to 28 by Skutella and Williamson [37]. Both algorithms are based on the same time-indexed linear program, but use different rounding techniques, namely standard randomized rounding [4] and α-points [37], respectively.

The currently best-known approximation ratio for GMSSC is 12.4, due to Im, Sviridenko and Zwaan [23]. However, Im et al. [23] conjecture that GMSSC admits a 4-approximation. By adapting the proof of Theorem 1, we obtain a 4-approximation for GMSSC if $\kappa(b) = \max\{|\mathcal{P}(b)| - 1, 1\}$ for all $b \in B$. To the best of the authors' knowledge, this case, which we call *all-but-one MSSC*, was not considered before. Here, each job (with more than one predecessor) needs at least all but one of them to be completed before it can start. This is a natural special case inbetween MSSC and AND-precedence constrained scheduling (where $\kappa(b) = 1$ and $\kappa(b) = |\mathcal{P}(b)|$, respectively). Note that all-but-one MSSC generalizes MSVC. The algorithm of Theorem 2 is described in Sect. 3.

Theorem 2. *There is a 4-approximation algorithm for all-but-one MSSC.*

Related Work on Scheduling Problems. The first polynomial-time algorithm for scheduling jobs on a single machine to minimize the sum of weighted completion times is due to Smith [38]. Once there are AND-precedence constraints, the problem becomes strongly NP-hard [27]. The first constant-factor approximation for AND-precedence constraints was proposed by Hall, Shmoys and Wein [20] with an approximation factor of $4+\varepsilon$. Their algorithm is based on a time-indexed linear program and α-point scheduling with a fixed value of α. Subsequently, various 2-approximations based on linear programs [10,19,34] as well as purely combinatorial algorithms [8,29] were derived. Assuming a variant of the Unique Games Conjecture of Khot [26], Bansal and Khot [5] showed that the approximation ratio of 2 is essentially best possible.

If the precedence constraints are of AND/OR-structure, then the problem does not admit constant-factor approximations anymore. Let $0 < c < \frac{1}{2}$ and $\gamma = (\log\log n)^{-c}$. It is NP-hard to approximate the sum of weighted completion times of unit processing time jobs on a single machine within a factor of $2^{\log^{1-\gamma} n}$, if AND/OR-precedence constraints are involved [13]. The precedence graph in the reduction consists of four layers with an OR/AND/OR/AND-structure. Erlebach, Kääb and Möhring [13] also showed that scheduling the jobs in order of non-decreasing processing times (among the available jobs) yields an n-approximation for general weights and a $\sqrt{n}$-approximation for unit weights, respectively. It can easily be verified that $1 \mid ao\text{-}prec = A \dot{\vee} B \mid \sum w_j C_j$ is a special case of the problem considered in [13].

Scheduling unit processing time jobs with OR-precedence constraints only on parallel machines to minimize the sum of completion times can be solved in polynomial time [24]. However, once we want to minimize the sum of *weighted* completion times, already the single-machine problem with unit processing times becomes strongly NP-hard [24]. We strengthen this result and show that even more restricted special cases are strongly NP-hard already.

Theorem 3. *$1 \mid ao\text{-}prec = A \dot{\vee} B,\ p_j \in \{0,1\} \mid \sum C_j$ and $1 \mid ao\text{-}prec = A \dot{\vee} B,\ p_j = 1 \mid \sum w_j C_j$ with $w_j \in \{0,1\}$ are NP-hard, even if $E_\wedge = \emptyset$.*

Note that the latter problem is a special case of the problem considered in [24], where it was denoted by $1 \mid or\text{-}prec, p_j = 1 \mid \sum w_j C_j$, and that $1 \mid ao\text{-}prec = A \dot{\vee} B,\ p_j = 1 \mid \sum C_j$ is trivial.

Our Techniques and LP Relaxations. The algorithms that lead to Theorems 1 and 2 are based on time-indexed linear programs and the concept of random α-point scheduling, similar to, e.g., [9,16,17,19,20,35]. One new element here is to not use a global value for α, but to use different values of α for the jobs in A and B, respectively. This is crucial in order to obtain feasible schedules. We focus on time-indexed linear programs, since other standard LP formulations fail in the presence of OR-precedence constraints; see Sect. 4.

More specifically, we show that these relaxations have an integrality gap that is linear in the number of jobs, even on instances with $\Delta = 2$ and $E_\wedge = \emptyset$. In Sect. 4.1, we discuss a formulation in linear ordering variables that was introduced by Potts [32]. We present a class of constraints that is facet-defining for the integer hull (Theorem 4), and show that the integrality gap remains linear, even if we add these inequalities. In Sect. 4.2, we consider an LP relaxation in completion time variables, which was proposed by Wolsey [41] and Queyranne [33]. We first generalize the well-known parallel inequalities [33,41], that fully describe the polytope in the absence of precedence constraints, to OR-precedence constraints (Theorem 5). Then we show that, even though we add an exponential number of tight valid inequalities, the corresponding LP relaxation still exhibits a linear integrality gap.

2 A New Generalization of Min-Sum Set Cover

Consider an instance of $1 \mid ao\text{-}prec = A \dot{\vee} B \mid \sum w_j C_j$. W.l.o.g., we may assume that $w_a = 0$ for all $a \in A$. Otherwise, we can shift a positive weight of a job in A to an additional successor in B with zero processing time. Further, we may assume that all data is integer and $p_j \geq 1$ for every job $j \in N$ that has no predecessors (otherwise such a job can be disregarded). So no job can complete at time 0 in a feasible schedule.

Suppose that $p_j \in \{0,1\}$ for all $j \in N$, and let $T = \sum_{j \in N} p_j$ be the time horizon. We consider the time-indexed linear programming formulation of Sousa and Wolsey [39] with AND-precedence constraints [20]. The binary variable x_{jt}

indicates whether job $j \in N$ completes at time $t \in [T]$ or not. Additionally, we introduce constraints corresponding to $E_\vee$. The resulting linear relaxation is

$$\min \quad \sum_{b \in B} \sum_{t=1}^{T} w_b \cdot t \cdot x_{bt} \tag{1a}$$

$$\text{s.t.} \quad \sum_{t=1}^{T} x_{jt} = 1 \quad \forall j \in N, \tag{1b}$$

$$\sum_{j \in N} \sum_{s=t-p_j+1}^{t} x_{js} \leq 1 \quad \forall t \in [T], \tag{1c}$$

$$\sum_{s=1}^{t+p_b} x_{bs} - \sum_{a \in \mathcal{P}(b)} \sum_{s=1}^{t} x_{as} \leq 0 \quad \forall b \in B : \mathcal{P}(b) \neq \emptyset, \ \forall t \in [T - p_b], \tag{1d}$$

$$\sum_{s=1}^{t+p_j} x_{js} - \sum_{s=1}^{t} x_{is} \leq 0 \quad \forall (i,j) \in E_\wedge, \ \forall t \in [T - p_j], \tag{1e}$$

$$x_{jt} \geq 0 \quad \forall j \in N, \ \forall t \in [T]. \tag{1f}$$

Constraints (1b) and (1c) ensure that each job is executed and no jobs overlap, respectively. Note that only jobs with $p_j = 1$ appear in (1c). Constraints (1d) and (1e) ensure OR- and AND-precedence constraints, respectively. Note that we can solve LP (1) in polynomial time, since $T \leq n$.

Let $\overline{x}$ be an optimal fractional solution of LP (1). For $j \in N$, we call $\overline{C}_j = \sum_t t \cdot \overline{x}_{jt}$ its *fractional completion time*. Note that $\sum_j w_j \overline{C}_j$ is a lower bound on the objective value of an optimal integer solution, which corresponds to an optimal schedule. For $0 < \alpha \leq 1$ and $j \in N$, we define its α-point, $t_j^\alpha := \min\{t \mid \sum_{s=1}^{t} \overline{x}_{js} \geq \alpha\}$, to be the first integer point in time when an α-fraction of j is completed [20].

The algorithm, hereafter called Algorithm 1, works as follows. First, solve LP (1) to optimality, and let $\overline{x}$ be an optimal fractional solution. Then, draw β at random from the interval $(0, 1]$ with density function $f(\beta) = 2\beta$, and set $\alpha = \frac{\beta}{\Delta}$. (Choosing α as a function of β is crucial in order to obtain a feasible schedule in the end. This together with (1d) ensures that at least one OR-predecessor of a job $b \in B$ completes early enough in the constructed schedule. The density function $f(\beta) = 2\beta$ is chosen to cancel out an unbounded term of $\frac{1}{\beta}$ in the expected value of the completion time of job b, as in [16,35].) Now, compute t_a^α and t_b^β for all jobs $a \in A$ and $b \in B$, respectively. Sort the jobs in order of non-decreasing values t_a^α ($a \in A$) and t_b^β ($b \in B$), and denote this total order by $\prec$. If there is $b \in B$ and $a \in \mathcal{P}(b)$ with $t_a^\alpha = t_b^\beta$, then set $a \prec b$. Similarly, set $i \prec j$, if $(i,j) \in E_\wedge$ and $t_i^\alpha = t_j^\alpha$ (for $i, j \in A$) or $t_i^\beta = t_j^\beta$ (for $i, j \in B$). (Recall that $E_\wedge \subseteq (A \times A) \cup (B \times B)$, so $(i,j) \in E_\wedge$ implies $i, j \in A$ or $i, j \in B$.) Break all other ties arbitrarily. Our main result shows that ordering jobs according to $\prec$ yields a feasible schedule and that the expected objective value of this schedule is at most 2Δ times the optimum.

Lemma 1. *Algorithm 1 is a randomized 2Δ-approximation for $1 \mid ao\text{-}prec = A \dot{\vee} B, p_j \in \{0,1\} \mid \sum w_j C_j$.*

For fixed $\overline{x}$ and $0 < \beta \leq 1$ we call the schedule that orders the jobs according to $\prec$ the *β-schedule of $\overline{x}$*. Given $\overline{x}$ and $0 < \beta \leq 1$, we can construct the β-schedule in time $\mathcal{O}(n)$. We derandomize Algorithm 1 by a simple observation similar to [9,17]. List all possible schedules that occur as β goes from 0 to 1, and pick the best one. The next lemma shows that the number of different β-schedules is not too large.

Lemma 2. *For every $\overline{x}$ there are $\mathcal{O}(n^2)$ different β-schedules.*

Lemmas 1 and 2 together prove the first part of Theorem 1. Note that for scheduling instances that are equivalent to MSVC, $\Delta = 2$. Hence, we immediately obtain a 4-approximation for these instances.

Corollary 1. *There is a 4-approximation algorithm for precedence-constrained MSVC.*

If we use an interval-indexed LP instead of a time-indexed LP, see e.g. [19,20], then Algorithm 1 can be generalized to arbitrary processing times. This will prove the second part of Theorem 1. Let $\varepsilon' > 0$, and recall that all processing times are non-negative integers. Let $T = \sum_{j \in N} p_j$ be the time horizon and L be minimal such that $(1+\varepsilon')^{L-1} \geq T$. Set $\tau_0 := 1$, and let $\tau_l = (1+\varepsilon')^{l-1}$ for every $l \in [L]$. We call $(\tau_{l-1}, \tau_l]$ the l-th interval for $l \in [L]$. (The first interval is the singleton $(1,1] := \{1\}$.) We introduce a binary variable x_{jl} for every $j \in N$ and for every $l \in [L]$ that indicates whether or not job j completes in the l-th interval. If we relax the integrality constraints on the variables we obtain the following relaxation:

$$\min \quad \sum_{b \in B} \sum_{l=1}^{L} w_b \cdot \tau_{l-1} \cdot x_{bl} \tag{2a}$$

$$\text{s.t.} \quad \sum_{l=1}^{L} x_{jl} = 1 \quad \forall j \in N, \tag{2b}$$

$$\sum_{j \in N} \sum_{k=1}^{l} p_j \, x_{jk} \leq \tau_l \quad \forall l \in [L], \tag{2c}$$

$$\sum_{k=1}^{l} x_{bk} - \sum_{a \in \mathcal{P}(b)} \sum_{k=1}^{l} x_{ak} \leq 0 \quad \forall b \in B : \mathcal{P}(b) \neq \emptyset, \ \forall l \in [L], \tag{2d}$$

$$\sum_{k=1}^{l} x_{jk} - \sum_{k=1}^{l} x_{ik} \leq 0 \quad \forall (i,j) \in E_\wedge, \ \forall l \in [L], \tag{2e}$$

$$x_{jl} \geq 0 \quad \forall j \in N, \ \forall l \in [L] : \tau_{l-1} \geq p_j. \tag{2f}$$

Given ε', the size of LP (2) is polynomial, so we can solve it in polynomial time. Again (2b) ensures that every job is executed. Constraints (2c) are valid for any feasible schedule, since the total processing time of all jobs that complete within the first l intervals cannot exceed τ_l. Constraints (2d) and (2e) model OR- and AND-precedence constraints, respectively.

Let $\overline{x}$ be an optimal fractional solution of LP (2), and let $\overline{C}_j = \sum_l \tau_{l-1} \overline{x}_{jl}$. Note that $\sum_j w_j \overline{C}_j$ is a lower bound on the optimal objective value of an integer solution, which is a lower bound on the optimal value of a feasible schedule. Let $l_j^\alpha = \min\{l \mid \sum_{k=1}^{l} \overline{x}_{jk} \geq \alpha\}$ be the α-interval of job $j \in N$. This generalizes the notion of α-points from before.

The algorithm for arbitrary processing times is similar to Algorithm 1. We call it Algorithm 2 and it works as follows. In order to achieve a $(2\Delta + \varepsilon)$-approximation, solve LP (2) with $\varepsilon' = \frac{\varepsilon}{2\Delta}$ and let $\overline{x}$ be an optimal solution. Then, draw β at random from the interval $(0, 1]$ with density function $f(\beta) = 2\beta$, and set $\alpha = \frac{\beta}{\Delta}$. Compute l_a^α and l_b^β for all jobs $a \in A$ and $b \in B$, respectively. Sort the jobs in order of non-decreasing values l_a^α $(a \in A)$ and l_b^β $(b \in B)$ and denote this total order by $\prec$. If $l_a^\alpha = l_b^\beta$ for some $b \in B$ and $a \in \mathcal{P}(b)$, set $a \prec b$. Similarly, set $i \prec j$, if $(i, j) \in E_\wedge$ and $l_i^\alpha = l_j^\alpha$ (for $i, j \in A$) or $l_i^\beta = l_j^\beta$ (for $i, j \in B$). Break all other ties arbitrarily. Finally, schedule the jobs in the order of $\prec$. Note that $\prec$ extends the order for α-points from Algorithm 1 to α-intervals.

Lemma 3. *For any $\varepsilon > 0$, Algorithm 2 is a randomized $(2\Delta+\varepsilon)$-approximation for $1 \mid ao\text{-}prec = A \dot{\vee} B \mid \sum w_j C_j$.*

We can derandomize Algorithm 2 similar to Lemma 2. Algorithms 1 and 2 can be further extended to release dates. To do so, we need to add constraints to LP (1) and LP (2) that ensure that no job completes too early. More precisely, fix $x_{jt} = 0$ for all $j \in N$ and $t < r_j + p_j$ in LP (1) and $x_{jl} = 0$ for all $j \in N$ and $\tau_{l-1} < r_j + p_j$ in LP (2), respectively. When scheduling the jobs according to $\prec$, we might have to add idle time in order to respect the release dates. This increases the approximation factor slightly.

Lemma 4. *There is a $(2\Delta + 2)$- and $(2\Delta + 2 + \varepsilon)$-approximation algorithm for $1 \mid r_j, ao\text{-}prec = A \dot{\vee} B, p_j \in \{0,1\} \mid \sum w_j C_j$ and $1 \mid r_j, ao\text{-}prec = A \dot{\vee} B \mid \sum w_j C_j$, respectively.*

3 The Generalized Min-Sum Set Cover Problem

Recall that we can model GMSSC as a single-machine scheduling problem to minimize the sum of weighted completion times with job set $N = A \dot{\cup} B$, processing times $p_j \in \{0, 1\}$, and certain precedence requirements $\kappa(b)$ for each job $b \in B$.

In this section, we prove Theorem 2. That is, we give a 4-approximation algorithm for the special case of GMSSC where $\kappa(b) = \max\{d(b) - 1, 1\}$ with $d(b) := |\mathcal{P}(b)|$ for all $b \in B$. So each job in B requires all but one of its predecessors to be completed before it can start, unless it has only one predecessor (*all-but-one MSSC*). Suppose we want to schedule a job $b \in B$ with $d(b) \geq 2$ at time $t \geq 0$. Then we need at least $d(b) - 1$ of its predecessors to be completed before t. Equivalently, for each pair of distinct $i, j \in \mathcal{P}(b)$ at most one of the two jobs i, j may complete after t. This gives the following linear relaxation with the same time-indexed variables as before and time horizon $T = \sum_{j \in N} p_j \leq n$.

$$\min \quad \sum_{b \in B} \sum_{t=1}^{T} w_b \cdot t \cdot x_{bt} \tag{3a}$$

$$\text{s.t.} \quad \sum_{t=1}^{T} x_{jt} = 1 \quad \forall j \in N, \tag{3b}$$

$$\sum_{j \in N} \sum_{s=t-p_j+1}^{t} x_{js} \leq 1 \quad \forall t \in [T], \tag{3c}$$

$$\sum_{s=1}^{t+p_b} x_{bs} - \sum_{s=1}^{t} (x_{is} + x_{js}) \leq 0 \quad \forall b \in B, \forall i, j \in \mathcal{P}(b), \forall t \in [T - p_b], \tag{3d}$$

$$\sum_{s=1}^{t+p_b} x_{bs} - \sum_{s=1}^{t} x_{is} \leq 0 \quad \forall b \in B : \mathcal{P}(b) = \{i\}, \forall t \in [T - p_b], \tag{3e}$$

$$x_{jt} \geq 0 \quad \forall j \in N, \forall t \in [T]. \tag{3f}$$

Constraints (3b) and (3c) again ensure that each job is processed and no jobs overlap, respectively. Note that only jobs with non-zero processing time contribute to (3c). If $d(b) = 1$, then (3e) dominates (3d). It ensures that the unique predecessor of $b \in B$ is completed before b starts. Note that this is a classical AND-precedence constraint which will not affect the approximation factor.

If $d(b) \geq 2$, then (3d) models the above observation. Suppose at most $d(b) - 2$ predecessors of b complete before time t. Then there are $i, j \in \mathcal{P}(b)$ such that $\sum_{s=1}^{t} (x_{is} + x_{js}) = 0 \geq \sum_{s=1}^{t} x_{bs}$, so b cannot complete by time t.

Note that we can solve LP (3) in polynomial time. Similar to the algorithms of Sect. 2, we first solve LP (3) and let $\bar{x}$ be an optimal fractional solution. We then draw β randomly from $(0, 1]$ with density function $f(\beta) = 2\beta$, and schedule the jobs in A and B in order of non-decreasing $\frac{\beta}{2}$-points and β-points, respectively. Again, we break ties consistently with precedence constraints. (Choosing $\frac{\beta}{2}$ for the jobs in A ensures that at most one of the predecessors of a job $b \in B$ is scheduled after b in the constructed schedule.) This algorithm is called Algorithm 3.

Lemma 5. *Algorithm 3 is a randomized 4-approximation for all-but-one MSSC.*

One can derandomize Algorithm 3 similar to Lemma 2, which proves Theorem 2. Note that Algorithm 3 also works if jobs in B have unit processing time. It can be generalized to release dates and arbitrary processing times, if we use an interval-indexed formulation similar to LP (2). If we choose $\varepsilon' = \frac{\varepsilon}{4}$ and solve the corresponding interval-indexed formulation instead of LP (3), then Algorithm 3 is a $(4 + \varepsilon)$-approximation for any $\varepsilon > 0$. Again, AND-precedence constraints do not affect the approximation factor, similar to Lemmas 1 and 3.

4 Integrality Gaps for Other LP Relaxations

In this section, we analyze other standard linear programming relaxations that have been useful for various scheduling problems, and show that they fail in the presence of OR-precedence constraints. More precisely, we show that the natural LPs in linear ordering variables (Sect. 4.1) and completion time variables (Sect. 4.2) both exhibit integrality gaps that are linear in the number of jobs, even on instances where $E_\wedge = \emptyset$ and $\Delta = 2$.

4.1 Linear Ordering Formulation

The following relaxation for single-machine scheduling problems was proposed by Potts [32]. It is based on linear ordering variables δ_{ij}, which indicate whether job i precedes job j ($\delta_{ij} = 1$) or not ($\delta_{ij} = 0$). This LP has played an important role in better understanding Sidney's decomposition [11,36], and in uncovering the connection between AND-scheduling and vertex cover [1,2,10,11]. A nice feature of this formulation is that we can model OR-precedence constraints in a very intuitive way with constraints $\sum_{a\in\mathcal{P}(b)} \delta_{ab} \geq 1$ for all $b \in B$. Together with the total ordering constraints ($\delta_{ij} + \delta_{ji} = 1$), standard transitivity constraints ($\delta_{ij} + \delta_{jk} + \delta_{ki} \geq 1$) and AND-precedence constraints ($\delta_{ij} = 1$) we thus obtain a polynomial size integer program for $1 \,|\, \textit{ao-prec} = A \dot\vee B \,|\, \sum w_j C_j$. The LP-relaxation is obtained by relaxing the integrality constraints to $\delta_{ij} \geq 0$.

$$
\begin{aligned}
\min \quad & \sum_{j\in N}\sum_{i\in N} w_j p_i \delta_{ij} && \text{(4a)}\\
\text{s.t.} \quad & \delta_{ij} + \delta_{ji} = 1 \quad \forall\, i,j \in N : i \neq j && \text{(4b)}\\
& \delta_{ij} + \delta_{jk} + \delta_{ki} \geq 1 \quad \forall\, i,j,k \in N && \text{(4c)}\\
& \sum_{a\in\mathcal{P}(b)} \delta_{ab} \geq 1 \quad \forall\, b \in B : \mathcal{P}(b) \neq \emptyset, && \text{(4d)}\\
& \delta_{ij} = 1 \quad \forall\, (i,j) \in E_\wedge, && \text{(4e)}\\
& \delta_{ii} = 1 \quad \forall\, i \in N, && \text{(4f)}\\
& \delta_{ij} \geq 0 \quad \forall\, i,j \in N. && \text{(4g)}
\end{aligned}
$$

We set $\delta_{ii} = 1$ in (4f) so the completion time of job j is $C_j = \sum_i p_i \delta_{ij}$. Note that every feasible single-machine schedule without idle time corresponds to a feasible integer solution of LP (4), and vice versa. If $E_\vee = \emptyset$, i.e., $\mathcal{P}(b) = \emptyset$ for all $b \in B$, then this relaxation has an integrality gap of 2 (lower and upper bound of 2 due to [8] and [34], respectively). However, in the presence of OR-precedence constraints, the gap of LP (4) grows linearly in the number of jobs, even if $E_\wedge = \emptyset$ and $\Delta = 2$.

Lemma 6. *There is a family of instances such that the integrality gap of LP (4) is $\Omega(n)$.*

The instances of Lemma 6 consist of copies of an instance on three jobs, as illustrated in Fig. 2. Note that these satisfy $|\mathcal{P}(b)| \leq 2$ for all $b \in B$. For this

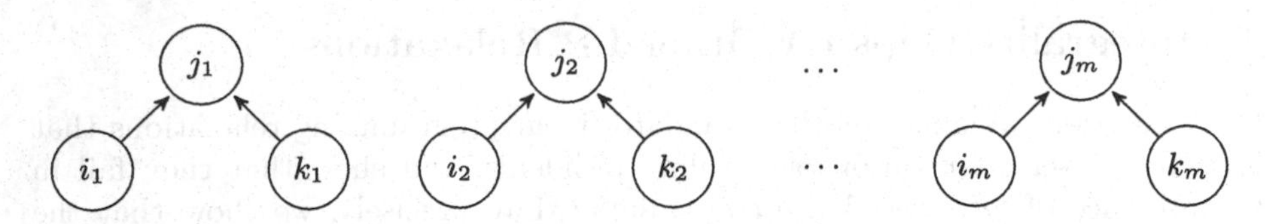

Fig. 2. Instance for which LP (4) exhibits an integrality gap that is linear in the number of jobs. The processing times and weights are $p_{j_q} = 0$, $p_{i_q} = p_{k_q} = 1$ and $w_{j_q} = 1$, $w_{i_q} = w_{k_q} = 0$ for all $q \in [m]$.

special case, we present facet-defining inequalities. If $|\mathcal{P}(b)| \leq 2$ for all $b \in B$, then LP (4) can be reformulated as

$$\begin{aligned}
\min \quad & \sum_{j \in N} \sum_{i \in N} w_j p_i \delta_{ij} && && \text{(5a)}\\
\text{s.t.} \quad & \delta_{ij} + \delta_{ji} = 1 && \forall i, j \in N : i \neq j, && \text{(5b)}\\
& \delta_{ij} + \delta_{jk} + \delta_{ki} \geq 1 && \forall i, j, k \in N, && \text{(5c)}\\
& \delta_{ab} + \delta_{a'b} \geq 1 && \forall b \in B : \mathcal{P}(b) = \{a, a'\}, && \text{(5d)}\\
& \delta_{ij} = 1 && \forall (i, j) \in E_\wedge, \text{ or } \mathcal{P}(j) = \{i\}, && \text{(5e)}\\
& \delta_{ii} = 1 && \forall i \in N, && \text{(5f)}\\
& \delta_{ij} \geq 0 && \forall i, j \in N. && \text{(5g)}
\end{aligned}$$

Note that constraints (5b), (5c), and (5f) coincide with the corresponding constraints in LP (4). Constraints (5d) model the OR-precedence constraints for jobs $b \in B$ with $|\mathcal{P}(b)| = 2$. For $b \in B$ with $|\mathcal{P}(b)| = 1$, the corresponding OR-precedence constraint is equivalent to an AND-constraint and is included in (5e).

Theorem 4. *For all $b \in B$ and $\mathcal{P}(b) = \{a, a'\}$, the constraints*

$$\delta_{aa'} + \delta_{a'b} \geq 1 \tag{6}$$

are valid for the integer hull of LP (5). Moreover, if they are tight, then they are either facet-defining or equality holds for all feasible integer solutions of LP (5).

One can verify that the integrality gap of LP (5) remains linear for the example in Fig. 2, even if we add constraints (6). Recall that in GMSSC each job $b \in B$ requires at least $\kappa(b) \in [|\mathcal{P}(b)|]$ of its predecessors to be completed before it can start. This can also be easily modeled with linear ordering variables by introducing a constraint $\sum_{a \in \mathcal{P}(b)} \delta_{ab} \geq \kappa(b)$. However, note that the instance in Fig. 2 is an instance of MSVC (which is a special case of MSSC and all-but-one MSSC). So already for $\kappa(b) = 1$ or $\kappa(b) = \max\{|\mathcal{P}(b)| - 1, 1\}$ and $\Delta = 2$ this formulation has an unbounded integrality gap.

4.2 Completion Time Formulation

The LP relaxation examined in this section contains one variable C_j for every job $j \in N$ that indicates the completion time of this job. In the absence of precedence constraints, the convex hull of all feasible completion time vectors can be fully described by the set of vectors $\{C \in \mathbb{R}^n \mid \sum_{j\in S} p_j C_j \geq f(S)\ \forall S \subseteq N\}$, where $f(S) := \frac{1}{2}\left(\sum_{j\in S} p_j\right)^2 + \frac{1}{2}\sum_{j\in S} p_j^2$ is a supermodular function [33,41]. One should note that, although there is an exponential number of constraints, one can separate them efficiently [33]. That is, one can solve $\min \sum_{j\in N} w_j C_j$ subject to $\sum_{j\in S} p_j C_j \geq f(S)$ for all $S \subseteq N$ in polynomial time. In the presence of AND-precedence constraints, Schulz [34] proposed the first 2-approximation algorithm. The algorithm solves the corresponding linear program with additional constraints $C_j \geq C_i + p_j$ for $(i,j) \in E_\wedge$ and schedules the jobs in non-decreasing order of their LP-values.

For OR-precedence constraints, we use the concept of *minimal chains*, see e.g. [21], to generalize the parallel inequalities of [33,41]. More specifically, we present a class of inequalities that are valid for all feasible completion time vectors of an instance of $1 \mid ao\text{-}prec = A \dot\vee B \mid \sum w_j C_j$, and that, in the absence of precedence constraints, coincide with the parallel inequalities. We add inequalities for AND-precedence constraints in the obvious way, $C_j \geq C_i + p_j$ for $(i,j) \in E_\wedge$, so we assume $E_\wedge = \emptyset$ for the moment.

We call $S \subseteq N$ a *feasible starting set*, if we can schedule the jobs in S without violating any OR-precedence constraints. The set of feasible starting sets is denoted by $\mathcal{S}$. That is, $S \in \mathcal{S}$, if $j \in B \cap S$ implies that $\mathcal{P}(j) \cap S \neq \emptyset$. The length of a minimal chain of a job k w.r.t. a set $S \subseteq N$ is defined as

$$mc(S,k) := \min\{\sum_{j\in T} p_j \mid T \subseteq N : \exists U \subseteq S \cup T \text{ with } k \in U \in \mathcal{S}\}. \tag{7}$$

Intuitively, the value $mc(S,k)$ is the minimal amount of time that we need to schedule job k in a feasible way, if we can schedule the jobs in S for free, i.e. if we assume all jobs in S have zero processing time. Let 2^N be the power set of N. For all $k \in N$, we define a set function $f_k(S) : 2^N \to \mathbb{R}_{\geq 0}$ via

$$f_k(S) := \frac{1}{2}\left(\sum_{j\in S} p_j + mc(S,k)\right)^2 + \frac{1}{2}\left(\sum_{j\in S} p_j^2 + mc(S,k)^2\right). \tag{8}$$

Note that if $k \in S \in \mathcal{S}$, then $mc(S,k) = 0$, so $f_k(S) = f(S)$. In particular, (8) generalizes the function $f : 2^N \to \mathbb{R}_{\geq 0}$ of [33,41] to OR-precedence constraints.[2]

Theorem 5. *For any $k \in N$ and $S \subseteq N$ the inequality*

$$\sum_{j\in S} p_j C_j + mc(S,k)\, C_k \geq f_k(S) \tag{9}$$

is valid for all feasible completion time vectors. If there is $T \in \operatorname{argmin}(mc(S,k))$ such that $S \cup T \in \mathcal{S}$ is a feasible starting set, then (9) is tight.

[2] One can also show that $mc(\cdot,k)$ and $f_k(\cdot)$ are supermodular for any k.

Theorem 5 suggests the following natural LP-relaxation for $1\,|\,\textit{ao-prec} = A \dot{\vee} B \,|\, \sum w_j C_j$:

$$\min \quad \sum_{j \in N} w_j C_j \tag{10a}$$

$$\text{s.t.} \quad \sum_{j \in S} p_j C_j + mc(S,k)\, C_k \geq f_k(S) \qquad \forall\, k \in N,\ \forall\, S \subseteq N, \tag{10b}$$

$$C_j - C_i \geq p_j \qquad \forall\, (i,j) \in E_\wedge. \tag{10c}$$

Note that it is not clear how to separate constraints (10b). The gap of LP (10) can grow linearly in the number of jobs, even for instances of $1\,|\,\textit{ao-prec} = A \dot{\vee} B \,|\, \sum w_j C_j$ with $E_\wedge = \emptyset$ and $\Delta = 2$, see Fig. 3.

Lemma 7. *There is a family of instances such that the gap between an optimal solution for LP (10) and an optimal schedule is $\Omega(n)$.*

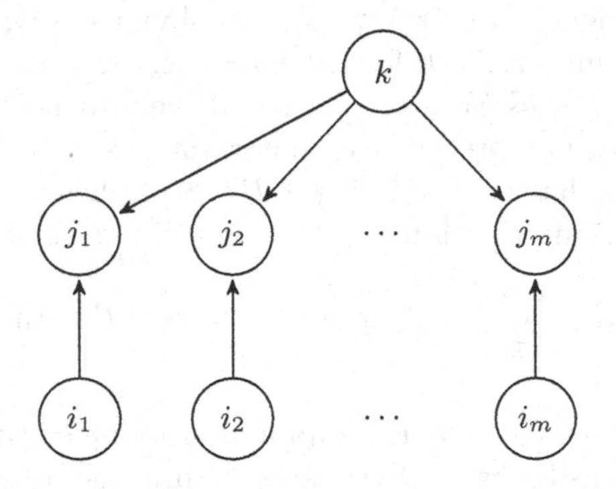

Fig. 3. Sketch of instance for which LP (10) exhibits a gap that is linear in the number of jobs. The processing times and weights are $p_k = \frac{m}{2}$, $w_k = w_{i_q} = p_{j_q} = 0$, and $w_{j_q} = p_{i_q} = 1$ for all $q \in [m]$.

5 Conclusion

In this extended abstract, we analyze single-machine scheduling problems with certain AND/OR-precedence constraints that are extensions of min-sum set cover, precedence-constrained min-sum set cover, pipelined set cover, minimum latency set cover, and set cover. Using machinery from the scheduling context, we derive new approximation algorithms that rely on solving time-indexed linear programming relaxations and scheduling jobs according to random α-points. In a nutshell, one may say that the new key technique is to choose the value of α for jobs in A dependent on the corresponding β-value of jobs in B. This observation allows us also to derive the best constant-factor approximation algorithm for an interesting special case of the generalized min-sum set cover problem—the all-but-one MSSC problem—which in itself is a generalization of min-sum vertex cover. This 4-approximation algorithm may further support the conjecture of Im et al. [23], namely that GMSSC is 4-approximable.

It is easy to see that one can also include AND-precedence constraints between jobs in A and B, i.e., allow $E_\wedge \subseteq (A \times N) \cup (B \times B)$. This does not affect the approximation guarantees or feasibility of the constructed schedules, since $\alpha \leq \beta$ and constraints (1e) imply $t_a^\alpha \leq t_b^\beta$ for $(a,b) \in E_\wedge$. Similarly, $l_a^\alpha \leq l_b^\beta$ for $(a,b) \in E_\wedge$ follows from (2e). Note that it is not clear whether the analysis of the algorithms in Sects. 2 and 3 are tight.

Besides deriving approximation algorithms based on time-indexed LPs, we analyze other standard LP relaxations, namely linear ordering and completion time formulations. These relaxations facilitated research on scheduling with AND-precedence constraints, see e.g. [1,2,10,11,19,32–34]. For the integer hull of the linear ordering relaxation we present a class of facet-defining valid inequalities and we generalize the well-known inequalities of [33,41] for the completion time relaxation. We show that, despite these additional constraints, both relaxations exhibit linear integrality gaps, even if $\Delta = 2$ and $E_\wedge = \emptyset$. Thus, unless one identifies stronger valid inequalities, these formulations seem to fail as soon as OR-precedence constraints are incorporated. In view of the integrality gaps in Sect. 4, it would be interesting to obtain stronger bounds on the integrality gap of the time-indexed formulation considered in Sects. 2 and 3.

References

1. Ambühl, C., Mastrolilli, M.: Single machine precedence constrained scheduling is a vertex cover problem. Algorithmica **53**(4), 488–503 (2009). https://doi.org/10.1007/s00453-008-9251-6
2. Ambühl, C., Mastrolilli, M., Mutsanas, N., Svensson, O.: On the approximability of single-machine scheduling with precedence constraints. Math. Oper. Res. **36**(4), 653–669 (2011). https://doi.org/10.1287/moor.1110.0512
3. Azar, Y., Gamzu, I., Yin, X.: Multiple intents re-ranking. In: Proceedings of the 41st Annual ACM Symposium on Theory of Computing, pp. 669–678. ACM (2009). https://doi.org/10.1145/1536414.1536505
4. Bansal, N., Gupta, A., Krishnaswamy, R.: A constant factor approximation algorithm for generalized min-sum set cover. In: Proceedings of the 21st Annual ACM-SIAM Symposium on Discrete Algorithms, pp. 1539–1545. SIAM (2010). https://doi.org/10.1137/1.9781611973075.125
5. Bansal, N., Khot, S.: Optimal long code test with one free bit. In: Proceedings of the 50th Annual IEEE Symposium on Foundations of Computer Science, pp. 453–462. IEEE (2009). https://doi.org/10.1109/FOCS.2009.23
6. Bar-Noy, A., Bellare, M., Halldórsson, M.M., Shachnai, H., Tamir, T.: On chromatic sums and distributed resource allocation. Inf. Comput. **140**(2), 183–202 (1998). https://doi.org/10.1006/inco.1997.2677
7. Charikar, M., Naamad, Y., Wirth, A.: On approximating target set selection. In: Approximation, Randomization, and Combinatorial Optimization. Algorithms and Techniques. Leibniz International Proceedings in Informatics (LIPIcs), vol. 60, pp. 4:1–4:16 (2016). https://doi.org/10.4230/LIPIcs.APPROX-RANDOM.2016.4

8. Chekuri, C., Motwani, R.: Precedence constrained scheduling to minimize sum of weighted completion times on a single machine. Discrete Appl. Math. **98**(1–2), 29–38 (1999). https://doi.org/10.1016/S0166-218X(98)00143-7
9. Chekuri, C., Motwani, R., Natarajan, B., Stein, C.: Approximation techniques for average completion time scheduling. SIAM J. Comput. **31**(1), 146–166 (2001). https://doi.org/10.1137/S0097539797327180
10. Chudak, F.A., Hochbaum, D.S.: A half-integral linear programming relaxation for scheduling precedence-constrained jobs on a single machine. Oper. Res. Lett. **25**(5), 199–204 (1999). https://doi.org/10.1016/S0167-6377(99)00056-5
11. Correa, J.R., Schulz, A.S.: Single-machine scheduling with precedence constraints. Math. Oper. Res. **30**(4), 1005–1021 (2005). https://doi.org/10.1287/moor.1050.0158
12. Dinur, I., Steurer, D.: Analytical approach to parallel repetition. In: Proceedings of the 46th Annual ACM Symposium on Theory of Computing, pp. 624–633. ACM (2014). https://doi.org/10.1145/2591796.2591884
13. Erlebach, T., Kääb, V., Möhring, R.H.: Scheduling AND/OR-networks on identical parallel machines. In: Solis-Oba, R., Jansen, K. (eds.) WAOA 2003. LNCS, vol. 2909, pp. 123–136. Springer, Heidelberg (2004). https://doi.org/10.1007/978-3-540-24592-6_10
14. Feige, U., Lovász, L., Tetali, P.: Approximating min-sum set cover. In: Jansen, K., Leonardi, S., Vazirani, V. (eds.) APPROX 2002. LNCS, vol. 2462, pp. 94–107. Springer, Heidelberg (2002). https://doi.org/10.1007/3-540-45753-4_10
15. Feige, U., Lovász, L., Tetali, P.: Approximating min sum set cover. Algorithmica **40**(4), 219–234 (2004). https://doi.org/10.1007/s00453-004-1110-5
16. Goemans, M.X.: Cited as personal communication in [35] (1996)
17. Goemans, M.X., Queyranne, M., Schulz, A.S., Skutella, M., Wang, Y.: Single machine scheduling with release dates. SIAM J. Discrete Math. **15**(2), 165–192 (2002). https://doi.org/10.1137/S089548019936223X
18. Graham, R.L., Lawler, E.L., Lenstra, J.K., Rinnooy Kan, A.H.G.: Optimization and approximation in deterministic sequencing and scheduling: a survey. In: Annals of Discrete Mathematics, vol. 5, pp. 287–326. Elsevier (1979). https://doi.org/10.1016/S0167-5060(08)70356-X
19. Hall, L.A., Schulz, A.S., Shmoys, D.B., Wein, J.: Scheduling to minimize average completion time: off-line and on-line approximation algorithms. Math. Oper. Res. **22**(3), 513–544 (1997). https://doi.org/10.1287/moor.22.3.513
20. Hall, L.A., Shmoys, D.B., Wein, J.: Scheduling to minimize average completion time: off-line and on-line algorithms. In: Proceedings of the 7th Annual ACM-SIAM Symposium on Discrete Algorithms, pp. 142–151. SIAM (1996)
21. Happach, F.: Makespan minimization with OR-precedence constraints. arXiv preprint arXiv:1907.08111 (2019)
22. Hochbaum, D.S.: Approximation algorithms for the set covering and vertex cover problems. SIAM J. Comput. **11**(3), 555–556 (1982). https://doi.org/10.1137/0211045
23. Im, S., Sviridenko, M., van der Zwaan, R.: Preemptive and non-preemptive generalized min sum set cover. Math. Program. **145**(1–2), 377–401 (2014). https://doi.org/10.1007/s10107-013-0651-2
24. Johannes, B.: On the complexity of scheduling unit-time jobs with OR-precedence constraints. Oper. Res. Lett. **33**(6), 587–596 (2005). https://doi.org/10.1016/j.orl.2004.11.009
25. Johnson, D.S.: Approximation algorithms for combinatorial problems. J. Comput. Syst. Sci. **9**(3), 256–278 (1974). https://doi.org/10.1016/S0022-0000(74)80044-9

26. Khot, S.: On the power of unique 2-prover 1-round games. In: Proceedings of the 34th Annual ACM Symposium on Theory of Computing, pp. 767–775. ACM (2002). https://doi.org/10.1145/509907.510017
27. Lenstra, J.K., Rinnooy Kan, A.H.G.: Complexity of scheduling under precedence constraints. Oper. Res. **26**(1), 22–35 (1978). https://doi.org/10.1287/opre.26.1.22
28. Lovász, L.: On the ratio of optimal integral and fractional covers. Discrete Math. **13**(4), 383–390 (1975). https://doi.org/10.1016/0012-365X(75)90058-8
29. Margot, F., Queyranne, M., Wang, Y.: Decompositions, network flows, and a precedence constrained single-machine scheduling problem. Oper. Res. **51**(6), 981–992 (2003). https://doi.org/10.1287/opre.51.6.981.24912
30. McClintock, J., Mestre, J., Wirth, A.: Precedence-constrained min sum set cover. In: 28th International Symposium on Algorithms and Computation. Leibniz International Proceedings in Informatics (LIPIcs), vol. 92, pp. 55:1–55:12 (2017). https://doi.org/10.4230/LIPIcs.ISAAC.2017.55
31. Munagala, K., Babu, S., Motwani, R., Widom, J.: The pipelined set cover problem. In: Eiter, T., Libkin, L. (eds.) ICDT 2005. LNCS, vol. 3363, pp. 83–98. Springer, Heidelberg (2004). https://doi.org/10.1007/978-3-540-30570-5_6
32. Potts, C.N.: An algorithm for the single machine sequencing problem with precedence constraints. Math. Program. Study **13**, 78–87 (1980)
33. Queyranne, M.: Structure of a simple scheduling polyhedron. Math. Program. **58**(1–3), 263–285 (1993). https://doi.org/10.1007/BF01581271
34. Schulz, A.S.: Scheduling to minimize total weighted completion time: performance guarantees of LP-based heuristics and lower bounds. In: Cunningham, W.H., McCormick, S.T., Queyranne, M. (eds.) IPCO 1996. LNCS, vol. 1084, pp. 301–315. Springer, Heidelberg (1996). https://doi.org/10.1007/3-540-61310-2_23
35. Schulz, A.S., Skutella, M.: Random-based scheduling new approximations and LP lower bounds. In: Rolim, J. (ed.) RANDOM 1997. LNCS, vol. 1269, pp. 119–133. Springer, Heidelberg (1997). https://doi.org/10.1007/3-540-63248-4_11
36. Sidney, J.B.: Decomposition algorithms for single-machine sequencing with precedence relations and deferral costs. Oper. Res. **23**(2), 283–298 (1975). https://doi.org/10.1287/opre.23.2.283
37. Skutella, M., Williamson, D.P.: A note on the generalized min-sum set cover problem. Oper. Res. Lett. **39**(6), 433–436 (2011). https://doi.org/10.1016/j.orl.2011.08.002
38. Smith, W.E.: Various optimizers for single-stage production. Nav. Res. Logist. Q. **3**(1–2), 59–66 (1956). https://doi.org/10.1002/nav.3800030106
39. Sousa, J.P., Wolsey, L.A.: A time indexed formulation of non-preemptive single machine scheduling problems. Math. Program. **54**(1–3), 353–367 (1992). https://doi.org/10.1007/BF01586059
40. Woeginger, G.J.: On the approximability of average completion time scheduling under precedence constraints. Discrete Appl. Math. **131**(1), 237–252 (2003). https://doi.org/10.1016/S0166-218X(02)00427-4
41. Wolsey, L.A.: Mixed integer programming formulations for production planning and scheduling problems. Invited Talk at the 12th International Symposium on Mathematical Programming (1985)

Fault Tolerant Clustering with Outliers

Tanmay Inamdar(✉) and Kasturi Varadarajan

Department of Computer Science, The University of Iowa, Iowa City, USA
{tanmay-inamdar,kasturi-varadarajan}@uiowa.edu

Abstract. In a clustering with outliers problem, we are required to cluster all but a specified number of points, called outliers. In a fault tolerant clustering problem, the objective function incorporates the distance of a point to its f-th closest center chosen in the solution. We combine these two orthogonal generalizations, and consider Fault Tolerant Clustering with Outliers problems for various clustering objectives, such as k-center, k-median, and sum of radii. We essentially reduce the Fault Tolerant Clustering with Outliers problem, to the corresponding (non Fault Tolerant) Clustering with Outliers problem, for which constant approximations are known. This can be seen as a generalization of the framework of Kumar and Raichel [20] to handle the presence of outliers. This reduction comes at a loss in the approximation guarantee; however, we show that it is bounded by $O(1)$ for the k-center objective, whereas it is $O(f)$ for k-median and sum of radii objectives, where f is the degree of Fault Tolerance required in the solution. This implies $O(1)$ and $O(f)$ approximations for these generalizations respectively.

1 Introduction

We take the k-center problem as a running example for the initial part of this discussion. In the k-center problem, we are given a set X of n points, and a distance function d that satisfies the metric properties. The objective is to select a set $C \subseteq X$ of at most k $(1 \leq k \leq n)$ *centers*, that minimizes the maximum distance of any point in X to its nearest center in C.

A small number of *outliers* may skew the performance of clustering algorithms. However, if one is willing to ignore these outliers, the resulting clustering is much better; which is reflected by much smaller cost of the solution. In the context of k-center, this issue is captured by a generalization called the k-center with outliers problem. Charikar et al. [8] introduced the outlier generalizations of various clustering problems.[1] Here, we are given an additional parameter p, which specifies the coverage requirement. Now, the objective function is defined

[1] Charikar et al. [8] use the adjective "robust" for the generalization described here, and use the term "clustering with outliers" for the prize-collecting versions. Nevertheless, we adopt the aforementioned convention, which is otherwise standard in the literature.

Supported by the National Science Foundation under grant CCF-1615845.

E. Bampis and N. Megow (Eds.): WAOA 2019, LNCS 11926, pp. 188–201, 2020.
https://doi.org/10.1007/978-3-030-39479-0_13

with respect to p $(1 \leq p \leq n)$ closest points to the set of chosen centers. Formally, we want to choose a set $X' \subseteq X$ of size (at least) p, and a set $C \subseteq X$ of at most k centers, such that the maximum distance of any point in X' to its nearest center in C is minimized. Here, the set $X \setminus X'$ of size at most $n-p$ can be thought of as *outliers*. It is easy to generalize other clustering objectives in a similar way.

Another orthogonal issue in the domain of clustering problems is that of Fault Tolerance. In a given clustering problem (without outliers), the "service cost" of each point is the distance to its closest center among the set of chosen centers. However, if some centers undergo failure, there may be a large increase in the service costs for some of the points. This issue is modeled by the fault tolerant k-center problem ([20]). Suppose, the solution is required to tolerate failure of at most $f-1$ centers in the solution, where $1 \leq f \leq k$ is a parameter. We define the service cost of a point in X as its distance to the f-th closest center in C. In the Fault Tolerant k-center problem, we want to find a set C of at most k centers that minimizes the maximum service cost of any point in X. Note that now we have a bound on the service cost of any point, even if any set of at most $f-1$ centers fail.

We consider a common generalization that we call Fault Tolerant Clustering with Outliers (FTCwO). Here, the goal is to find a Fault Tolerant clustering for all the points in X, excluding a specified number of outliers. In the Fault Tolerant k-center with Outliers problem, we are required to find a set $X' \subseteq X$ of at least p points and a set $C \subseteq X$ of at most k centers, such that the maximum distance of any point in X' to its f-th closest center in C is minimized. The formal definitions of the FTCwO versions for various clustering objectives are given in the next section. But first, we discuss the related work.

1.1 Related Work

It is unsurprising that the clustering problems considered in this paper—even without Fault Tolerance and Outliers—are NP-hard. We start with the k-center problem. There are simple 2-approximations for this problem [12], and it is well-known that obtaining an approximation ratio of $2-\varepsilon$ is NP-hard from a simple reduction from the Minimum Dominating Set problem [15]. Charikar et al. [8], who introduced the k-center with outliers problem, gave a 3-approximation via a simple greedy algorithm. This was recently improved to the optimal approximation ratio of 2 [6,14]. Several constant factor approximations are known for the Fault Tolerant k-center problem [17,20], and closely related variants thereof [9,19].

k-median is another popular clustering objective function, where we seek to minimize the sum of distances from each point to its closest center. The k-median problem is also widely studied: the current best approximation ratio is $2.675+\varepsilon$ [5], and there is a known lower bound of 1.736 [16]. The first constant approximation for the k-median with outliers was given by Chen [10], but the approximation ratio was a large unspecified constant. This was recently improved by Krishnaswamy et al. [18] to $53.004+\varepsilon$ using an iterative rounding algorithm, and they also extended it to k-means with outliers and other related problems. Constant factor approximations for Fault Tolerant k-median problem are also known ([20] and [13,24]). However, we note that the notion of Fault Tolerance in [13,24] is more general compared to that of ours—they allow varying degrees of Fault Tolerance for each client.

Sum of radii is yet another well-studied clustering objective. Here, we have a set of clients and a set of servers. The goal is to open some balls centered at (a subset of) servers, such that each client belongs to at least one opened ball, while minimizing the sum of radii of the opened balls. A slight generalization asks to minimize the sum of α-th powers of radii, for some fixed $\alpha \geq 1$. Another related problem, called k-clustering, imposes a restriction of using at most k balls in the solution. Constant approximations for these problems were given by Charikar and Panigrahy [7]. There are subsequent results that give better (possibly bicriteria) guarantees in quasi-polynomial time for these problems [2, 11]. A constant approximation for the k-clustering with outliers problem was given by Ahmadian and Swamy [1][2]. Constant approximations for the Fault Tolerant generalizations of these problems were given by Bhowmick et al. [4].

Finally, we note that there is a generalization of the well-known Set Cover problem, called Partial Set Multi-Cover, which is a covering analog of the problems considered here. There has been some recent work on this problem [22,23]; however their approximation guarantees can be polynomial in n in general, where n is the number of elements in the set system. This is unsurprising, since even the very restricted case of 2-Fault Tolerant Covering of edges with Outliers in an undirected graph is closely related to the Densest k-subgraph problem. Assuming ETH, there is an almost-polynomial lower bound on the approximation ratio for this problem [21], and the best known upper bound is $O(n^{1/4+\varepsilon})$ [3]. On the other hand, clustering problems can be thought of as "soft" covering problems, which intuitively explains why we can obtain constant approximation guarantees in many cases.

In the next section, we set up notation, define the problems considered in the paper, and give a brief overview of our general technique. In the subsequent sections, we give our approximation algorithms for various fault tolerant clustering with outliers objectives.

2 Problem Statements and Results

Notation. Let (X, d) be a metric space. For a point $x \in X$, a set $Z \subseteq X$, and an integer $1 \leq i \leq |Z|$, let $\mathsf{nn}_i(x, Z, d)$ denote the i-th point $z \in Z$, in the non-decreasing order of distance from x. Here, we assume the ties are broken according to an arbitrary global ordering. Furthermore, let $\mathsf{NN}_i(x, Z, d) := \bigcup_{j=1}^{i} \mathsf{nn}_j(x, Z, d)$. Finally, let $d_i(x, Z) := d(x, \mathsf{nn}_i(x, Z, d))$. The ball centered at $x \in X$ of radius r is $B_d(x, r) := \{y \in X \mid d(x, y) \leq r\}$. Finally, when the distance function d is obvious from the context, we may omit it from these notations.

[2] In fact, Ahmadian and Swamy [1] give a constant factor approximation for a generalization of k-clustering with lower bounds and outliers.

2.1 Problem Statements

We denote a generic FTCwO instance by $\mathcal{I} = (X, Y, d, k, p, f)$. Here, X is a set of points (or clients), Y is a set of centers (or servers), in a metric space $(X \cup Y, d)$. We also have $1 \leq p \leq |X|$, and $1 \leq f \leq k \leq |Y|$. The objective of a generic FTCwO problem pr is to find a solution $\sigma_{\mathsf{pr}}(\mathcal{I})$ consisting of $X' \subseteq X$ and $Y' \subseteq Y$ such that:

1. $|X'| \geq p$,
2. $|Y'| \leq k$,
3. A problem-specific objective function $\mu_{\mathsf{pr}}(\cdot)$ is minimized, over all feasible solutions σ_{pr}.

The problem objective $\mu_{\mathsf{pr}}(\cdot)$ can be defined differently to model different FTCwO problems. A solution is said to be feasible if it satisfies Conditions 1 and 2.

For a problem pr, let $\mathsf{opt}_{\mathsf{pr}}(\mathcal{I})$ denote an optimal solution for an instance $\mathcal{I}$. A solution $\sigma_{\mathsf{pr}}(\mathcal{I})$ is an α-approximate solution if $\mu_{\mathsf{pr}}(\sigma_{\mathsf{pr}}(\mathcal{I})) \leq \alpha \cdot \mu_{\mathsf{pr}}(\mathsf{opt}_{\mathsf{pr}}(\mathcal{I}))$. We will omit the subscript pr in the subsequent sections, since the problem will be clear from the context.

We now specify the objective functions μ of the FTCwO versions of the problems in the notation defined above.

- k-center: $\max_{x \in X'} d_f(x, Y')$
- k-median (v1): $\sum_{x \in X'} d_f(x, Y')$
- k-median (v2): $\sum_{x \in X'} \sum_{i=1}^{f} d_i(x, Y')$
- k-clustering: $\sum_{y \in Y'} r(y)$
  Here, $r : Y' \to \mathbb{R}^+$ is a "radius function" such that for any $x \in X'$, $|\{y \in Y' \mid d(x, y) \leq r(y)\}| \geq f$. That is, every point in X' belongs to at least f balls.

We note that two different objectives have been studied under the name *Fault Tolerant k-median*. The first version (v1) was studied by [20], whereas (a generalization of) the second one (v2) was studied by [13,24]. We also note that for technical reasons, for the first three problems, we require that $Y = X$, i.e., the set of clients and servers be the same. For the k-clustering problem, however, our algorithm also works in the general case where Y may not be the same as X.

2.2 Our Results and Technique

For all the problems we consider, we know efficient constant factor approximation algorithms when $f = 1$, i.e., the solutions are not required to be Fault Tolerant. Our contribution is to show that with simple pre- and post-processing steps, we can essentially reduce FTCwO problems, to the corresponding Clustering with Outliers problems, at the expense of increase in the approximation factors. For the k-center and the k-median (v1) objective functions, the increase in the approximation factor is constant. On the other hand, for k-median (v2) and k-clustering objective functions, the increase in the approximation factor is $O(f)$. We summarize our results in the following table. These results are described in Sects. 3 to 6 respectively.

Table 1. Known constant factor approximations for Clustering with Outliers problems (second column) and our approximations for the corresponding FTCwO version (third column).

Problem	Outliers approx.	Fault tolerance with outliers approx.
k-center	2 ([6,14])	6
k-median (v1)	7.081 ([18])	35.405
k-median (v2)	7.081 ([18])	$28.324 \cdot f + 7.081$
k-clustering	12.365 ([1])	$49.46 \cdot f + 12.365$

Our general strategy for finding an approximate solution for a FTCwO instance $\mathcal{I} = (X, Y, d, k, p, f)$ is as follows. Let $m := \lfloor k/f \rfloor$, and let $\mathcal{I}' = (X, Y, d', m, p, 1)$ be a related instance. Notice that in the instance $\mathcal{I}'$, the cardinality constraint has been reduced from k to $\lfloor k/f \rfloor$, and the distance function d' may be different from d. Furthermore, observe that a solution for $\mathcal{I}'$ is not required to be Fault Tolerant, i.e., $\mathcal{I}'$ is an instance of a Clustering with Outliers problem, for which different constant factor approximation algorithms are known (see Table 1). We use such an algorithm as a black box to obtain an approximate solution (P, Q) such that: (1) $P \subseteq X$ with $|P| \geq p$, and $Q \subseteq Y$ with $|Q| \leq m$. Now, let $Y' = \bigcup_{y \in Q} \mathsf{NN}_f(y, Y, d)$. That is, Y' consists of f nearest centers from Y for all centers in the approximate set of centers Q. It is easy to see that (P, Y') satisfies Conditions 1 and 2, and therefore it is feasible[3]. We return this solution as an approximate solution for $\mathcal{I}$. We note that, at a high level, our approach can be seen as a generalization of the approach in [20], who showed similar reductions from Fault Tolerant k-center and k-median (without outliers) to their respective non Fault Tolerant versions. Our contribution here is to handle the presence of outliers.

3 Fault Tolerant k-Center with Outliers

We consider the Fault Tolerant k-center with Outliers instance $\mathcal{I} = (X, X, d, k, p, f)$, i.e., the set of points to be covered, X, is the same as the

[3] For the k-clustering objective, the last step takes some more work, since we also have to compute a radius assignment $r : Y' \to \mathbb{R}^+$. However, at a high level, the strategy is similar.

set of centers. Recall that p is the number of points to be clustered, and f is the degree of fault-tolerance required in the solution. Also, recall that the cost of a feasible solution (X', Y'), is defined as $\max_{x \in X'} d_f(x, Y')$.

Algorithm. Notice that the cost of the optimal solution $\mathsf{opt}(\mathcal{I}) = (X^*, Y^*)$ must be an inter-point distance between some $x, y \in X$. By parametric search, we assume that we know the cost of the optimal solution, denoted by D. Let $X_p = \{x \in X : d_f(x, X) \leq \mathsf{D}\}$ be the set of points that have at least f points within a ball of radius D. It is easy to see that since D is the correct guess for the optimal cost, the set X_p has size at least p, since $X^* \subseteq X_p$. Now, we will focus on the D-neighborhoods of X_p, i.e., on the set $T = \bigcup_{x \in X_p} B_d(x, \mathsf{D})$. Let $m := \lfloor k/f \rfloor$ and let $\mathcal{I}' = (T, T, d, m, p, 1)$ be an instance of the m-center with outliers problem. Note that $\mathcal{I}'$ has fault-tolerance of 1. Let $\sigma(\mathcal{I}') = (P, Q)$ be the solution returned by an α-approximate algorithm $\mathcal{A}$. Let $Y' = \bigcup_{y \in Q} \mathsf{NN}_f(y, X, d)$. We return $\sigma(\mathcal{I}) = (P, Y')$ as the solution.

Before we analyze the cost of this solution, we need the following lemma (due to Kumar and Raichel [20]) that will be useful in this as well as the subsequent sections.

Lemma 1. *Given a finite metric space* (X, d), *let* $Y \subseteq X$ *be any non-empty subset. Then for any integer parameter* $1 \leq h \leq |Y|$, *and any subset* $Z \subseteq X$, *there exists a subset* $S \subseteq Z$ *such that*

- $|S| \leq \lfloor |Y|/h \rfloor$
- *For any* $v \in Z$, $d(v, S) \leq 2 \cdot d_h(v, Y)$.

In other words, the set S is an approximation to the h-th nearest neighbors from the set Y, and its size is bounded by $|Y|/h$. We note here that even though we will only use this lemma in the analyses, such a set S can be found efficiently using a simple iterative algorithm [20]. In the following lemma, we analyze the cost of the solution $\sigma(\mathcal{I}) = (P, Y')$.

Lemma 2.

$$\mu(\sigma(\mathcal{I})) \leq (2\alpha + 2) \cdot \mu(\mathsf{opt}(\mathcal{I}))$$

Proof. Fix a point $x \in P$. We first upper bound the connection cost of x to its f-th closest center in Y'. Let $y \in Q$ denote a closest center of x in Q. Then,

$$d_f(x, Y') \leq d(x, y) + d_f(y, Y') = d(x, y) + d_f(y, X) \quad (1)$$

The equality follows from the fact that $\mathsf{NN}_f(y, X) \subseteq Y'$. Now, let $u \in X_p$ denote a closest point to y from the set X_p. Then,

$$d_f(y, X) \leq d(y, u) + d_f(u, X) = d(y, X_p) + d_f(u, X).$$

Since, $y \in Q \subseteq T$, we have that $d(y, X_p) \leq \mathsf{D}$. Similarly, since $u \in X_p$, $d_f(u, X) \leq \mathsf{D}$. Combining this with (1), we get that $d_f(x, Y') \leq d(x, y) + 2 \cdot \mathsf{D}$. Taking maximum over all $x \in P$, we get that,

$$\mu(\sigma(\mathcal{I})) = \max_{x \in P} d_f(x, Y') \leq \max_{x \in P} (d(x, Q) + 2 \cdot \mathsf{D}) \leq \alpha \cdot \mu(\mathsf{opt}(\mathcal{I}')) + 2 \cdot \mathsf{D}. \quad (2)$$

Where $\mathsf{opt}(\mathcal{I}') = (P^*, Q^*)$ is an optimal solution for the instance $\mathcal{I}'$. Recall that $\mathsf{opt}(\mathcal{I}) = (X^*, Y^*)$ is an optimal solution for the original instance $\mathcal{I}$. Let $S \subseteq T$ be the set guaranteed by Lemma 1, by setting $Y \leftarrow Y^* \subseteq T$, $h \leftarrow f$, and $Z \leftarrow X^* \subseteq T$. Therefore, we have that $|S| \leq m$, and for any $x \in X^*$ we have that, $d(x, S) \leq 2 \cdot d_f(x, Y^*) \leq 2 \cdot \mathsf{D}$. That is, (X^*, S) is a feasible solution for the instance $\mathcal{I}'$, of cost at most $2\mathsf{D}$, which implies that $\mu(\mathsf{opt}(\mathcal{I}')) \leq 2\mathsf{D}$. Combining this with (2), yields the desired bound in the lemma. □

Finally, we note that using an $\alpha = 2$ approximation for k-center with Outliers [6,14], we get a 6-approximation for the Fault Tolerant k-center with Outliers problem.

4 Fault Tolerant k-Median with Outliers (v1)

The Fault Tolerant k-median With Outliers instance that we consider is $\mathcal{I} = (X, X, d, k, p, f)$, i.e. the set of points X is the same as the set from which the centers can be chosen. However, it is convenient to imagine the set of potential centers, as a distinct copy of X, which we denote by Y. Recall that p is the number of points to be clustered, and f is the degree of fault-tolerance required in the solution. Also, recall that the cost of a feasible solution is (X', Y') is $\sum_{x \in X'} d_f(x, Y')$.

Given the metric space $(X \cup Y, d)$, we create a new distance function d' as follows.

$$d'(x,y) = d'(y,x) = \begin{cases} d(x,y) & \text{if } x, y \in X \\ d(x,y) + d_f(y,Y) & \text{if } x \in X, y \in Y \\ d_f(x,Y) + d(x,y) + d_f(y,Y) & \text{if } x, y \in Y \end{cases}$$

It is easy to see that the distance function satisfies the metric properties, and hence $(X \cup Y, d')$ is a metric space. We use a known α-approximation algorithm on the instance $\mathcal{I}' = (X, Y, d', m, p, 1)$, where $m := \lfloor k/f \rfloor$. Notice that this is an m-median with outlier instance, and here the fault-tolerance is 1. Let $\sigma(\mathcal{I}') = (P, Q)$ be the α-approximate solution obtained, such that $\mu(\sigma(\mathcal{I}')) = \sum_{x \in P} d'(x, Q) \leq \alpha \cdot \sum_{x \in P*} d'(x, Q^*)$, where $\mathsf{opt}(\mathcal{I}') = (P^*, Q^*)$ is an optimal solution for $\mathcal{I}'$. We return $\sigma(\mathcal{I}) = (P, Y')$ as a solution, where $Y' = \bigcup_{y \in Q} \mathsf{NN}_f(y, Y, d)$.

Lemma 3.

$$\mu(\sigma(\mathcal{I})) \leq 5\alpha \cdot \mu(\mathsf{opt}(\mathcal{I}))$$

Proof. Recall that $\sigma(\mathcal{I}) = (P, Y')$. We fix a point $x \in P$ and first analyze its connection cost $d_f(x, Y')$. Let $y = \mathsf{nn}_1(x, Q, d')$ denote a closest center to x from the set Q, according to the new distance function d'. Then, we have the following.

$$d_f(x, Y') \leq d(x,y) + d_f(y, Y') = d(x,y) + d_f(y, Y) = d'(x,y) \tag{3}$$

Where the first equality follows from the fact that $\mathsf{NN}_f(y, Y, d) \subseteq Y'$, i.e., Y' contains the f nearest neighbors of y from the set Y. The second equality follows from the definition of $d'(x, y)$. Now, summing (3) over all $x \in P$, we get, $\mu(\sigma(\mathcal{I})) = \sum_{x \in P} d_f(x, Y') \leq \sum_{x \in P} d'(x, y) = \sum_{x \in P} d'(x, Q)$.

Observe that $\sum_{x \in P} d'(x, y)$ is the cost of the α-approximate solution $\sigma(\mathcal{I}')$. Therefore,

$$\mu(\sigma(\mathcal{I})) \leq \mu(\sigma(\mathcal{I}')) \leq \alpha \cdot \mu(\mathsf{opt}(\mathcal{I}')) \tag{4}$$

Recalling that, $\mathsf{opt}(\mathcal{I}') = (P^*, Q^*)$ is an optimal solution for the instance $\mathcal{I}'$. Now, let $\mathsf{opt}(\mathcal{I}) = (X^*, Y^*)$ be an optimal solution for the original instance $\mathcal{I}$. Let S be the set guaranteed by Lemma 1, by setting $Y \leftarrow Y^*$, $h \leftarrow f$, and $Z \leftarrow X^*$. Therefore, we have that $|S| \leq m$, and that for any $x \in X^*$,

$$d(x, S) \leq 2 \cdot d_f(x, Y^*) \tag{5}$$

Let S' denote the set of centers in Y that are co-located with the points in S. Notice that (X^*, S') is a feasible solution for the instance $\mathcal{I}'$. Therefore, $\mu(\mathsf{opt}(\mathcal{I}')) \leq \sum_{x \in X^*} d'(x, S')$. We first bound the connection cost of a fixed $x \in X^*$. Let $s' = \mathsf{nn}_1(x, S', d)$, i.e., s' is a nearest neighbor of x from the set S', according to the original distance function d. Then,

$$d'(x, S') \leq d'(x, s') = d(x, s') + d_f(s', Y) \leq d(x, s') + d(x, s') + d_f(x, Y) \tag{6}$$

Where the equality follows from the definition of d'. Now, let $s \in S$ be the point co-located with $s' \in S'$. Then, we get that, $d'(x, S') \leq 2 \cdot d(x, s) + d_f(x, Y) \leq 4 \cdot d_f(x, Y^*) + d_f(x, Y)$. Here the second inequality follows from (5).

Since $d_f(x, Y) \leq d_f(x, Y^*)$, we get that, $d'(x, S') \leq 5 \cdot d_f(x, Y^*)$. Summing this over all $x \in X^*$, we get,

$$\mu(\mathsf{opt}(\mathcal{I}')) \leq \sum_{x \in X^*} d'(x, S') \leq 5 \cdot \sum_{x \in X^*} d_f(x, Y^*) = 5 \cdot \mu(\mathsf{opt}(\mathcal{I})) \tag{7}$$

Combining inequalities (4) and (7) yields the lemma. □

Using the $\alpha \approx 7.081$ approximation for k-median with Outliers [18], we get a 35.405-approximation for the Fault-Tolerant k-median with Outliers problem (v1).

5 Fault Tolerant k-Median with Outliers (v2)

Recall that the cost of a feasible solution of this version of Fault Tolerant k-median with Outliers is $\mu(\sigma(\mathcal{I})) = \sum_{x \in X'} \sum_{i=1}^{f} d_i(x, Y')$. Despite many similarities between the current and previous sections, we give all the details for the sake of completeness.

Given the metric space $(X \cup Y, d)$, we create a new distance function d' as follows.

$$d'(x,y) = d'(y,x) = \begin{cases} \sum_{i=1}^{f} d(x,y) & \text{if } x, y \in X \\ \sum_{i=1}^{f} (d(x,y) + d_i(y,Y)) & \text{if } x \in X, y \in Y \\ \sum_{i=1}^{f} (d(x,y) + d_i(x,Y) + d_i(y,Y)) & \text{if } x, y \in Y \end{cases}$$

Similar to the previous section, let $\sigma(\mathcal{I}') = (P, Q)$ be an α-approximate solution for the instance $\mathcal{I}' = (X, Y, d', m, p, 1)$. We return $\sigma(\mathcal{I}) = (P, Y')$ as a solution, where $Y' = \bigcup_{y \in Q} \mathsf{NN}_f(y, Y, d)$. Now we analyze the cost of this solution.

Lemma 4.

$$\mu(\sigma(\mathcal{I})) \leq (4f+1) \cdot \alpha \cdot \mu(\mathsf{opt}(\mathcal{I}))$$

Proof. Recall that $\sigma(\mathcal{I}) = (P, Y')$. We fix a point $x \in P$ and first analyze its connection cost $\sum_{i=1}^{f} d_i(x, Y')$. Let $y = \mathsf{nn}_1(x, Q, d')$ denote a closest center to x from the set Q, according to the new distance function d'. We have the following.

$$\sum_{i=1}^{f} d_i(x, Y') \leq \sum_{i=1}^{f} (d(x,y) + d_i(y, Y')) = \sum_{i=1}^{f} (d(x,y) + d_i(y, Y)) = d'(x,y). \tag{8}$$

Where the first equality follows from the fact that $\mathsf{NN}_f(y, Y, d) \subseteq Y'$, i.e., Y' contains the f nearest neighbors of y from the set Y. The second equality follows from the definition of $d'(x,y)$. Now, summing (8) over all $x \in P$, we get,

$$\mu(\sigma(\mathcal{I})) = \sum_{x \in P} \sum_{i=1}^{f} d_i(x, Y') \leq \sum_{x \in P} d'(x,y) = \sum_{x \in P} d'(x, Q).$$

Observe that $\sum_{x \in P} d'(x,y)$ is the cost of the α-approximate solution $\sigma(\mathcal{I}')$. Therefore,

$$\mu(\sigma(\mathcal{I})) \leq \mu(\sigma(\mathcal{I}')) \leq \alpha \cdot \mu(\mathsf{opt}(\mathcal{I}')). \tag{9}$$

Recall that $\mathsf{opt}(\mathcal{I}') = (P^*, Q^*)$ is an optimal solution for the instance $\mathcal{I}'$. Now, let $\mathsf{opt}(\mathcal{I}) = (X^*, Y^*)$ be an optimal solution for the original instance $\mathcal{I}$. Let S be the set guaranteed by Lemma 1, by setting $Y \leftarrow Y^*$, $h \leftarrow f$, and $Z \leftarrow X^*$. Therefore, we have that $|S| \leq m$, and that for any $x \in X^*$,

$$d(x, S) \leq 2 \cdot d_f(x, Y^*). \tag{10}$$

Let S' denote the set of centers in Y that are co-located with the points in S. Notice that (X^*, S') is a feasible solution for the instance $\mathcal{I}'$. Therefore, $\mu(\mathsf{opt}(\mathcal{I}')) \leq \sum_{x \in X^*} d'(x, S')$. We first bound the connection cost of a fixed $x \in X^*$. Let $s' = \mathsf{nn}_1(x, S', d)$, i.e., s' is a nearest neighbor of x from the set S' according to the original distance function d. Then,

$$d'(x, S') \leq d'(x, s') = \sum_{i=1}^{f} (d(x, s') + d_i(s', Y)) \leq \sum_{i=1}^{f} (2 \cdot d(x, s') + d_i(x, Y)).$$

Let $s \in S$ be a point co-located with $s' \in S'$. Using (10), we get, $\sum_{i=1}^{f} d(x,s) = \sum_{i=1}^{f} d(x,s) \leq 2f \cdot d_f(x, Y^*)$. Therefore, we have,

$$d'(x, S') \leq 4f \cdot d_f(x, Y^*) + \sum_{i=1}^{f} d_i(x, Y).$$

Adding this over $x \in X^*$, we get,

$$\mu(\mathsf{opt}(\mathcal{I}')) \leq 4f \sum_{x \in X^*} d_f(x, Y^*) + \sum_{x \in X^*} \sum_{i=1}^{f} d_i(x, Y).$$

Now we claim that each of the sums on the right hand side is a lower bound on $\mu(\mathsf{opt}(\mathcal{I}))$. The cost of connecting a point $x \in X^*$ to f nearest centers in Y^* is at least $d_f(x, Y^*)$. At the same time, the same cost is also at least $\sum_{i=1}^{f} d_i(x, Y)$. Therefore, we get the following bound.

$$\mu(\mathsf{opt}(\mathcal{I}')) \leq (4f + 1) \cdot \mu(\mathsf{opt}(\mathcal{I})) \tag{11}$$

Finally, combining inequalities (9) and (11) yields the lemma.

Using the $\alpha \approx 7.081$ approximation for k-median with Outliers ([18]), we get an $28.324 \cdot f + 7.081 = O(f)$-approximation for the Fault Tolerant k-median with Outliers problem (v2).

6 Fault Tolerant k-Clustering with Outliers

In this section, we consider the Fault Tolerant k-Clustering with Outliers problem. Even though the details are somewhat involved, the high level ideas remain similar to the ones employed in the previous section for the Fault Tolerant k-median with Outliers problem (v2).

We are given an instance $\mathcal{I} = (X, Y, d, k, p, f)$. In this section, we do not require the set of clients X to be the same as the set of servers Y. A feasible solution is (X', Y', r'), where $|X'| \geq p$, $|Y'| \leq k$. Furthermore, we need to compute a radius assignment $r' : Y' \to \mathbb{R}^+$, such that for any $x \in X'$, $|\{y \in Y' : d(x, y) \leq r'(y)\}| \geq f$. In other words, if $\mathcal{B}_{r'} = \{B_d(y, r'(y)) : y \in Y'\}$ denotes the set of balls defined by the radius function r', then there are at least f balls in $\mathcal{B}_{r'}$ that contain each point $x \in X'$. The cost of this solution is $\sum_{y \in Y'} r'(y)$. Without loss of generality, we can assume that in any feasible solution (X', Y', r'), for any $y \in Y'$, there is at least one client on the boundary of the ball $B_d(y, r'(y))$ – for otherwise, we can obtain a new feasible solution of a smaller cost by shrinking some of the balls. For $y \in Y'$, we denote such a client by $x_{r'}(y)$.

Recall that $m := \lfloor k/f \rfloor$. First, we prove an analogue of Lemma 1 for the k-clustering objective. Intuitively, this lemma lets us relate the costs of solutions to an f-Fault Tolerant k-Clustering with Outliers instance to an instance of m-Clustering with Outliers.

Lemma 5. *Let $\mathcal{I} = (X, Y, d, k, p, f)$, and $\overline{\mathcal{I}} = (X, Y, d, m, p, 1)$ be two instances. Given any feasible solution $\sigma(\mathcal{I}) = (X^*, Y^*, r^*)$, it is possible to obtain a feasible solution $\sigma(\overline{\mathcal{I}}) = (X^*, S, r_S)$, such that,*

1. *$S \subseteq Y^*$,*
2. *$\sum_{y \in S} d_f(y, Y) \leq \sum_{y \in S} r_S(y) \leq 3 \cdot \sum_{y \in Y^*} r^*(y)$,*
3. *Furthermore, for each $s \in S$, we identify a unique point $u(s) \in B(s, r^*(s)) \cap X^*$ such that, $\sum_{s \in S} \sum_{i=1}^{f} d_i(u(s), Y^*) \leq \mu(\sigma(\mathcal{I}))$.*

Proof. We construct the solution $\sigma(\overline{\mathcal{I}})$ iteratively from the solution $\sigma(\mathcal{I})$. Let $\mathcal{B}_{r'}$ be the set of balls corresponding to the solution $\sigma(\mathcal{I})$. Initially, all points in X^*, and all balls in $\mathcal{B}_{r^*}$ are unmarked. The set S is initialized to be the empty set.

The algorithm proceeds in iterations, while there exists an unmarked point in X^*. In each iteration, we select an unmarked ball $B(y, r^*(y)) \in \mathcal{B}_{r^*}$ of the largest radius (ties are broken arbitrarily) containing an unmarked point of X^*. We denote this point of X^* by $u(y)$. Let $\mathcal{B}' \subseteq \mathcal{B}_{r^*}$ be the set of balls that intersect $B(y, r^*(y))$ (including itself). We mark all the balls of $\mathcal{B}'$, and we also mark points from X^* that belong to any ball from $\mathcal{B}'$. Finally, we add y to the set S, and we set $r_S(y) := 3r^*(y)$. This completes the description of one iteration of the algorithm.

It is easy to see that $S \subseteq Y^*$, which establishes the first property. Note that a point $x \in X^*$ is marked once any ball in $\mathcal{B}_{r^*}$ that contains x is marked. Therefore, in the iteration when we select a ball $B(y, r^*(y))$ that contains a (previously) unmarked point $u(y) \in X^*$, there are at least f unmarked balls in $\mathcal{B}_{r^*}$ that contain $u(y)$. That is, we mark at least f balls in each iteration. From this, it follows that $|S| \leq m$. Also from the same argument, it is easy to see that for any $y \in S$, $\mathsf{NN}_f(y, Y, d) \subseteq B(y, r_S(y))$. Thus, the first inequality in the second property follows.

Recall that $\mathcal{B}_{r_S}$ is the set of balls corresponding to the radius function r_S. Since in each iteration, we select an unmarked ball of the largest radius, and expand it by 3, it is easy to see that the balls in $\mathcal{B}_{r_S}$ cover the points in X^*. From the same argument, the second inequality of the second property also follows. This shows that $\sigma(\overline{\mathcal{I}})$ is a feasible solution.

Now we prove the third property. Consider an iteration of algorithm when we add $s \in Y^*$ to the set S. As argued earlier, we mark at least f balls containing the point $u(s)$ in this iteration. The cost of these f balls is lower bounded by $\sum_{i=1}^{f} d_i(u(s), Y^*)$. Furthermore, for distinct $s, s' \in S$, the two sets of f balls covering $u(s)$ and $u(s')$ are disjoint. Combining these two facts yields the desired lower bound on the cost of the solution $\mu(\sigma(\mathcal{I}))$. □

Now, we discuss our algorithm. First, we define a new distance function d' from the original distance function d – the definition is exactly the same as in the previous section. Let $\sigma(\mathcal{I}') = (P, Q, \lambda)$ be an α-approximate solution for the instance $\mathcal{I}' = (X, Y, d', m, p, 1)$. Define $Y' := \bigcup_{q \in Q} \mathsf{NN}_f(q, Y, d)$, which is the set of servers in the solution to be returned. We compute the radius assignment $r : Y' \to \mathbb{R}^+$, such that, if $y \in Y'$ is one of the f nearest neighbors of $q \in Q$, then $B_{d'}(q, \lambda) \subseteq B_d(y, r)$. Formally, for any $y \in Y'$, its radius $r(y)$ is defined as follows.

$$r(y) := \max_{q \in Q: y \in N_f(q,Y,d)} \Big[d(x_\lambda(q), q) + d(q, y) \Big] \tag{12}$$

This ensures that each point of P is contained in at least f balls in the solution $\sigma(\mathcal{I}) = (P, Y', r)$. Now we analyze the cost of this solution.

Lemma 6.

$$\mu(\sigma(\mathcal{I})) \leq (4f+1) \cdot \alpha \cdot \mu(\mathsf{opt}(\mathcal{I}))$$

Proof. Recall that $\sigma(\mathcal{I}) = (P, Y', r)$.

$$\begin{aligned}
\mu(\sigma(\mathcal{I})) &= \sum_{y \in Y'} r(y) \\
&= \sum_{y \in Y'} \max_{q \in Q} \left(d(x_\lambda(q), q) + d(q, y)\right) \\
&\qquad\qquad (\text{max is taken over } q \in Q \text{ s.t. } y \in \mathsf{NN}_f(q, Y, d)) \\
&\leq \sum_{q \in Q} \sum_{i=1}^{f} \left(d(x_\lambda(q), q) + d(q, y_i)\right) \qquad (\text{Where } y_i = \mathsf{nn}_i(q, Y, d)) \\
&= \sum_{q \in Q} d'(q, x_\lambda(q)) \\
&= \mu(\sigma(\mathcal{I}')) \\
&\leq \alpha \cdot \mu(\mathsf{opt}(\mathcal{I}'))
\end{aligned} \tag{13}$$

Recall that $\mathsf{opt}(\mathcal{I}') = (P^*, Q^*, \lambda^*)$ is an optimal solution for the instance $\mathcal{I}'$. Now, let $\mathsf{opt}(\mathcal{I}) = (X^*, Y^*, r^*)$ be an optimal solution for the original instance $\mathcal{I}$. Let (X^*, S, r_S) be the solution guaranteed by Lemma 5, by setting $\sigma(\mathcal{I}) \leftarrow \mathsf{opt}(\mathcal{I})$. Note that the distance function used for this solution is d. Consider the analogue of this solution in the distance function d'. More formally, for any $y \in S$, let $x(y) := x_{r_S}(y)$ be a point on the boundary according to the radius function r_S. Now, for $y \in S$, let $r'_S(y) := d'(y, x(y))$. Note that (X^*, S, r'_S) is a feasible solution for the instance $\mathcal{I}'$. Therefore,

$$\begin{aligned}
\mu(\mathsf{opt}(\mathcal{I}')) &\leq \sum_{y \in S} r'_S(y) \\
&= \sum_{y \in S} d'(y, x(y)) \\
&\leq \sum_{y \in S} \sum_{i=1}^{f} (d(y, x(y)) + d_i(y, Y^*)) \qquad (\text{Since } Y^* \subseteq Y) \\
&\leq \sum_{y \in S} \sum_{i=1}^{f} d(y, x(y)) + \sum_{y \in S} \sum_{i=1}^{f} d(y, u(y)) + \sum_{y \in S} \sum_{i=1}^{f} d_i(u(y), Y^*)
\end{aligned}$$

Where, for any $y \in S$, $u(y) \in X^*$ is the point guaranteed in the third property of Lemma 5. Now, we bound each of the three terms separately.

We use the second property of Lemma 5 to bound the first term by $3f \cdot \sum_{y \in Y^*} r^*(y)$.

Since $d(y, u(y)) \leq r^*(y)$, we bound the second term by $f \cdot \sum_{y \in S} r^*(y) \leq f \cdot \sum_{y \in Y^*} r^*(y)$.

We use the third property of Lemma 5 to bound the third term by $\sum_{y \in Y^*} r^*(y)$.

Combining the bounds on the three terms, we get that $\mu(\mathsf{opt}(\mathcal{I}')) \leq (4f+1) \cdot \mu(\mathsf{opt}(\mathcal{I}))$. Finally, combining this inequality with (13) gives the claimed bound in the Lemma. □

By using $\alpha \approx 12.365$ approximation for k-clustering with Outliers ([1]), we get a $49.46 \cdot f + 12.365 = O(f)$-approximation for the Fault Tolerant k-clustering with Outliers problem.

References

1. Ahmadian, S., Swamy, C.: Approximation algorithms for clustering problems with lower bounds and outliers. In: 43rd International Colloquium on Automata, Languages, and Programming, ICALP 2016, Rome, Italy, 11–15 July 2016, pp. 69:1–69:15 (2016)
2. Bandyapadhyay, S., Varadarajan, K.R.: Approximate clustering via metric partitioning. In: 27th International Symposium on Algorithms and Computation, ISAAC 2016, Sydney, Australia, 12–14 December 2016, pp. 15:1–15:13 (2016)
3. Bhaskara, A., Charikar, M., Chlamtac, E., Feige, U., Vijayaraghavan, A.: Detecting high log-densities: an $O(n^{1/4})$ approximation for densest k-subgraph. In: Proceedings of the 42nd ACM Symposium on Theory of Computing, STOC 2010, 5–8 June 2010, Cambridge, Massachusetts, USA, pp. 201–210 (2010)
4. Bhowmick, S., Inamdar, T., Varadarajan, K.R.: Improved approximation for metric multi-cover. CoRR, abs/1602.04152 (2016)
5. Byrka, J., Pensyl, T., Rybicki, B., Srinivasan, A., Trinh, K.: An improved approximation for k-median and positive correlation in budgeted optimization, vol. 13, pp. 23:1–23:31 (2017)
6. Chakrabarty, D., Goyal, P., Krishnaswamy, R.: The non-uniform k-center problem. In: 43rd International Colloquium on Automata, Languages, and Programming, ICALP 2016, Rome, Italy, 11–15 July 2016, pp. 67:1–67:15 (2016)
7. Charikar, M., Panigrahy, R.: Clustering to minimize the sum of cluster diameters. J. Comput. Syst. Sci. **68**(2), 417–441 (2004)
8. Charikar, M., Khuller, S., Mount, D.M., Narasimhan, G.: Algorithms for facility location problems with outliers. In: Proceedings of the Twelfth Annual Symposium on Discrete Algorithms, Washington, DC, USA, 7–9 January 2001, pp. 642–651 (2001)
9. Chaudhuri, S., Garg, N., Ravi, R.: The p-neighbor k-center problem. Inf. Process. Lett. **65**(3), 131–134 (1998)

10. Chen, K.: A constant factor approximation algorithm for k-median clustering with outliers. In: Proceedings of the Nineteenth Annual ACM-SIAM Symposium on Discrete Algorithms, SODA 2008, San Francisco, California, USA, 20–22 January 2008, pp. 826–835 (2008)
11. Gibson, M., Kanade, G., Krohn, E., Pirwani, I.A., Varadarajan, K.R.: On metric clustering to minimize the sum of radii. Algorithmica **57**(3), 484–498 (2010)
12. Gonzalez, T.F.: Clustering to minimize the maximum intercluster distance. Theor. Comput. Sci. **38**, 293–306 (1985)
13. Hajiaghayi, M.T., Hu, W., Li, J., Li, S., Saha, B.: A constant factor approximation algorithm for fault-tolerant k-median. ACM Trans. Algorithms **12**(3), 36:1–36:19 (2016)
14. Harris, D.G., Pensyl, T., Srinivasan, A., Trinh, K.: A lottery model for center-type problems with outliers. ACM Trans. Algorithms (TALG) **15**, 36:1–36:25 (2019)
15. Hsu, W.-L., Nemhauser, G.L.: Easy and hard bottleneck location problems. Discrete Appl. Math. **1**(3), 209–215 (1979)
16. Jain, K., Mahdian, M., Markakis, E., Saberi, A., Vazirani, V.V.: Greedy facility location algorithms analyzed using dual fitting with factor-revealing LP. J. ACM **50**(6), 795–824 (2003)
17. Khuller, S., Pless, R., Sussmann, Y.J.: Fault tolerant k-center problems. Theor. Comput. Sci. **242**(1–2), 237–245 (2000)
18. Krishnaswamy, R., Li, S., Sandeep, S.: Constant approximation for k-median and k-means with outliers via iterative rounding. In: Proceedings of the 50th Annual ACM SIGACT Symposium on Theory of Computing, STOC 2018, Los Angeles, CA, USA, 25–29 June 2018, pp. 646–659 (2018)
19. Krumke, S.O.: On a generalization of the p-center problem. Inf. Process. Lett. **56**(2), 67–71 (1995)
20. Kumar, N., Raichel, B.: Fault tolerant clustering revisited. In: Proceedings of the 25th Canadian Conference on Computational Geometry, CCCG 2013, Waterloo, Ontario, Canada, 8–10 August 2013 (2013)
21. Manurangsi, P.: Almost-polynomial ratio eth-hardness of approximating densest k-subgraph. In: Proceedings of the 49th Annual ACM SIGACT Symposium on Theory of Computing, STOC 2017, Montreal, QC, Canada, 19–23 June 2017, pp. 954–961 (2017)
22. Ran, Y., Shi, Y., Zhang, Z.: Local ratio method on partial set multi-cover. J. Comb. Optim. **34**(1), 302–313 (2017)
23. Ran, Y., Shi, Y., Zhang, Z.: Primal dual algorithm for partial set multi-cover. In: Kim, D., Uma, R.N., Zelikovsky, A. (eds.) COCOA 2018. LNCS, vol. 11346, pp. 372–385. Springer, Cham (2018). https://doi.org/10.1007/978-3-030-04651-4_25
24. Swamy, C., Shmoys, D.B.: Fault-tolerant facility location. ACM Trans. Algorithms **4**(4), 51:1–51:27 (2008)

Improved (In-)Approximability Bounds for d-Scattered Set

Ioannis Katsikarelis(✉), Michael Lampis, and Vangelis Th. Paschos

Université Paris-Dauphine, PSL Research University, CNRS, UMR 7243 LAMSADE, 75016 Paris, France
{ioannis.katsikarelis,michail.lampis,paschos}@lamsade.dauphine.fr

Abstract. In the d-Scattered Set problem we are asked to select at least k vertices of a given graph, so that the distance between any pair is at least d. We study the problem's (in-)approximability and offer improvements and extensions of known results for Independent Set, of which the problem is a generalization. Specifically, we show:

- A lower bound of $\Delta^{\lfloor d/2 \rfloor - \epsilon}$ on the approximation ratio of any polynomial-time algorithm for graphs of maximum degree Δ and an improved upper bound of $O(\Delta^{\lfloor d/2 \rfloor})$ on the approximation ratio of any greedy scheme for this problem.
- A polynomial-time $2\sqrt{n}$-approximation for bipartite graphs and even values of d, that matches the known lower bound by considering the only remaining case.
- A lower bound on the complexity of any ρ-approximation algorithm of (roughly) $2^{\frac{n^{1-\epsilon}}{\rho d}}$ for even d and $2^{\frac{n^{1-\epsilon}}{\rho(d+\rho)}}$ for odd d (under the randomized ETH), complemented by ρ-approximation algorithms of running times that (almost) match these bounds.

1 Introduction

In this paper we study the d-Scattered Set problem: given graph $G = (V, E)$, we are asked if there exists a set K of at least k *selections* from V, such that the distance between any pair $v, u \in K$ is at least $d(v, u) \geq d$, where $d(v, u)$ denotes the shortest-path distance from v to u. The problem can already be seen to be hard as it generalizes Independent Set (for $d = 2$) and thus the optimal k cannot be approximated to $n^{1-\epsilon}$ in polynomial time [15] (under standard complexity assumptions), while an alternative name is Distance-d Independent Set [9,20]. The problem has been well-studied, also from the parameterized point of view [18,21], while approximability in polynomial time has already been considered for bipartite, regular and degree-bounded graphs [9,10], perhaps the natural candidate for the next intractability frontier.

This paper aims to advance our understanding in this direction by providing the first lower bound on the approximation ratio of any polynomial-time algorithm as a function of the maximum degree of any vertex in the input graph, while also improving upon the known ratios to match this lower bound. On bipartite graphs, our aim is to complete the picture by considering the only remaining

E. Bampis and N. Megow (Eds.): WAOA 2019, LNCS 11926, pp. 202–216, 2020.
https://doi.org/10.1007/978-3-030-39479-0_14

open case for this class with an approximation algorithm whose ratio matches the problem's known inapproximability, before we turn our attention to super-polynomial running times with the purpose of extending known upper/lower bounds for INDEPENDENT SET, so as to also take into account the range of values for d: we observe that d acts as a scaling factor for the size of the instance, whereby the problem becomes easier when vertices are required to be much further apart.

Before moving on to describe our results in detail, we note that these may be dependent on the parity of our distance parameter d as being even or odd. Both our running times and ratios can be affected by this peculiarity of the problem that, intuitively, arises due to the (non)existence of a *middle* vertex on a path of length d between two endpoints: if d is even then such a vertex can exist at equal distance $d/2$ from any number of vertices in the solution, while if d is odd there can be no vertex at equal distance from any pair of vertices in the solution. This idiosyncrasy can change the way in which both our algorithms and hardness constructions work and in some cases even entirely alters the problem's complexity (e.g. [9]).

Our Contribution: Our results are also summarized in Table 1 below. Section 3 concerns itself with strictly polynomial running times. We first show that there is no polynomial-time approximation algorithm for d-SCATTERED SET with ratio $\Delta^{\lfloor d/2 \rfloor - \epsilon}$ in graphs of maximum degree Δ. Our complexity assumption is NP $\not\subseteq$ BPP due to our use as a starting point of a randomized construction for INDEPENDENT SET from [6], that we then build upon to produce highly efficient (in terms of maximum vertex degree and diameter) instances of d-SCATTERED SET. This is the first lower bound that considers Δ and generalizes the known $\Delta^{1-\epsilon}$-inapproximability of INDEPENDENT SET (see Theorem 5.2 of [6], restated here as Theorem 2.1, as well as [1]). Maximum vertex degree Δ plays an important role in the context of independence (e.g. [2,8,14]) and was specifically studied for d-SCATTERED SET in [10], where polynomial-time $O(\Delta^{d-1})$- and $O(\Delta^{d-2}/d)$-approximations are given. We improve upon these upper bounds by showing that any degree-based greedy approximation algorithm in fact achieves a ratio of $O(\Delta^{\lfloor d/2 \rfloor})$, also matching our lower bound. We then turn our attention to bipartite graphs and show that d-SCATTERED SET can be approximated within a factor of $2\sqrt{n}$ in polynomial time also for even values of d, matching its known $n^{1/2-\epsilon}$-inapproximability from [9] and complementing the known $\sqrt{n}$-approximation for odd d [13].

Section 4 follows this up by considering super-polynomial running times, presenting first an exact exponential-time algorithm for d-SCATTERED SET of complexity $O^*((ed)^{\frac{2n}{d}})$ based on a straightforward upper bound on the size of any solution and then considering the inapproximability of the problem in the same complexity range. We show that no ρ-approximation algorithm can take time (roughly) $2^{\frac{n^{1-\epsilon}}{\rho d}}$ for even d and $2^{\frac{n^{1-\epsilon}}{\rho(d+\rho)}}$ for odd d, under the (randomized) ETH. This is complemented by (almost) matching ρ-approximation algorithms of running times $O^*((e\rho d)^{\frac{2n}{\rho d}})$ for even d and $O^*((e\rho d)^{\frac{2n}{\rho(d+\rho)}})$ for odd d. We note that

Table 1. A summary of our results (theorem numbers), for even/odd values of d.

	Inapproximability	Approximation
Super-polynomial	$2^{\frac{n^{1-\epsilon}}{\rho d}}$ (4.3)/$2^{\frac{n^{1-\epsilon}}{\rho(d+\rho)}}$ (4.4)	$O^*((e\rho d)^{\frac{2n}{\rho d}})$ (4.5)/$O^*((e\rho d)^{\frac{2n}{\rho(d+\rho)}})$ (4.6)
Polynomial	$\Delta^{\lfloor d/2 \rfloor - \epsilon}$ (3.1)	$O(\Delta^{\lfloor d/2 \rfloor})$ (3.14)
Bipartite graphs	$n^{1/2-\epsilon}$ [9]	$2\sqrt{n}$ (3.17)

the current state-of-the-art PCPs are unable to distinguish between optimal running times of the form $2^{n/\rho}$ and $\rho^{n/\rho}$ for ρ-approximation algorithms, due to the poly-logarithmic factors added by even the most efficient constructions and we thus do not focus on such factors differentiating our upper and lower bounds. These results provide a complete characterization of the optimal relationship between the worst-case approximation ratio ρ achievable for d-Scattered Set by any algorithm, its running time and the distance parameter d, *for any point in the trade-off curve*, in a similar manner as was done for Independent Set in [6,7] (see also [4,5]). Due to space restrictions, some of our proofs (marked with a $\star$) are omitted in this extended abstract.

Related Work: Eto et al. [10] showed that on r-regular graphs the problem is APX-hard for $r, d \geq 3$, while also providing polynomial-time $O(r^{d-1})$-approximations. They also show a polynomial-time 2-approximation on cubic graphs and a polynomial-time approximation scheme (PTAS) for planar graphs and every fixed constant $d \geq 3$. For a class of graphs with at most a polynomial (in n) number of minimal separators, d-Scattered Set can be solved in polynomial time for even d, while it remains NP-hard on chordal graphs and any odd $d \geq 3$ [20]. For bipartite graphs, the problem is NP-hard to approximate within a factor of $n^{1/2-\epsilon}$ and W[1]-hard for any fixed $d \geq 3$. Further, for any odd $d \geq 3$, it remains NP-complete, inapproximable and W[1]-hard [9]. It is NP-hard even for planar bipartite graphs of maximum degree 3, yet a 1.875-approximation is available on cubic graphs [11]. Furthermore, [12] shows the problem admits an EPTAS on (apex)-minor-free graphs, based on the theory of bidimensionality, while on a related result [19] offers an $n^{O(\sqrt{n})}$-time algorithm for planar graphs, making use of Voronoi diagrams and based on ideas previously used to obtain geometric QPTASs. Finally, [18] presents tight upper/lower bounds on the structurally parameterized complexity of the problem, while [21] shows that it admits an almost linear kernel on every nowhere dense graph class.

2 Definitions and Preliminaries

We use standard graph-theoretic notation. For a graph $G = (V, E)$, we let $V(G) := V$ and $E(G) := E$, an edge $e \in E$ between $u, v \in V$ is denoted by (u, v) and for a subset $X \subseteq V$, $G[X]$ denotes the graph induced by X. We let $d_G(v, u)$ denote the shortest-path distance (i.e. the number of edges) from v to

u in G. We may omit subscript G if it is clear from the context. The maximum distance between vertices is the *diameter* of the graph, while the minimum among all the maximum distances between a vertex to all other vertices (their *eccentricities*) is considered as the *radius* of the graph. For a vertex v, we let $N_G^d(v)$ denote the (open) *d-neighborhood* of v in G, i.e. the set of vertices at distance $\leq d$ from v in G (without v), while for a subset $U \subseteq V$, $N_G^d(U)$ denotes the union of the d-neighborhoods of vertices $u \in U$. For an integer q, the *q-th power graph of G*, denoted by G^q, is defined as the graph obtained from G by adding to $E(G)$ all edges between vertices $v, u \in V(G)$ for which $d_G(v, u) \leq q$. Furthermore, we let $OPT_d(G)$ denote the maximum size of a d-scattered set in G and $\alpha(G) = OPT_2(G)$ denote the size of the largest independent set. The Exponential Time Hypothesis (ETH) [16,17] implies that 3-SAT cannot be solved in time $2^{o(n)}$ on instances with n variables (a slightly weaker statement), while the definition can also refer to *randomized* algorithms. Finally, we recall here the following result by [6] that some of our reductions will be relying on (slightly paraphrased, see also [4]), that can furthermore be seen as implying the $\Delta^{1-\epsilon}$-inapproximability of Independent Set:

Theorem 2.1 ([6], Theorem 5.2). *For any sufficiently small $\epsilon > 0$ and any $r \leq N^{5+O(\epsilon)}$, there is a randomized polynomial reduction that builds from a formula ϕ of* SAT *on N variables a graph G of size $n = N^{1+\epsilon}r^{1+\epsilon}$ and maximum degree r, such that with high probability:*

- *If ϕ is satisfiable, then $\alpha(G) \geq N^{1+\epsilon}r$.*
- *If ϕ is not satisfiable, then $\alpha(G) \leq N^{1+\epsilon}r^{2\epsilon}$.*

3 Polynomial Time

3.1 Inapproximability

We show that for sufficiently large Δ and any $\epsilon_1 > 0, d \geq 4$, the d-Scattered Set problem is inapproximable to $\Delta^{\lfloor d/2 \rfloor - \epsilon_1}$ on graphs of degree bounded by Δ, unless NP $\subseteq$ BPP. Let us first summarize our reduction. Starting from an instance of Independent Set of bounded degree, we create an instance of d-Scattered Set where the degree is (roughly) the $d/2$-th square root of that of the original instance. As we are able to maintain a direct correspondence of solutions in both instances, the $\Delta^{1-\epsilon'}$-inapproximability of IS implies the $\Delta^{\lfloor d/2 \rfloor - \epsilon_1}$-inapproximability of d-Scattered Set.

The technical part of our reduction involves preserving the adjacency between vertices of the original graph without increasing the maximum degree (too far) beyond $\Delta^{2/d}$. We are able to construct a regular tree as a gadget for each vertex and let the edges of the leaves (their total number being equal to Δ) represent the edges of the original graph. To ensure that our gadget has some useful properties (i.e. small diameter), we overlay a number of extra edges on each level of the tree (i.e. between vertices at equal distance from the root), only sacrificing a small increase in maximum degree. Our complexity assumption is NP$\not\subseteq$BPP, since for the $\Delta^{1-\epsilon'}$-inapproximability of IS we use the randomized reduction from SAT of [6] (Theorem 2.1 above). In particular, we will prove the following theorem:

Theorem 3.1. *For sufficiently large Δ and any $d \geq 4, \epsilon \in (0, \lfloor d/2 \rfloor)$, there is no polynomial-time approximation algorithm for d-Scattered Set with ratio $\Delta^{\lfloor d/2 \rfloor - \epsilon}$ for graphs of maximum degree Δ, unless $NP \subseteq BPP$.*

Construction: Let $\delta = \left\lceil \sqrt[\lfloor d/2 \rfloor]{\Delta} \right\rceil$. Given $\epsilon_1 \in (0, \lfloor d/2 \rfloor)$ and an instance of Independent Set $G = (V, E)$, where the degree of any vertex is bounded by Δ, we will construct an instance $G' = (V', E')$ of d-Scattered Set, where the degree is bounded[1] by $\delta^{1+\epsilon_2} = 6\delta^{1+2\epsilon_1/d}$, for $\epsilon_2 = 2\epsilon_1/d + \log_\delta 3 > 2\epsilon_1/d$, while $OPT_2(G) = OPT_d(G')$. We assume Δ is sufficiently large for $\epsilon_1 \geq \frac{d(\log(\log(\Delta)) + c)}{4\log(\Delta)/d}$, where $c \leq 10$ is a small constant, for reasons that become apparent in the following.

Our construction for G' builds a gadget $T(v)$ for each vertex $v \in V$. For even d, each gadget $T(v)$ is composed of a $(\delta+1)$-regular tree of height $d/2-1$ and we refer to vertices of $T(v)$ at distance exactly i from the root t_v as being in the i-th *height-level* of $T(v)$, letting each such subset be denoted by $T_i(v)$. That is, every vertex of $T_i(v)$ has one neighbor in $T_{i-1}(v)$ (its parent) and δ neighbors in $T_{i+1}(v)$ (its children). For odd values of d, the difference is in the height of each tree being $\lfloor d/2 \rfloor$ instead of $d/2-1$. Since for even d the number of leaves of $T(v)$ is $\delta^{d/2-1} = (\Delta^{2/d})^{d/2-1} = \Delta^{1-2/d}$ and each such leaf also has $\delta = \Delta^{2/d}$ edges, the number of edges leading outside each gadget is $\delta^{d/2} = \Delta$ and we let each of them correspond to one edge of the original vertex v in G, i.e. we add an edge between a leaf x_v of $T(v)$ and a leaf y_u of $T(u)$, if $(v, u) \in E$. For odd d, the number of leaves is $\left(\left\lceil \sqrt[\lfloor d/2 \rfloor]{\Delta} \right\rceil\right)^{\lfloor d/2 \rfloor}$ (i.e. at least Δ) and we let each leaf correspond to an edge of the original vertex v in G, i.e. we *identify* two such leaves x_v, y_u of two gadgets $T(v), T(u)$, if $(v, u) \in E$ in G. In this way, the gadgets $T(v), T(u)$ share a common "leaf" of degree 2, that is at distance $\lfloor d/2 \rfloor$ from both roots $t_v \in T(v), t_u \in T(u)$. See Fig. 1 for an illustration.

Next, in order to make the diameter of our gadgets at most equal to their height, we will add a number of edges between the vertices of each height-level i of each gadget $T(v)$, for every $v \in V$. We will first add the edges of a cycle plus a random matching (using a technique from [3]) and then the edges of an appropriately chosen power graph of this subgraph containing the edges of the cycle plus the matching. These edges will be overlaid on each height-level, meaning our final construction will contain the edges of the tree, the cycle, the matching, as well as the power graph.

For even d and each gadget $T(v)$, we first make all vertices $T_i(v)$ at each height-level $i < 1 + \epsilon_2$ into a clique. For larger height-levels $i \in [1 + \epsilon_2, d/2 - 1]$, we first make the vertices into a cycle (arbitrarily ordered) and then also add

[1] We note that this value of ϵ_2 is for odd values of d. For d even, the correct value is such that we have the (slightly lower) bound $\delta^{1+\epsilon_2} = \delta + 3\delta^{1+2\epsilon_1/d}$, but we write ϵ_2 for both cases to simplify notation.

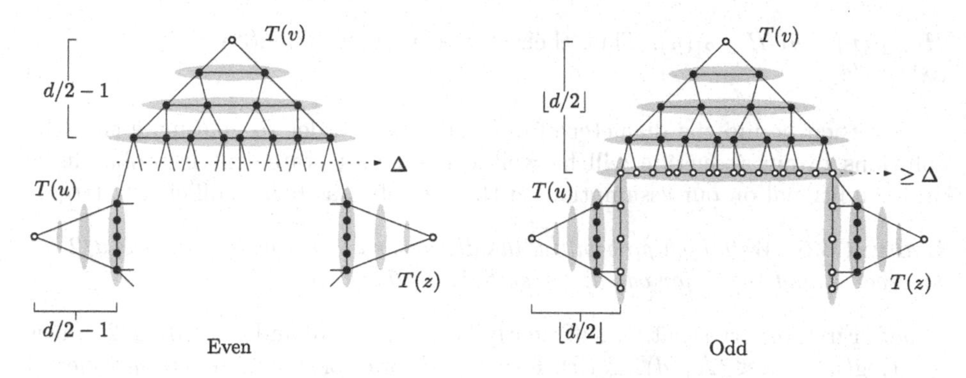

Fig. 1. Our constructions for an example subgraph consisting of a path on three vertices (u, v, z) and even/odd d. Ellipses in grey designate the overlaid edges on each height-level.

a random matching, i.e. the edges between each pair of a random partition of $T_i(v)$ into disjoint pairs (plus a singleton if $|T_i(v)|$ is odd). Letting $P_i(v)$ denote these edges of the cycle plus the matching for each $T_i(v)$, we define the subgraph $H_i(v) = (T_i(v), P_i(v))$ and compute the $\lceil((1 + 2\epsilon_1/d)\log_2(\delta))\rceil$-power graph of $H_i(v)$, finally also adding its edges to G'. For odd d and each gadget $T(v)$, we again make the vertices $T_i(v)$ at each height-level $i < 1 + \epsilon_2$ into a clique and for larger height-levels $i \in [1 + \epsilon_2, \lfloor d/2 \rfloor]$, we follow the same process.

This concludes our construction, while to prove our claims on the diameter of our gadgets we also make use of the following statements:

Theorem 3.2 ([3], Theorem 1). *Let G be a graph formed by adding a random matching to an n-cycle. Then with probability tending to 1 as n goes to infinity, G has diameter upper-bounded by $\log_2(n) + \log_2(\log(n)) + c$, where c is a small constant (at most 10).*

Lemma 3.3 ($\star$). *Let G be a graph of diameter $\leq a$. Then the diameter of the b-power graph G^b is $\leq \lceil a/b \rceil$, for any integer $b < a$.*

We are now ready to argue about the maximum degree of any vertex in the instances built by our construction.

Lemma 3.4. *The maximum degree of any vertex in G' is $\leq \delta + 3\delta^{1+2\epsilon_1/d}$ for even d and $\leq 6\delta^{1+2\epsilon_1/d}$ for odd d.*

Proof. Observe that for d even, the degree of any vertex is bounded by the sum of the $\delta + 1$ edges of the tree plus the number of edges added by the power graph (including the three edges of the cycle and matching): $\sum_{k=0}^{\lceil(1+2\epsilon_1/d)\log_2(\delta)\rceil}(3 \cdot 2^k) = 3 \cdot 2^{\lceil(1+2\epsilon_1/d)\log_2(\delta)-1\rceil} - 1 \leq 3 \cdot 2^{(1+2\epsilon_1/d)\log_2(\delta)} - 1 = 3 \cdot \delta^{1+2\epsilon_1/d} - 1$, for a total of $\delta + 3\delta^{1+2\epsilon_1/d}$.

For d odd, we note that the degree of all other vertices is strictly lower than that of the "shared" leaves between gadgets, since each leaf between two gadgets $T(v), T(u)$ (representing the edge (v, u) of G) will belong to two subgraphs

$H_{\lfloor d/2 \rfloor}(v)$ and $H_{\lfloor d/2 \rfloor}(u)$. Thus their degree will be $2 + 2(3 \cdot \delta^{1+2\epsilon_1/d} - 1) = 6\delta^{1+2\epsilon_1/d}$. □

We then bound the diameter of our gadgets in order to guarantee that the solutions in our reduction will be well-formed. Our statement is probabilistic and conditional on our assumption on the size of Δ as being sufficiently large.

Lemma 3.5. *With high probability, the diameter of each gadget $T(v)$ is $d/2-1$ for even d and $\lfloor d/2 \rfloor$ for odd d, for sufficiently large Δ.*

Proof. First, observe that for sufficiently large n, $c \leq 10$ and $\epsilon_1 \in (0, \lfloor d/2 \rfloor)$, it is $\log_2(\log(n)) + c < (2\epsilon_1/d)\log_2(n)$. For even d, our construction uses n-cycles of length $n = \delta^i$ for each $i \in [1+\epsilon_2, d/2-1]$, meaning that Δ must be sufficiently large for $\epsilon_1 \geq \frac{d(\log(2i\log(\Delta)/d)+c)}{4i\log(\Delta)/d}$, while for odd d it is $i \in [1+\epsilon_2, \lfloor d/2 \rfloor]$. As noted above, our assumption for Δ requires that it is sufficiently large for $\epsilon_1 \geq \frac{d(\log(\log(\Delta))+c)}{4\log(\Delta)/d}$, which is $> \frac{d(\log(2i\log(\Delta)/d)+c)}{4i\log(\Delta)/d}$ for the required range of i in both cases.

By Theorem 3.2, the distance between any pair of vertices at height-level i after adding the edges of $P_i(v)$ is at most $\log_2(\delta^i) + \log_2(\log(\delta^i)) + c$ (with high probability). This is $< (1+2\epsilon_1/d)\log_2(\delta^i)$ for sufficiently large Δ. By Lemma 3.3, taking the $\lceil((1+2\epsilon_1/d)\log_2(\delta))\rceil$-power of $H_i(v)$ shortens the distance to at most $\frac{(1+2\epsilon_1/d)\log_2(\delta^i)}{\lceil(1+2\epsilon_1/d)\log_2(\delta)\rceil} \leq i$, for each height-level $i \in [1+\epsilon_2, d/2-1]$. For smaller values of i, the vertices of each height-level form a clique and the distance between any pair of them is thus at most 1.

For odd values of d, the size n of the cycles we use is again δ^i, with $i \in [1+\epsilon_2, \lfloor d/2 \rfloor]$ and we thus have once more that for sufficiently large Δ the distance between any pair of vertices after adding the edges of $P_i(v)$ to each height-level i of each gadget $T(v)$ is at most $(1+2\epsilon_1/d)\log_2(\delta^i)$ (with high probability) and at most i after taking the $\lceil((1+2\epsilon_1/d)\log_2(\delta))\rceil$-power of $H_i(v)$. Again, for smaller $i < 1+\epsilon_2$, $T_i(v)$ is a clique.

Since at each height-level i, no pair of vertices is at distance $> i$ with $i \leq d/2-1$ for even d and $i \leq \lfloor d/2 \rfloor$ for odd d, the distance between any vertex x at some height-level i_x to another vertex y at height-level $i_y > i_x$ will be at most i_x from x to the root of the subtree of $T(v)$ (at level i_x) that contains y. From there to y it will be at most $d/2-1-i_x$ for even d and at most $\lfloor d/2 \rfloor - i_x$ for odd d. Furthermore, the distance from the root of $T(v)$ to a leaf is exactly $d/2-1$ for even d and exactly $\lfloor d/2 \rfloor$ for odd d. □

We finalize our argument with a series of lemmas leading to the proof of Theorem 3.1, that detail the behaviour of solutions that can form in our construction, relative to the independence of vertices in the original graph.

Lemma 3.6. *No d-*SCATTERED SET *in G' can contain a vertex from gadget $T(v)$ and a vertex from gadget $T(u)$, if $(u,v) \in E$.*

Proof. Since $(u,v) \in E$, there is an edge $(x_v, y_u) \in E'$ between a leaf $x_v \in T(v)$ and $y_u \in T(u)$ for even d, while for odd d the leaf x belongs to both $T(v), T(u)$

and is at distance $\lfloor d/2 \rfloor$ from each of their roots. Thus for even d the maximum distance from any vertex of $T(v)$ to $y_u \in T(u)$ is $d/2-1+1 = d/2$, by Lemma 3.5, and for odd d this is $\lfloor d/2 \rfloor$. Since, by the same lemma, the diameter of $T(u)$ is $d/2-1$ for even d and $\lfloor d/2 \rfloor$ for odd d, there is no vertex of $T(u)$ that can be in any d-Scattered Set along with any vertex of $T(v)$, as the maximum distance is $\leq d/2 + d/2 - 1 = d - 1$ for even d and $\leq \lfloor d/2 \rfloor + \lfloor d/2 \rfloor = d - 1$ for odd d. □

Lemma 3.7. *If $(u, v) \notin E$, then the distance between the root t_v of $T(v)$ and the root t_u of $T(u)$ is at least d.*

Proof. Since $(u, v) \notin E$, then there is no edge between any pair of leaves x_u of $T(u)$ and y_v of $T(v)$ for even d. Thus the shortest possible distance between any such pair of leaves is 2 for even d, through a third leaf z_w of another gadget $T(w)$ corresponding to a vertex w adjacent to both u and v in G. The distance from t_v to any leaf of $T(v)$ is $d/2-1$ and the distance from t_u to any leaf of $T(u)$ is also $d/2-1$. Thus the distance from t_u to t_v must be at least $d/2 - 1 + d/2 - 1 + 2 = d$.

For odd d, there is no shared leaf x between the two gadgets, i.e. at distance $\lfloor d/2 \rfloor$ from both roots. Thus the distance between two leaves $x_u \in T(u)$ and $y_v \in T(v)$ is at least 1, if each of these is shared with a third gadget $T(w)$ corresponding to a vertex w that is adjacent to both u and v in G. The distance from t_v to any leaf of $T(v)$ is $\lfloor d/2 \rfloor$ and the distance from t_u to any leaf of $T(u)$ is also $\lfloor d/2 \rfloor$. Thus the distance from t_u to t_v is at least $2\lfloor d/2 \rfloor + 1 = d$. □

Lemma 3.8. *For any independent set S in G, there is a d-Scattered Set K in G', with $|S| = |K|$.*

Proof. Given an independent set S in G, we let K include the root vertex $t_v \in T(v)$ for each $v \in S$. Clearly, $|S| = |K|$. Since S is independent, there is no edge (u, v) between any pair $u, v \in S$ and thus, by Lemma 3.7, vertices t_v and t_u are at distance at least d. □

Lemma 3.9. *For any d-Scattered Set K in G', there is an independent set S in G, with $|K| = |S|$.*

Proof. Given a d-Scattered Set K in G, we know there is at most one vertex from each gadget $T(v)$ in K, since its diameter is $d/2-1$ for even d and $\lfloor d/2 \rfloor$ for odd d, by Lemma 3.5. Furthermore, for any two vertices $x, y \in K$, we know by Lemma 3.6 that if $x \in T(u)$ and $y \in T(v)$ for gadgets corresponding to vertices $u, v \in V$, then $(u, v) \notin E$ and thus u, v are independent in G. We let set S contain each vertex $v \in V$ whose corresponding gadget $T(v)$ contains a vertex of K. These vertices are all independent and also $|K| = |S|$. □

Proof of Theorem 3.1. We suppose the existence of a polynomial-time approximation algorithm for d-Scattered Set with ratio $\Delta^{\lfloor d/2 \rfloor - \epsilon_1}$ for graphs of maximum degree Δ and some $0 < \epsilon_1 < d/2$. We assume Δ is sufficiently large for $\epsilon_1 \geq \frac{d(\log(\log(\Delta)) + 10)}{4\log(\Delta)/d}$.

Starting from a formula ϕ of SAT on N variables, where N is also sufficiently large, i.e $N > \Delta^{1/(5+O(\epsilon'))}$ (where ϵ' is defined below), we use Theorem 2.1 to produce an instance $G = (V, E)$ of Independent Set on $|V| = N^{1+\epsilon'}\Delta^{1+\epsilon'}$ vertices and of maximum degree Δ, such that with high probability: if ϕ is satisfiable, then $\alpha(G) \geq N^{1+\epsilon'}\Delta$; if ϕ is not satisfiable, then $\alpha(G) \leq N^{1+\epsilon'}\Delta^{2\epsilon'}$. Thus approximating Independent Set in polynomial time on G within a factor of $\Delta^{1-2\epsilon'}$, for $\epsilon' > 0$, would permit us to decide if ϕ is satisfiable, with high probability.

We next use the above construction to create an instance G' of d-Scattered Set where the degree is bounded by $6\delta^{1+2\epsilon_1/d} = \delta^{1+\epsilon_2}$, for $\epsilon_2 > 2\epsilon_1/d$, by Lemma 3.4. Slightly overloading notation, we let $\epsilon_3 \geq \epsilon_2$ be such that $\delta^{1+\epsilon_2} = (\lceil \Delta^{\frac{1}{\lfloor d/2 \rfloor}} \rceil)^{1+\epsilon_2} = (\Delta^{\frac{1}{\lfloor d/2 \rfloor}})^{1+\epsilon_3}$. We now let $\epsilon' = \frac{\epsilon_1(1+\epsilon_3) - \epsilon_3\lfloor d/2 \rfloor}{2\lfloor d/2 \rfloor}$. Note that $\epsilon' > 0$, since $\epsilon_3 \geq \epsilon_2 > 2\epsilon_1/d$.

We then apply the supposed approximation for d-Scattered Set on G'. This returns a solution at most $(\delta^{1+\epsilon_2})^{\lfloor d/2 \rfloor - \epsilon_1} = \Delta^{(1-\frac{\epsilon_1}{\lfloor d/2 \rfloor})(1+\epsilon_3)} = \Delta^{1-\frac{\epsilon_1(1+\epsilon_3)-\epsilon_3\lfloor d/2 \rfloor}{\lfloor d/2 \rfloor}} = \Delta^{1-2\epsilon'}$ from the optimum. By Lemma 3.9 we can find a solution for Independent Set in G of the same size, i.e. we can approximate $\alpha(G)$ within a factor of $\Delta^{1-2\epsilon'}$, again, with high probability (as Lemma 3.5 is also randomized). This would allow us to decide if ϕ is satisfiable and thus solve SAT in polynomial time with two-sided bounded errors, implying NP $\subseteq$ BPP. □

3.2 Approximation

We next show that any (degree-based) greedy polynomial-time approximation algorithm for d-Scattered Set achieves a ratio of $O(\Delta^{\lfloor d/2 \rfloor})$, thus improving upon the analysis of [10] and the $O(\Delta^{d-1})$- and $O(\Delta^{d-2}/d)$-approximations given therein.

Our strategy is to bound the size of the largest d-scattered set in any graph of maximum degree at most Δ and radius at most $d-1$, centered on some vertex v. The idea is that in one of its iterations our greedy algorithm would select v and thus exclude all other vertices within distance $d-1$ from v, yet an upper bound on the size of the largest possible d-scattered set can guarantee that the ratio of our algorithm will not be too large.

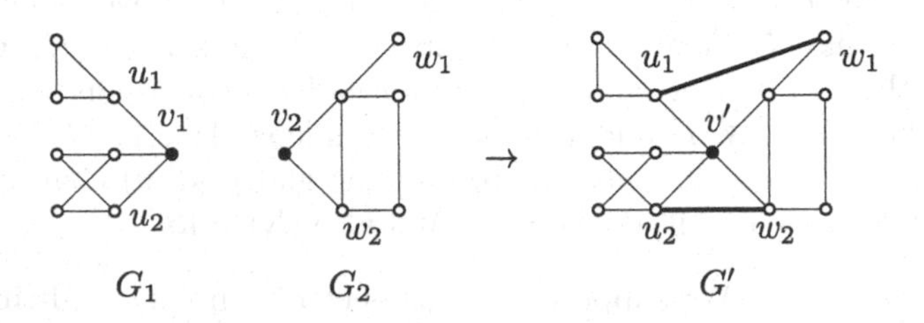

Fig. 2. An example graph $G' = M_{G_1}^{G_2}(v_1, v_2, [u_1, u_2], [w_1, w_2])$ for G_1, G_2 shown on the left, with edges added by the third merge operation shown in bold.

The following definition of our "merge" operation (see also Fig. 2) will allow us to consider all possible graphs of a given radius and degree and provide upper bounds on the size of the optimal solution in such graphs. These bounds on the size of the optimal are then used to compare it to those solutions produced by our greedy scheme.

Definition 3.10 (Merge operation). *For two connected graphs* $G_1 = (V_1, E_1), G_2 = (V_2, E_2)$*, the* merged *graph* $M_{G_1}^{G_2}(v_1, v_2, \mathbf{U}, \mathbf{W})$*, where* $\mathbf{U} = [u_1, \dots, u_{k_1}]$*,* $\mathbf{W} = [w_1, \dots, w_{k_2}]$ *are ordered (possibly empty and with repetitions allowed) sequences of vertices from* V_1 *and* V_2*, respectively, is defined as the graph* $G' = (V', E')$ *obtained by: (1) Identification of vertex* $v_1 \in G_1$ *and vertex* $v_2 \in G_2$*, i.e.* V' *is composed of the union of* V_1, V_2 *after removal of vertices* v_1, v_2 *and addition of a new vertex* v'*. (2) Replacement of all edges of* v_1, v_2 *by new edges with* v' *as the new endpoint (3)Addition of a number of edges between vertices of* G_1, G_2*, i.e.* E' *also contains an edge between every pair* (u_i, w_j) *from* $\mathbf{U}, \mathbf{W}$*, for* $i = j$*.*

Lemma 3.11. *The maximum size of a d-scattered set in any graph of maximum degree at most* Δ *and radius at most* $\lceil d/2 \rceil$ *centered on some vertex* v *is at most* Δ*.*

Proof. Consider a graph of maximum degree Δ and radius $\lceil d/2 \rceil$ centered on some vertex v: the only pairs of vertices at distance d from each other must be at distance $\geq \lceil d/2 \rceil$ from v (or $\lfloor d/2 \rfloor$ in one side for odd d), as any vertex u at distance $< \lfloor d/2 \rfloor$ from v will be at distance $< d$ from any other vertex z in the graph, since z is at distance $\leq \lceil d/2 \rceil$ from v (due to the graph's radius). Furthermore, for every vertex in the d-scattered set, there must be an edge-disjoint path of length at least $\lceil d/2 \rceil$ to v that is not shared with any other such vertex, i.e. these paths can only share vertex v at distance $\lceil d/2 \rceil$ from their endpoints (the vertices that can be in a d-scattered set). As the degree of v is bounded by Δ, the number of such disjoint paths also cannot be more than Δ. □

Lemma 3.12. *Given two graphs* G_1, G_2*, for* $G' = M_{G_1}^{G_2}(v_1, v_2, \mathbf{U}, \mathbf{W})$ *and any* $\mathbf{U}, \mathbf{W}$*, it is* $OPT_d(G') \leq OPT_d(G_1) + OPT_d(G_2)$*.*

Proof. Assume for the sake of contradiction that $OPT_d(G') > OPT_d(G_1) + OPT_d(G_2)$ and let S denote an optimum d-scattered set in G' of this size, with $S_1 = S \cap \{V_1 \setminus \{v_1\} \cup \{v'\}\}$ and $S_2 = S \cap \{V_2 \setminus \{v_2\} \cup \{v'\}\}$ denoting the parts in G_1, G_2, respectively. Since $|S| > OPT_d(G_1) + OPT_d(G_2)$, then for any pair of optimal $K_1 \subseteq V_1$, $K_2 \subseteq V_2$ in G_1, G_2, there must be at least one vertex s in S for which $s \notin K_1$ and $s \notin K_2$, but it must be $s \in S_1$ or $s \in S_2$ (or both, if $s = v'$).

Observe that the distance between any pair of vertices in G_1 (the same holds for G_2) cannot increase in G' after the merge operation, since identification of a pair of vertices between two graphs and addition of any number of edges between the two can only decrease their distance Thus if $s \in V_1$, then S_1 is also a d-scattered set in G_1 (potentially substituting v' for v_1) and so, if $|S_1| > |K_1|$ then

K_1 was not optimal for G_1. If $|S_1| \leq |K_1|$, it must be $|S_2| > |K_2|$ contradicting the optimality of K_2 for G_2. Similarly, if $u \in V_2$ we have either $|S_1| > |K_1|$ or $|S_2| > |K_2|$. If u is the merged vertex v' then there must be at least two other vertices added from V_1, V_2 for $|S| > |K_1| + |K_2|$, since S can only contain v' in the place of $v_1 \in K_1$ and $v_2 \in K_2$. In this case the same argument as above gives the contradiction. □

Lemma 3.13. *For any graph $G = (V, E)$ of maximum degree at most Δ and radius at most $d-1$ centered on some vertex v, it is $OPT_d(G) \leq O(\Delta^{\lfloor d/2 \rfloor})$.*

Proof. Any graph G of maximum degree at most Δ and radius at most $d-1$ centered on a vertex v can be obtained by the following process: we begin with a graph H of radius at most $\lfloor d/2 \rfloor - 1$ and maximum degree Δ. Let $\{v_1, \ldots, v_k\} \in H$ be the set of vertices at maximum distance from v, i.e. $d_H(v, v_i) = \lfloor d/2 \rfloor - 1$. Since the degree of H is bounded by Δ, it must be $k \leq \Delta^{\lfloor d/2 \rfloor - 1}$. We now let H_i, for each $i \leq k$, denote a series of at most k graphs of radius at most $\lceil d/2 \rceil$ centered on a vertex v_i and maximum degree Δ.

Repeatedly applying the merge operation $M_H^{H_i}(v_1, v_i, \mathbf{U}, \mathbf{W})$ between graph H (or the result of the previous operation) and such a graph H_i we can obtain any graph G of radius at most $d-1$: identifying a vertex $v_j \in H$ (for $j \in [1, k]$) at maximum distance from v with the central vertex v_i of H_i and then adding any number of edges between the vertices of H and H_i (while respecting the maximum degree of Δ), we can produce any graph of radius $\leq d-1$, since the distance from v to each v_j is at most $\lfloor d/2 \rfloor - 1$ and from there to any vertex of H_i it is at most $\lceil d/2 \rceil$. The remaining structure of G can be constructed by the chosen structures of the graphs H, H_i and the added edges between them, i.e. the sequences $\mathbf{U}, \mathbf{W}$. By Lemma 3.11 it is $OPT_d(H_i) \leq \Delta$ and by Lemma 3.12, it must be $OPT_d(G) \leq OPT_d(H) + \sum_{i=1}^{k} OPT_d(H_i) \leq 1 + \Delta \cdot \Delta^{\lfloor d/2 \rfloor - 1} \leq 1 + \Delta^{\lfloor d/2 \rfloor}$. □

Theorem 3.14. *Any degree-based greedy approximation algorithm for d-*Scattered Set *achieves a ratio of $O(\Delta^{\lfloor d/2 \rfloor})$ on graphs of degree bounded by Δ.*

Proof. Let $G = (V, E)$ be the input graph and consider the process of our supposed greedy algorithm: it picks a vertex v_i, removes it from consideration along with the set $V_i \subseteq V$ of vertices at distance at most $d-1$ from v_i and continues the process until there are no vertices left to consider. The sets $V_1, \ldots, V_{ALG}$ thus form a partition of G. By Lemma 3.13, the optimum size of a d-scattered set in any such V_i is at most $O(\Delta^{\lfloor d/2 \rfloor})$ and thus $OPT_d(G) \leq ALG \cdot O(\Delta^{\lfloor d/2 \rfloor})$, by Lemma 3.12, since G can be seen as the merged graph of $G[V_1], \ldots, G[V_{ALG}]$. □

3.3 Bipartite Graphs

Here we consider bipartite graphs and show that d-Scattered Set is approximable to $2\sqrt{n}$ in polynomial time also for even values of d. Our algorithm will be applied on both sides of the bipartition and each time it will only consider vertices from one side as candidates for inclusion in the solution. Appropriate sub-instances of SET PACKING are then defined and solved using the known $\sqrt{n}$-approximation for that problem.

Definition 3.15. *For a bipartite graph $G = (A \cup B, E)$, let $1OPT_d(G)$ denote the size of the largest* one-sided *d-scattered set of G, i.e. a set that only includes vertices from the same side of the bipartition A or B, but not both.*

Lemma 3.16 (⋆). *For a bipartite graph $G = (A \cup B, E)$, it is $1OPT_d(G) \geq OPT_d(G)/2$.*

Theorem 3.17. *For any bipartite graph $G = (A \cup B, E)$ of size n and d even, the d-SCATTERED SET problem can be approximated within a factor of $2\sqrt{n}$ in polynomial time.*

Proof. We will consider two cases based on the parity of $d/2$ and define appropriate SET PACKING instances whose solutions are in a one-to-one correspondence with one-sided d-scattered sets in G. We will then be able to apply the $\sqrt{n}$-approximation for SET PACKING of [13]. We will repeat this process for both sides A, B of the bipartition and retain the best solution found. Thus we will be able to approximate $1OPT_d(G)$ within a factor of $\sqrt{n}$ and then rely on Lemma 3.16 to obtain the claimed bound. Our SET PACKING instances are defined as follows: for $d/2$ even, we make a set c_i for every vertex a_i of A (i.e. from one side) and an element e_j for every vertex b_j of B (i.e. from the other side). For $d/2$ odd, we make a set c_i for every vertex a_i of A (again from one side) and an element e_j for every vertex b_j of B *and* an element r_i for every vertex $a_i \in A$ (i.e. from both sides). Note that $i, j \leq n$. In both cases we include an element corresponding to vertex $x \in G$ in a set corresponding to a vertex $y \in G$, if $d_G(x, y) \leq d/2 - 1$. We then claim that for any collection C of compatible (i.e. non-overlapping) sets in the above definitions, we can always find a one-sided d-scattered set $S \subseteq A$ in G with $|C| = |S|$ and vice-versa.

First consider the case where $d/2$ is even. Given a one-sided d-scattered set $S \subseteq A$, we let C include all the sets that correspond to some vertex in S and suppose for a contradiction that there exists a pair of sets $c_1, c_2 \in C$ that are incompatible, i.e. that there exists some element e with $e \in c_1$ and $e \in c_2$. Let $a_1, a_2 \in A$ be the vertices corresponding to sets c_1, c_2 and $b \in B$ be the vertex corresponding to element e. Then it must be $d_G(a_1, b) \leq d/2 - 1$ since $e \in c_1$ and $d_G(b, a_2) \leq d/2 - 1$ since $e \in c_2$, that gives $d_G(v_1, v_2) \leq d - 2$, which contradicts S being a d-scattered set. On the other hand, given collection C of compatible sets we let $S \subseteq A$ include all the vertices corresponding to some set in C and suppose there exists a pair of vertices $a_1, a_2 \in S$ for which it is $d_G(a_1, a_2) < d$. Since d is even and $a_1, a_2 \in A$, if $d_G(a_1, a_2) < d$ it must be $d_G(a_1, a_2) \leq d - 2$, as any shortest path between two vertices on the same side of a bipartite graph must be of even length. Thus there must exist at least one vertex $b \in B$ on a shortest path between a_1, a_2 in G for which it is $d_G(a_1, b) \leq d/2 - 1$ and $d_G(b, a_2) \leq d/2 - 1$. This means that the element e corresponding to vertex

$b \in B$ must be included in both sets c_1, c_2 corresponding to vertices $a_1, a_2 \in A$, which contradicts the compatibility of sets in C.

We next consider the case where $d/2$ is odd. Given a one-sided d-scattered set $S \subseteq A$, we again let C include all sets that correspond to some vertex in S. If there exists a pair of sets $c_1, c_s \in C$ that contain the same element e corresponding to some vertex $b \in B$ or some element r that corresponds to a vertex $a \in A$, then by the same argument as in the even case we know that there must exist paths of length $\leq d/2-1$ from both vertices $a_1, a_2 \in A$ (corresponding to $c_1, c_2 \in C$) to vertex $b \in B$ or $a \in A$ and thus it must be $d_G(a_1, a_2) < d$. On the other hand, given a collection C of compatible sets we again let $S \subseteq A$ include all the vertices corresponding to sets in C. Supposing there exists a pair $a_1, a_2 \in S$ for which it is $d_G(a_1, a_2) < d$, then again as d is even it must be $d_G(a_1, a_2) \leq d - 2$. This means there must be a vertex $a \in A$ on a shortest path between a_1 and a_2 for which $d_G(a_1, a) \leq d/2 - 1$ and $d_G(a, a_2) \leq d/2 - 1$, which means the corresponding sets $c_1, c_2 \in C$ must both contain element r that corresponds to this vertex $a \in A$, giving a contradiction.

Our algorithm then is as follows. For a given bipartite graph $G = (A \cup B, E)$, we define an instance of SET PACKING as described above (depending on the parity of $d/2$) and apply the $\sqrt{n}$-approximation of [13]. Observe that $|A|, |B| \leq n$. We then exchange the sets A, B in the definitions of our instances and repeat the same process. This will return a solution S of size $|S| \geq \frac{1OPT_d(G)}{\sqrt{n}}$, which by Lemma 3.16 is $\geq \frac{OPT_d(G)}{2\sqrt{n}}$. □

4 Super-Polynomial Time

We begin with a straightforward upper bound on the size of the solution in any connected graph that is then employed in obtaining an exact exponential-time algorithm that simply tries all subsets of vertices up to the size bound.

Lemma 4.1 (⋆). *The maximum size of any d-Scattered Set in a connected graph is* $\left\lfloor \frac{n}{\lceil d/2 \rceil} \right\rfloor$.

Theorem 4.2 (⋆). *The d-*SCATTERED SET *problem can be solved in* $O^*((ed)^{\frac{2n}{d}})$ *time.*

4.1 Inapproximability

We now turn our attention to the problem's hardness of approximation in super-polynomial time. We use Theorem 2.1 in conjunction with standard reductions from INDEPENDENT SET to d-SCATTERED SET for the two cases, that depend on the parity of d (see Fig. 3).

Theorem 4.3 (⋆). *Under the randomized ETH, for any even* $d \geq 4$, $\epsilon > 0$ *and* $\rho \leq (2n/d)^{5/6}$, *no* ρ*-approximation for* d-SCATTERED SET *can take time* $2^{\left(\frac{n^{1-\epsilon}}{\rho^{1+\epsilon} d^{1-\epsilon}}\right)} \cdot n^{O(1)}$.

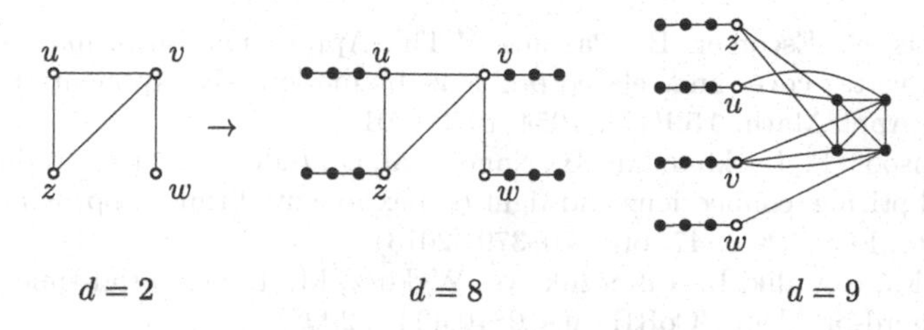

Fig. 3. Examples of the constructions for even (center) and odd (right) values of d. Note the existence of an edge "gadget" for the odd case. This necessity is responsible for the difference in running times and is due to the parity idiosyncrasies of the problem.

Theorem 4.4 ($\star$). *Under the randomized ETH, for any odd* $d \geq 5$, $\epsilon > 0$ *and* $\rho \leq (2n/d)^{5/6}$, *no* ρ*-approximation for* d-SCATTERED SET *can take time* $2^{\left(\frac{n^{1-\epsilon}}{\rho^{1+\epsilon}(d+\rho)^{1+\epsilon}}\right)} \cdot n^{O(1)}$.

4.2 Approximation

We complement the above hardness results with approximation algorithms of almost matching super-polynomial running times. Similarly to the exact algorithm of Theorem 4.2, the upper bound from the beginning of this section is used for even values of d, while for the odd values this idea is combined with a greedy scheme based on minimum vertex degree.

Theorem 4.5 ($\star$). *For any even* $d \geq 2$ *and any* $\rho \leq \frac{n}{\lfloor d/2 \rfloor}$, *there is a* ρ*-approximation algorithm for* d-SCATTERED SET *of running time* $O^*((e\rho d)^{\frac{2n}{\rho d}})$.

Theorem 4.6 ($\star$). *For any odd* $d \geq 3$ *and any* $\rho \leq \frac{n}{\lceil d/2 \rceil}$, *there is a* ρ*-approximation algorithm for* d-SCATTERED SET *of running time* $O^*((e\rho d)^{\frac{2n}{\rho(d+\rho)}})$.

References

1. Alon, N., Feige, U., Wigderson, A., Zuckerman, D.: Derandomized graph products. Comput. Complex. **5**(1), 60–75 (1995)
2. Berman, P., Karpinski, M.: On some tighter inapproximability results (Extended Abstract). In: Wiedermann, J., van Emde Boas, P., Nielsen, M. (eds.) ICALP 1999. LNCS, vol. 1644, pp. 200–209. Springer, Heidelberg (1999). https://doi.org/10.1007/3-540-48523-6_17
3. Bollobás, B., Chung, F.R.K.: The diameter of a cycle plus a random matching. SIAM J. Discret. Math. **1**(3), 328–333 (1988)
4. Bonnet, E., Lampis, M., Paschos, V.Th.: Time-approximation trade-offs for inapproximable problems. In: STACS, LIPIcs, vol. 47, pp. 22:1–22:14 (2016)

5. Bourgeois, N., Escoffier, B., Paschos, V.Th.: Approximation of max independent set, min vertex cover and related problems by moderately exponential algorithms. Discret. Appl. Math. **159**(17), 1954–1970 (2011)
6. Chalermsook, P., Laekhanukit, B., Nanongkai, D.: Independent set, induced matching, and pricing: connections and tight (subexponential time) approximation hardnesses. In: FOCS, vol. 47, pp. 370–379 (2013)
7. Cygan, M., Kowalik, L., Pilipczuk, M., Wykurz, M.: Exponential-time approximation of hard problems. CoRR, abs/0810.4934 (2008)
8. Demange, M., Paschos, V.Th.: Improved approximations for maximum independent set via approximation chains. Appl. Math. Lett. **10**(3), 105–110 (1997)
9. Eto, H., Guo, F., Miyano, E.: Distance- d independent set problems for bipartite and chordal graphs. J. Comb. Optim. **27**(1), 88–99 (2014)
10. Eto, H., Ito, T., Liu, Z., Miyano, E.: Approximability of the distance independent set problem on regular graphs and planar graphs. In: Chan, T.-H.H., Li, M., Wang, L. (eds.) COCOA 2016. LNCS, vol. 10043, pp. 270–284. Springer, Cham (2016). https://doi.org/10.1007/978-3-319-48749-6_20
11. Eto, H., Ito, T., Liu, Z., Miyano, E.: Approximation algorithm for the distance-3 independent set problem on cubic graphs. In: Poon, S.-H., Rahman, M.S., Yen, H.-C. (eds.) WALCOM 2017. LNCS, vol. 10167, pp. 228–240. Springer, Cham (2017). https://doi.org/10.1007/978-3-319-53925-6_18
12. Fomin, F.V., Lokshtanov, D., Raman, V., Saurabh, S.: Bidimensionality and EPTAS. In: SODA, pp. 748–759. SIAM (2011)
13. Halldórsson, M.M., Kratochvil, J., Telle, J.A.: Independent sets with domination constraints. Discret. Appl. Math. **99**(1–3), 39–54 (2000)
14. Halldórsson, M.M., Radhakrishnan, J.: Greed is good: approximating independent sets in sparse and bounded-degree graphs. Algorithmica **18**(1), 145–163 (1997)
15. Håstad, J.: Clique is hard to approximate within $n^{1-\epsilon}$. Acta Mathematica **182**, 105–142 (1999)
16. Impagliazzo, R., Paturi, R.: On the complexity of k-SAT. J. Comput. Syst. Sci. **62**(2), 367–375 (2001)
17. Impagliazzo, R., Paturi, R., Zane, F.: Which problems have strongly exponential complexity? J. Comput. Syst. Sci. **63**(4), 512–530 (2001)
18. Katsikarelis, I., Lampis, M., Paschos, V.Th.: Structurally parameterized d-scattered set. In: Brandstädt, A., Köhler, E., Meer, K. (eds.) WG. LNCS, pp. 292–305, vol. 11159. Springer, Heidelberg (2018). https://doi.org/10.1007/978-3-030-00256-5_24
19. Marx, D., Pilipczuk, M.: Optimal parameterized algorithms for planar facility location problems using Voronoi diagrams. In: Bansal, N., Finocchi, I. (eds.) ESA 2015. LNCS, vol. 9294, pp. 865–877. Springer, Heidelberg (2015). https://doi.org/10.1007/978-3-662-48350-3_72
20. Montealegre, P., Todinca, I.: On distance-d independent set and other problems in graphs with "few" minimal separators. In: Heggernes, P. (ed.) WG 2016. LNCS, vol. 9941, pp. 183–194. Springer, Heidelberg (2016). https://doi.org/10.1007/978-3-662-53536-3_16
21. Pilipczuk, M., Siebertz, S.: Kernelization and approximation of distance-r independent sets on nowhere dense graphs. CoRR, abs/1809.05675 (2018)

Greedy Is Optimal for Online Restricted Assignment and Smart Grid Scheduling for Unit Size Jobs

Fu-Hong Liu[1(✉)], Hsiang-Hsuan Liu[2,3], and Prudence W. H. Wong[4]

[1] Department of Computer Science, National Tsing Hua University, Hsinchu, Taiwan
fhliu@cs.nthu.edu.tw

[2] Department Information and Computing Sciences, Utrecht University, Utrecht, The Netherlands
H.H.Liu@uu.nl

[3] Institute of Computer Science, Wroclaw University, Wroclaw, Poland

[4] Department of Computer Science, University of Liverpool, Liverpool, UK
pwong@liverpool.ac.uk

Abstract. We study online scheduling of unit-sized jobs in two related problems, namely, restricted assignment problem and smart grid problem. The input to the two problems are in close analogy but the objective functions are different. We show that the greedy algorithm is an optimal online algorithm for both problems. Typically, an online algorithm is proved to be an optimal online algorithm through bounding its competitive ratio and showing a lower bound with matching competitive ratio. However, our analysis does not take this approach. Instead, we prove the optimality without giving the exact bounds on competitive ratio. Roughly speaking, given any online algorithm and a job instance, we show the existence of another job instance for greedy such that (i) the two instances admit the same optimal offline schedule; (ii) the cost of the online algorithm is at least that of the greedy algorithm on the respective job instance. With these properties, we can show that the competitive ratio of the greedy algorithm is the smallest possible.

Keywords: Optimal online algorithm · Restricted assignment · Smart grid scheduling

1 Introduction

In this paper, we study online scheduling of unit-sized jobs in two related problems, namely, restricted assignment problem and smart grid problem. The input

H.-H. Liu—Partially supported by Polish National Science Centre grant 2016/22/E/ST6/00499. This work was partially done when Hsiang-Hsuan Liu worked in Wroclaw University, Poland.

P.W.H. Wong—Supported by Networks Sciences & Technologies (NeST), School of EEECS, University of Liverpool.

E. Bampis and N. Megow (Eds.): WAOA 2019, LNCS 11926, pp. 217–231, 2020.
https://doi.org/10.1007/978-3-030-39479-0_15

to the two problems are in close analogy but the objective functions are different. We show that the greedy algorithm is an optimal online algorithm for both problems by showing that both objective functions have led to the same property of the greedy algorithm. The property is crucial for the optimality of the greedy algorithm.

Smart Grid Scheduling. The smart grid scheduling problem arises in demand response management in electrical smart grid [16,21,23,35,48] - one of the major challenges in the 21st century [15,44,45]. The smart grid [17,37] makes power generation, distribution and consumption more efficient through information and communication technologies. One of the main challenges is that peak demand hours happen only for a short duration, yet can make electrical grid very inefficient. For example, in the US power grid, 10% of generation assets and 25% of distribution infrastructure are required for the peak hours which is roughly 5% of the whole time [13,45]. Demand response management is to reduce peak load by shifting demand to non-peak hours [11,26,34,36,38,41] through technological advances in smart meters [27]. It is beneficial to both the power supplier and consumers. On one hand, it can bring down the cost of for the supplier operating the grid [34]. On the other hand, it can reduce electricity bill for consumers as it is common that suppliers charge according to generation cost [41]. Research initiatives in the area include [24,33,40,43].

We consider online scheduling of unit-sized requests with the following input. A consumer sends in a power request j with unit power requirement, unit duration of service, and feasible timeslots $F(j)$ that j can be served. The operator of the smart grid selects a timeslot from $F(j)$ for each request j. The *load* of the grid at each timeslot t is the number of requests allocated to t. The *energy cost* is modeled by a strictly increasing convex function $f(t)$ on load(t). The objective is to minimize the total energy cost over time, i.e., minimize $\sum_t f(\text{load}(t))$.

Restricted Assignment Problem. The assignment problem [19,20] and its variant restricted assignment problem [7] have been extensively studied. The assignment problem is concerned with a set of jobs and a set of machines in which each job specifies a vector of processing times (a.k.a. load) it takes to complete if it is assigned in the corresponding machine. The objective is to minimize over the machines the total load of jobs scheduled on each machine. For the restricted assignment problem, each job is associated with a processing time (a.k.a. size) and a subset of machines that the job can be scheduled on. As pointed out in [7], the restricted assignment problem can be applied to say a wireless communication network where customers arriving one-by-one each request a certain amount of service and must be assigned a base-station within range to service it. We consider online scheduling of unit size jobs. This means that a job increases the load of the assigned machine by one. The objective is to minimize the maximum number of jobs assigned to any machine while satisfying the assignment restriction constraints.

Our Contribution. Notice that with unit size, the input for the grid scheduling problem and the restricted assignment problem is indeed the same. Timeslots in grid scheduling is in analogy to machines in restricted assignment; feasible timeslots in analogy to subset of machines; load of timeslots in analogy to load of machines. The difference of the two problems lie in the objective functions. Our main contribution is the following theorem about both problems.

Theorem 1. *When the input to the grid scheduling problem and the restricted assignment problem is a set of unit-sized jobs, the greedy algorithm is an optimal online algorithm having the best possible competitive ratio.*

Typically, an online algorithm is proved to be an optimal online algorithm through bounding its competitive ratio and showing a lower bound with matching competitive ratio. However, our analysis does not take this approach. Instead, we prove the optimality without giving the exact bounds on competitive ratio.

Roughly speaking, given any online algorithm and a load configuration (to be defined precisely later), we show the existence of two job instances J_1 and J_2 such that (i) J_1 and J_2 admit the same optimal offline schedule represented by the given load configuration; (ii) the cost of the schedule produced by the given online algorithm on J_1 is at least the cost of the schedule produced by the greedy algorithm on J_2. This means that when we consider any job instance for the greedy algorithm, there is always another job instance such that the ratio versus the (same) optimal offline schedule of the greedy algorithm is not larger than any online algorithm. Hence, we can show that the competitive ratio of the greedy algorithm is the smallest possible. The existence of the two job sets relies on a property about the relative costs of two comparable schedules (see Theorem 2). We show that this property holds for both objective functions for the two problems in concern, hence, the optimality holds for both problems.

Related Work on Grid Scheduling. The offline version of the grid problem with unit power requirement and unit service duration can be solved optimally in polynomial time [10]. The solution iteratively assigns each request and rearranges the assignment to maintain optimality. However in the online setting where a request must be irrevocably scheduled, rearrangement is not allowed. It is interesting to study the performance of the greedy strategy without the rearrangement. A previous work [18] has studied the greedy strategy on the problem with unit power requirement, unit service duration and cost function $f(t) = \text{load}^2(t)$ and claimed that the algorithm is 2-competitive. However, as stated in [31], the greedy algorithm is indeed at least 3-competitive. Hence, it is still an open problem that how good or bad the greedy strategy is. Our results in this paper establish the optimality of the greedy algorithm.

For arbitrary power requirement and service duration, the problem becomes NP-hard [10,26]. Theoretical study on this problem mainly focuses on the cost function $f(t) = \text{load}^\alpha(t)$ [12,32]. In particular, Chau et al. [12] designed a greedy algorithm based on a primal-dual approach and improved the upper bound on the competitive ratio to $O(\alpha^\alpha)$, which is asymptotically optimal. Other work on demand response management can be found in [26,34,35,41].

Related Work on (Restricted) Assignment Problem. Online (restricted) assignment problem of jobs with arbitrary size has also been studied as the problem of load balancing. When jobs can be scheduled on any (unrestricted) machine, Graham [19,20] has showed that the greedy algorithm is $(2 - \frac{1}{m})$-competitive where m is the number of machines and this has been improved to $2 - \epsilon$ in [8]. For restricted assignment, Azar et al. [7] have shown that the greedy algorithm is $(\lceil \log m \rceil + 1)$-competitive and no online algorithm can do better than $\lceil \log(m+1) \rceil$-competitive. This implies that the greedy algorithm is very close to optimal. Our result indeed shows that the greedy algorithm is the best possible online algorithm for unit-sized jobs although the precise competitive ratio is yet to been established.

In the offline setting, the (unrestricted) assignment problem has also been studied as scheduling on unrelated machines in which Lenstra et al. [30] have shown a 2-approximation algorithm and that approximating the problem with approximation ratio 3/2 is NP-hard. For restricted assignment, a breakthrough was made by Svensson [42] who has shown that the integrality gap of the configuration LP for the restricted assignment problem is at most 1.942. Various special cases have been studied [14,22,25,29,39,46,47]. The (restricted and unrestricted) assignment problem has also been studied for temporary jobs that depart [2–4,6,28].

Organization of the Paper. We present some preliminaries in Sect. 2. We then present a framework of analysis in Sect. 3 and establish the optimality of the greedy algorithm in Sect. 4. Finally, we conclude in Sect. 5. Due to space limit, proofs are given in the full paper.

2 Preliminaries

Problem Definition. We unify the two problems as follows. We are given a set of machines. Each job j has unit size and a set of permitted machines P_j, which is a subset of machines where the job can be assigned to. A *job instance* J is a set of jobs together with their release order. Two job instances can contain the same set of jobs but with different release orders.

A schedule $\mathcal{S}(J)$ of a job instance J is an assignment assigning each job to a machine. We simply use $\mathcal{S}$ when the context is clear. We denote the machine where j is assigned to by the schedule $\mathcal{S}$ by $m_{\mathcal{S}}(j)$. A schedule $\mathcal{S}$ is *feasible* if each job $j \in J$ is assigned to one of the machines in P_j. That is, $\mathcal{S}$ is feasible if $m_{\mathcal{S}}(j) \in P_j$ for all j in the job instance. We denote by $\mathcal{A}(J)$ the schedule produced by a scheduling algorithm $\mathcal{A}$ on J. We denote the optimal offline algorithm by $\mathcal{O}$ and its schedule $\mathcal{O}(J)$.

In a schedule $\mathcal{S}$ of some job instance J, the *load* of machine i, $\text{load}_{\mathcal{S}}(i)$, is the number of jobs assigned to the machine i. That is, $\text{load}_{\mathcal{S}}(i) = |\{j : m_{\mathcal{S}}(j) = i\}|$. The *cost* of machine i, $\text{cost}_{\mathcal{S}}(i)$ is a strictly increasing convex function of the load of i and $\text{cost}_{\mathcal{S}}(i) = 0$ if the load of i is 0. We overload the notation and use $\mathcal{S}(J)$ to also denote the total cost of schedule $\mathcal{S}$ with instance J, which is the sum of $\text{cost}_{\mathcal{S}}(i)$ over the machines. The goal is to minimize the total cost $\mathcal{S}(J)$.

Online Model. We consider the online model. The jobs are released one by one, and the released job has to be scheduled before the next one is released. And any time, the online algorithm knows only the released jobs without any knowledge about the future. The decisions of an online algorithm are made irrevocably.

We measure the performance of online algorithms by *competitive ratio* [9], which is defined as the maximum ratio between the cost of the online algorithm and the cost of an optimal offline algorithm knowing the whole input.

The Greedy Algorithm $\mathcal{G}$. When a job arrives, it is assigned to the machine with the smallest number of jobs currently assigned.

A Critical Theorem. We first introduce a theorem which is useful when comparing the costs of two schedules.

Definition 1. *Consider an algorithm $\mathcal{A}$, the* level *of job j decided by $\mathcal{A}$, $level_{\mathcal{A}}(j)$, is the number of jobs on the machine $m_{\mathcal{A}}(j)$ right after the time when j is assigned to it. That is, a job with $level_{\mathcal{A}}(j)$ means that it is the $level_{\mathcal{A}}(j)$-th job assigned to $m_{\mathcal{A}}(j)$ by $\mathcal{A}$.*

Definition 2. *Given a schedule $\mathcal{S}$ produced by an algorithm $\mathcal{A}$ on job instance J, the* accumulated size at level k, *$L_{\mathcal{S}}^{(k)}$, is defined as the total number of jobs with level at most k. That is, $L_{\mathcal{S}}^{(k)} := |\{j : level_{\mathcal{S}}(j) \in [1, k]\}|$.*

Theorem 2. *Given two schedules $\mathcal{S}_1$ and $\mathcal{S}_2$ which have the same number of jobs (which are not necessary of the same job instance), if $L_{\mathcal{S}_1}^{(k)} \geq L_{\mathcal{S}_2}^{(k)}$ for all $k \geq 1$, then the cost of $\mathcal{S}_1$ is at most that of $\mathcal{S}_2$.*

Proof. Let $f(x)$ be the cost corresponding to load x. First of all, we observe that the cost of schedule $\mathcal{S}$ can be written as

$$\sum_{j \in J} \Big(f(\text{level}_{\mathcal{S}}(j)) - f(\text{level}_{\mathcal{S}}(j) - 1) \Big) . \tag{1}$$

We claim that we can map each job j in $\mathcal{S}_2$ to a unique job j' in $\mathcal{S}_1$ such that $\text{level}_{\mathcal{S}_2}(j) \geq \text{level}_{\mathcal{S}_1}(j')$. The claim can be proved inductively by first mapping jobs in $\mathcal{S}_2$ at level 1 and because of $L_{\mathcal{S}_1}^{(1)} \geq L_{\mathcal{S}_2}^{(1)}$, there are enough jobs in $\mathcal{S}_1$ at level 1 to have a unique mapping. Then we can map jobs in $\mathcal{S}_2$ at level 2 to unmapped jobs in $\mathcal{S}_1$ at level 1 and any jobs at level 2 because $L_{\mathcal{S}_1}^{(2)} \geq L_{\mathcal{S}_2}^{(2)}$. Since the number of jobs up to level i in $\mathcal{S}_1$ is always at least that in $\mathcal{S}_2$, we can repeat this mapping for each level. The claim then follows. Furthermore, as the cost function f is convex, we have $f(\text{level}_{\mathcal{S}_2}(j)) - f(\text{level}_{\mathcal{S}_2}(j) - 1) \geq f(\text{level}_{\mathcal{S}_1}(j')) - f(\text{level}_{\mathcal{S}_1}(j') - 1)$. Summing up over all pairs of mapped jobs using Eq. 1 concludes the theorem. □

Remark: Note that Theorem 2 also holds for the objective of minimizing the maximum load over machines. This objective is equivalent to ℓ_∞ norm by viewing the loads of machines as a vector. Since ℓ_p norm for any $p \geq 1$ is a valid total cost function for the problem, the proof of Theorem 2 applies to ℓ_p norm.

3 Framework of Analysis

In this section, we give a framework of the analysis and we then present the details of analysis in the next section. As proved in Theorem 2, we can compare schedules by looking at some aggregate property of the schedule instead of the precise allocation of which job in which machine. We further formalize this notion as configuration of a schedule.

Given an arbitrary schedule $\mathcal{S}$, the *configuration* of $\mathcal{S}$, config($\mathcal{S}$), is defined as the multi-set of loads of the machines. Two schedules are considered as having the same configuration if they have the same multi-set of machine loads even with different order. Moreover, we represent the configuration of a schedule as the sequence of loads sorted from low to high and we can compute the *cost* of a certain configuration.

Example. Consider a case with five machines and ten jobs, and two schedules $\mathcal{S}_1$ and $\mathcal{S}_2$. Let ℓ_i be the load on machine m_i. Suppose the load of $\mathcal{S}_1$ is $\ell_1 = 1$, $\ell_2 = 2$, $\ell_3 = 2$, $\ell_4 = 5$, and $\ell_5 = 0$; the load of $\mathcal{S}_2$ is $\ell_1 = 2$, $\ell_2 = 1$, $\ell_3 = 0$, $\ell_4 = 2$, $\ell_5 = 5$. The two schedules $\mathcal{S}_1$ and $\mathcal{S}_2$ have the same configuration $(0, 1, 2, 2, 5)$.

The high level idea of the analysis is roughly as follows. We attempt to find some "bad" instances for the greedy algorithm $\mathcal{G}$ and show that for each such bad instance we can always find another bad instance for every other online algorithm $\mathcal{A}$ such that the ratio of $\mathcal{G}$ to $\mathcal{O}$ on its bad instance is no more than the ratio of $\mathcal{A}$ to $\mathcal{O}$ on its own bad instance. We can then bound the competitive ratio of $\mathcal{G}$ by that of $\mathcal{A}$. We are going to find these bad instances through characterizing the job instances by the configuration of their optimal schedules.

Let $\mathcal{I}$ be the set of all possible job instances. We partition $\mathcal{I}$ according to the optimal configuration of job instances. Job instances J and J' are in the same partition $\mathcal{I}_\mathcal{C}$ if and only if they both have the optimal configuration the same as $\mathcal{C}$. That is, config($\mathcal{O}(J)$) = config($\mathcal{O}(J')$) = $\mathcal{C}$. The following are some properties of $\mathcal{I}_\mathcal{C}$.

Observation 1. *Consider a partition $\mathcal{I}_\mathcal{C}$ and the corresponding optimal configuration $\mathcal{C}$.*

(1) Since the cost function is strictly increasing and convex, any two different configurations have different cost. Hence, for each job instance J, there is exactly one $\mathcal{I}_\mathcal{C}$ such that $J \in \mathcal{I}_\mathcal{C}$, i.e., the partition is well defined.
(2) By definition, for any job instance $J \in \mathcal{I}_\mathcal{C}$, config($\mathcal{O}(J)$) = $\mathcal{C}$.
(3) For any two job instances $J_1, J_2 \in \mathcal{I}_\mathcal{C}$, consider their optimal schedules $\mathcal{O}_1$ and $\mathcal{O}_2$, respectively. Although config($\mathcal{O}_1$) = config($\mathcal{O}_2$), $\mathcal{O}_1$ may not be a feasible schedule for J_2, and neither the other way round.

With the above partition, we can express the competitive ratio of $\mathcal{G}$, denoted by $\mathcal{R}(\mathcal{G})$, as follows.

$$\mathcal{R}(\mathcal{G}) = \max_{J \in \mathcal{I}} \frac{\mathcal{G}(J)}{\mathcal{O}(J)} = \max_{\mathcal{I}_\mathcal{C}} \max_{J \in \mathcal{I}_\mathcal{C}} \frac{\mathcal{G}(J)}{\mathcal{O}(J)} = \max_{\mathcal{I}_\mathcal{C}} \max_{J \in \mathcal{I}_\mathcal{C}} \frac{\mathcal{G}(J)}{\mathcal{C}} .$$

This means that we can characterize the competitive ratio by considering the job instance in each $\mathcal{I}_\mathcal{C}$ with the highest greedy cost. We denote this job instance as $J_\mathcal{G}$, i.e., for a given $\mathcal{C}$, $J_\mathcal{G} = \arg\max_{J\in\mathcal{I}_\mathcal{C}} \mathcal{G}(J)$. It is not clear how to find such job instances directly and instead we try to find their counter parts (bad instances) for any online algorithm $\mathcal{A}$ which share the same $\mathcal{C}$. Precisely, for any online algorithm $\mathcal{A}$, we show the existence of a job instance $J_\mathcal{A} \in \mathcal{I}_\mathcal{C}$ such that $\mathcal{A}(J_\mathcal{A}) \geq \mathcal{G}(J_\mathcal{G})$. This implies $\frac{\mathcal{G}(J_\mathcal{G})}{\mathcal{C}} \leq \frac{\mathcal{A}(J_\mathcal{A})}{\mathcal{C}} = \frac{\mathcal{A}(J_\mathcal{A})}{\mathcal{O}(J_\mathcal{A})}$, where the last equality is because that $J_\mathcal{A} \in \mathcal{I}_\mathcal{C}$. We can then bound the competitive ratio of $\mathcal{G}$ by that of $\mathcal{A}$ as follows:

$$\mathcal{R}(\mathcal{G}) = \max_{\mathcal{I}_\mathcal{C}} \max_{J\in\mathcal{I}_\mathcal{C}} \frac{\mathcal{G}(J_\mathcal{G})}{\mathcal{C}} \leq \max_{\mathcal{I}_\mathcal{C}} \max_{J\in\mathcal{I}_\mathcal{C}} \frac{\mathcal{A}(J_\mathcal{A})}{\mathcal{O}(\mathcal{A})} = \mathcal{R}(\mathcal{A}) .$$

Then we can conclude Theorem 1.

4 Optimality of the Greedy Algorithm

In this section, we construct $J_\mathcal{G}$ and $J_\mathcal{A}$ as required in the framework in Sect. 3.

4.1 The Job Instance $J_\mathcal{G}$ for the greedy algorithm $\mathcal{G}$

Given an optimal configuration $\mathcal{C}$ and the corresponding set of job instances $\mathcal{I}_\mathcal{C}$, we aim to find a job instance $J_\mathcal{G} \in \mathcal{I}_\mathcal{C}$ such that $J_\mathcal{G}$ is the most troublesome job instance for $\mathcal{G}$ among all job instances in $\mathcal{I}_\mathcal{C}$. That is, for any job instance $J \in \mathcal{I}_\mathcal{C}$, $\mathcal{G}(J_\mathcal{G}) \geq \mathcal{G}(J)$.

We find $J_\mathcal{G}$ by artificially designing a job instance. More specifically, $J_\mathcal{G}$ has the same number of jobs as the given $\mathcal{C}$, and we design the set of permitted machines of each job and the release order of the jobs. First, we transform the given $\mathcal{C}$ to schedule $\mathcal{S}_\mathcal{G}$ by changing the configuration. We make sure that $\mathcal{C}$ is the optimal configuration of $J_\mathcal{G}$ (Lemma 2) and the schedule $\mathcal{S}_\mathcal{G}$ is the consequence of running a greedy algorithm on $J_\mathcal{G}$ (Lemma 1). To achieve this, we design the set of permitted machines of each job and choose the release order carefully.

Although the job instance $J_\mathcal{G}$ seems to be artificial, we can prove that $\mathcal{G}(J_\mathcal{G})$ is the highest among all job instances in $\mathcal{I}_\mathcal{C}$ (Corollary 1). That is, consider any job set and any release order, as long as the job set with the release order has optimal configuration $\mathcal{C}$, its greedy cost is no greater than the greedy cost of $J_\mathcal{G}$.

Construction of the Job Instance $J_\mathcal{G}$. We aim to construct a job instance with high greedy cost, i.e., we want the greedy schedule for the job instance to have as few jobs at each level as possible. This can be done by setting a small set of permitted machines. However, this may result in a high optimal cost as well and the ratio between the greedy cost and the optimal cost is still small. Hence we have to balance the greedy schedule and the feasibility of optimal configuration of the job instance.

First we explain how to transform the given optimal configuration $\mathcal{C}$ to schedule $\mathcal{S}_\mathcal{G}$. Assume that $\mathcal{C} = (v_1, v_2, \cdots, v_k)$, where $0 < v_1 \leq v_2 \leq \cdots \leq v_k$, we treat $\mathcal{C}$ as building blocks with k columns and each column i has v_i blocks (where each block corresponding to one job). The transformation runs in rounds, in each round, we choose certain number of blocks, remove them from $\mathcal{C}$ and put them in $\mathcal{S}_\mathcal{G}$ (which is initially empty) and produce another configuration. During the process, the configuration $\mathcal{C}$ changes to reflect the removing of blocks. Hence, the number of non-zero terms in the configuration changes over the process as well. The number of non-zero terms in the configuration in each round takes an important role in our construction. We denote the number of non-zero terms in the configuration at the beginning of round i by n_i. Note that n_i is also the number of non-zero terms in the configuration at the end of round $i-1$.

At the beginning of round i, let $m_1, m_2, \cdots, m_{n_i}$ be the non-empty columns in the configuration and $v_1 \leq v_2 \leq \cdots \leq v_{n_i}$ be the number of blocks in the corresponding non-empty columns. We remove the jobs one by one from the lowest load non-empty column m_1 and update the number of v_1 to reflect the moving. The removing procedure stops once the number of the set of removing blocks in this round, $J_\mathcal{G}^{(i)}$, is greater than or equal to the number of non-empty columns in the current configuration (that is, the configuration after removing the jobs). Notice that by the construction, $n_{i+1} \leq |J_\mathcal{G}^{(i)}| \leq n_{i+1} + 1$.

In round i, after removing the blocks from $\mathcal{C}$, we place them in $\mathcal{S}_\mathcal{G}$ (which is initially empty at the beginning of the first round). Recall that there are n_{i+1} non-empty columns in the (updated) $\mathcal{C}$ at the end of round i. Let K_i' denote the corresponding set of these n_{i+1} non-empty columns. In $\mathcal{S}_\mathcal{G}$, the blocks in $J_\mathcal{G}^{(i)}$ are evenly placed at columns with highest load and cover all columns corresponding to K_i' (Observation 2).

Now we design other parameters of the job instance $J_\mathcal{G}$. As mentioned before, each block is corresponding to one job. For each job j, its permitted machines are the machines corresponding to the columns the block was in $\mathcal{C}$ and $\mathcal{S}_\mathcal{G}$. That is, $P_j = \{m_\mathcal{C}(j)\} \cup \{m_{\mathcal{S}_\mathcal{G}}(j)\}$, where $m_\mathcal{C}(j)$ and $m_{\mathcal{S}_\mathcal{G}}(j)$ are the columns of block j in the configurations $\mathcal{C}$ and $\mathcal{S}_\mathcal{G}$, respectively. The release order of the jobs in $J_\mathcal{G}$ is exactly the order their corresponding blocks removed from $\mathcal{C}$. Algorithm 1 is a demonstration to find the job instance $J_\mathcal{G}$. Figure 1a gives an example configuration and Fig. 1b is the corresponding $J_\mathcal{G}$ of the configuration in Fig. 1a.

The Construction Guarantees that $\mathcal{C}$ and $\mathcal{S}_\mathcal{G}$ are Feasible for $J_\mathcal{G}$. We have to show that $J_\mathcal{G} \in \mathcal{I}_\mathcal{C}$. Moreover, we show that $\mathcal{S}_\mathcal{G}$ generated during the construction process is a greedy schedule for $J_\mathcal{G}$. That is, there is a greedy algorithm for the input job set and the release order generating the schedule $\mathcal{S}_\mathcal{G}$ (with a certain tie breaking).

Before showing the construction produces a desired $J_\mathcal{G}$, we first show that Algorithm 1 is valid. More specifically, we prove that at the end of round i, the level of each jobs $j \in J_\mathcal{G}^{(i)}$ is equal to i (that is, in Algorithm 1, Line 10 is achievable). This property of the construction is essential for proving that the resulting schedule $\mathcal{S}_\mathcal{G}$ is a greedy schedule for $J_\mathcal{G}$.

Algorithm 1. Find $J_{\mathcal{G}}$

Input: The given configuration $\mathcal{C} = (v_1, v_2, \cdots, v_k)$, where $0 < v_1 \leq v_2 \leq v_3 \leq \cdots \leq v_k$.

Output: Job instance $J_{\mathcal{G}}$ with job subsets $J_{\mathcal{G}}^{(1)}, J_{\mathcal{G}}^{(2)}, \cdots$
Schedule $\mathcal{S}_{\mathcal{G}}$ of $J_{\mathcal{G}}$

1: $\mathcal{C}' \leftarrow \mathcal{C}$ (we ignore all zero terms and only consider non-zero terms in $\mathcal{C}'$)
2: **while** there is at least one job in the updated configuration $\mathcal{C}'$ **do**
3: **for** round $i = 1, 2, 3, \cdots$ **do**
4: **while** $|J_{\mathcal{G}}^{(i)}| <$ the number of non-zero entries in $\mathcal{C}' = (v'_1, v'_2, \cdots, v'_{k'_i})$ **do**
5: Let $j \leftarrow$ be a job with lowest level at v'_1 (which is the non-zero vector with the smallest index in $\mathcal{C}'$)
6: Add j to $J_{\mathcal{G}}^{(i)}$
7: Remove j from $\mathcal{C}$; update $\mathcal{C}'$
8: **end while**
9: Let M_i be a set of $|J_{\mathcal{G}}^{(i)}|$ machines such that M_i covers all non-empty machines in $\mathcal{C}'$ (which is K'_i) and the machine $m_{\mathcal{C}}(j)$ of the last job $j \in J_{\mathcal{G}}^{(i)}$
10: Arrange the jobs in $J_{\mathcal{G}}^{(i)}$ evenly at the machines in M_i such that $\text{level}_{\mathcal{S}_{\mathcal{G}}}(j)$ are the same for all jobs $j \in J_{\mathcal{G}}^{(i)}$
11: The permitted machines of job j is $\{m_{\mathcal{C}}(j)\} \cup m_{\mathcal{S}_{\mathcal{G}}}(j)\}$, where $m_{\mathcal{C}}(j)$ and $m_{\mathcal{S}_{\mathcal{G}}}(j)$ are the machines j is assigned in $\mathcal{C}$ and $\mathcal{S}_{\mathcal{G}}$, respectively.
12: **end for**
13: The release order of jobs is the order they are removed from $\mathcal{C}$
14: **end while**

Let M_i be the subset of machines we choose to place jobs in $J_{\mathcal{G}}^{(i)}$, and K'_i be the subset of non-empty machines in the updated configuration $\mathcal{C}'$ at the end of round i. We first observe the relation of machines in M_i and K'_i, which then lead to the feasibility of $\mathcal{S}_{\mathcal{G}}$ (Lemma 1).

Observation 2. *There is a $\mathcal{G}$ with some tie breaking to choose a subset M_i of $|J_{\mathcal{G}}^{(i)}|$ machines such that $K'_i \subseteq M_i$.*

Proof. First we notice that $|M_i| = |J_{\mathcal{G}}^{(i)}|$ and $|J_{\mathcal{G}}^{(i)}| \geq |K'_i|$ by the construction (Line 4 in Algorithm 1). By the construction (Line 9), $M_i = K'_i$ or $M_i = K'_i \cup m_{\mathcal{C}}(j)$. The second case is the situation where $m_{\mathcal{C}}(j) \notin K'_i$, that is, the removing of j from $\mathcal{C}$ creates another empty machine. In this case, $|J_{\mathcal{G}}^{(i)}| = |K'_i| + 1$. □

Observation 3. *For all jobs $j \in J_{\mathcal{G}}^{(i)}$, $m_{\mathcal{C}}(j) \in K'_{i-1}$. ($K'_0$ is defined as the whole set of machines.)*

Proof. In the updated $\mathcal{C}'$, in the beginning of round i, the job $j \in J_{\mathcal{G}}^{(i)}$ is at one of the non-empty machines. That is, the position of job j in $\mathcal{C}$, $m_{\mathcal{C}}(j)$ is one of the machines in K'_{i-1}. □

Lemma 1. *$\mathcal{S}_{\mathcal{G}}$ is a greedy schedule for $J_{\mathcal{G}}$.*

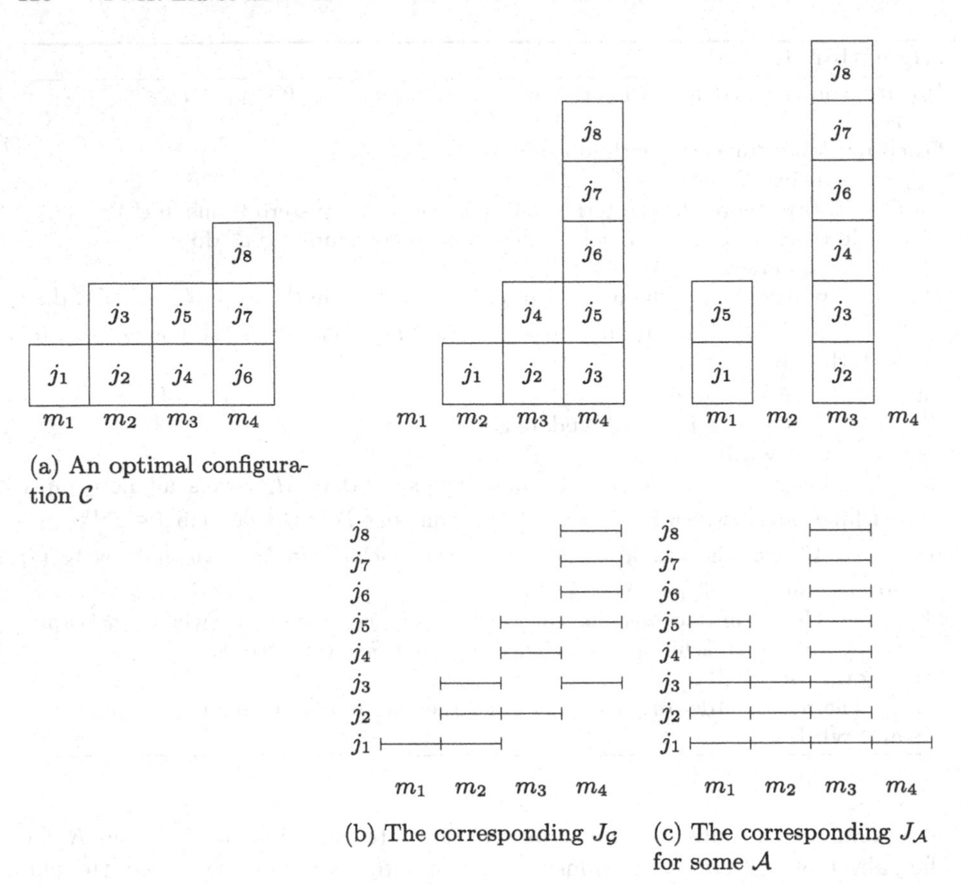

(a) An optimal configuration $\mathcal{C}$

(b) The corresponding $J_{\mathcal{G}}$

(c) The corresponding $J_{\mathcal{A}}$ for some $\mathcal{A}$

Fig. 1. An example of finding $J_{\mathcal{G}}$ and $J_{\mathcal{A}}$. (a) is a configuration with 8 jobs and 4 machines. To obtain $J_{\mathcal{G}}$, we first remove from (a) the jobs j_1, j_2 and j_3, which are in the lowest-load machines and the number of such jobs is at least the number of non-empty machines: m_3 and m_4. These 3 jobs are assigned to m_2, m_3 and m_4 in $J_{\mathcal{G}}$ respectively. Then we remove j_4 and j_5 from (a) and assign them evenly on the second level of $J_{\mathcal{G}}$. After that, we stack the remaining jobs onto $J_{\mathcal{G}}$ such that each job occupies a level since the current number of non-empty machines in (a) is at most 1. The bottom part of (b) shows the permitted machines of each job which is the union of machines the job assigned to in (a) and (b). On the other hand, for finding $J_{\mathcal{A}}$ for some $\mathcal{A}$, we first release j_1, which is at the lowest-load machine in (a), with all machines being permitted and get that $\mathcal{A}$ schedules j_1 to m_1. Then we release j_2 and j_3 with the permitted machines of the 3 highest-load machines in the current $J_{\mathcal{A}}$, which are m_1, m_2 and m_3. And then we release j_4 and j_5 with the permitted machines m_1 and m_3. Finally we release j_6, j_7 and j_8 with the current highest-load machine as their permitted machine.

Lemma 2. *$J_{\mathcal{G}}$ is a job instance in $\mathcal{I}_{\mathcal{C}}$. That is, the optimal configuration of $J_{\mathcal{G}}$ is $\mathcal{C}$.*

The Job Instance $J_{\mathcal{G}}$ is the most Troublesome Among $\mathcal{I}_{\mathcal{C}}$ for $\mathcal{G}$. Now we show that the job instance $J_{\mathcal{G}}$ has the highest greedy cost among all job instances in $\mathcal{I}_{\mathcal{C}}$. Recall that the schedule $\mathcal{S}_{\mathcal{G}}$ produced by Algorithm 1 is $\mathcal{G}(J_{\mathcal{G}})$. We compare $\mathcal{S}_{\mathcal{G}}$ with any greedy schedule of job instances in $\mathcal{I}_{\mathcal{C}}$.

Lemma 3. *Given any job instance J where $J, J_{\mathcal{G}} \in \mathcal{I}_{\mathcal{C}}$ for some $\mathcal{I}_{\mathcal{C}}$, $L^{(k)}_{\mathcal{G}(J)} \geq L^{(k)}_{\mathcal{S}_{\mathcal{G}}}$ for all k.*

By Lemma 3 and Theorem 2,

Corollary 1. *Given any job instance J where $J, J_{\mathcal{G}} \in \mathcal{I}_{\mathcal{C}}$ for some $\mathcal{I}_{\mathcal{C}}$, $\mathcal{G}(J) \leq \mathcal{G}(J_{\mathcal{G}})$*

4.2 A Job Instance $J_{\mathcal{A}}$ for an Online Algorithm $\mathcal{A}$

Given an arbitrary online algorithm $\mathcal{A}$ and an optimal configuration $\mathcal{C}$ with the corresponding set of job instances $\mathcal{I}_{\mathcal{C}}$, we prove that there is a bad instance $J_{\mathcal{A}} \in \mathcal{I}_{\mathcal{C}}$ for $\mathcal{A}$ such that $\mathcal{A}(J_{\mathcal{A}}) \geq \mathcal{G}(J_{\mathcal{G}})$.

Find a Job Instance for Any Online Algorithm $\mathcal{A}$. Similar to the construction of $J_{\mathcal{G}}$, we aim to construct a job instance which has a high cost for the online algorithm $\mathcal{A}$. However, unlike the greedy strategy, we have no knowledge about the behavior of $\mathcal{A}$. Hence we reference $\mathcal{A}$ as an oracle and design the job instance such that every decision made by $\mathcal{A}$ makes some trouble for itself in the future. Note that since $\mathcal{A}$ is an online algorithm, it is practicable for us to make use of the history of $\mathcal{A}$ and design the next group of released jobs such that the previous decisions of $\mathcal{A}$ become bad choices.

Given an optimal configuration $\mathcal{C} = (v_1, v_2, \cdots, v_k)$ where $0 < v_1 \leq v_2 \leq \cdots \leq v_k$. In each round i, we release the set of jobs at column corresponding to v_i as $J^{(i)}_{\mathcal{A}}$. The permitted machines of jobs $j \in J^{(i)}_{\mathcal{A}}$ is decided by the simulation of $\mathcal{A}$ on jobs released in previous rounds. The number of these permitted machines is $k - i + 1$. Note that we can make the simulation since $\mathcal{A}$ is an online algorithm. Algorithm 2 is a detailed instruction of finding $J_{\mathcal{A}}$. In the end, let $\mathcal{S}_{\mathcal{A}}$ be the schedule returned by running $\mathcal{A}$ on $J_{\mathcal{A}}$.

Figure 1 is a demonstration to find the job instance $J_{\mathcal{A}}$. Figure 1c is the corresponding $J_{\mathcal{A}}$ for some online algorithm $\mathcal{A}$ of the configuration in Fig. 1b.

By the construction, the output schedule $\mathcal{S}_{\mathcal{A}}$ is the result of running $\mathcal{A}$ on job set $J_{\mathcal{A}}$ (Line 4 in Algorithm 2). Now we need to prove that the job instance $J_{\mathcal{A}}$ satisfied the requirement that $J_{\mathcal{A}} \in \mathcal{I}_{\mathcal{C}}$.

Lemma 4. *$\mathcal{O}(J_{\mathcal{A}})$ and $\mathcal{C}$ have the same configuration.*

The Property that $\mathcal{A}(J_{\mathcal{A}})$ is at Least $\mathcal{G}(J_{\mathcal{G}})$. Recall that the schedule $\mathcal{S}_{\mathcal{A}}$ produced by Algorithm 2 is $\mathcal{A}(J_{\mathcal{A}})$. We compare $\mathcal{S}_{\mathcal{A}}$ with the greedy schedule $\mathcal{S}_{\mathcal{G}} = \mathcal{G}(J_{\mathcal{G}})$ produced by Algorithm 1.

Algorithm 2. Find $J_{\mathcal{A}}$

Input: The given configuration $\mathcal{C} = (v_1, v_2, \cdots, v_k)$, where $0 < v_1 \le v_2 \le v_3 \le \cdots \le v_k$.
Online scheduling algorithm $\mathcal{A}$

Output: Job instance $J_{\mathcal{A}}$ with job subsets $J_{\mathcal{A}}^{(1)}, J_{\mathcal{A}}^{(2)}, \cdots$
Schedule $\mathcal{S}_{\mathcal{A}}$ of $J_{\mathcal{A}}$

1: **for** round $i = 1, 2, \cdots, k$ **do**
2: Let $t_1, t_2, \cdots, t_{k-i+1}$ be the first $k - (i-1)$ machines with highest load in $\mathcal{A}(\bigcup_{j=1}^{i-1} J_{\mathcal{A}}^{(j)})$.
3: $J_{\mathcal{A}}^{(i)} \leftarrow$ jobs at machine v_i in $\mathcal{C}$, where for each job $j \in J_{\mathcal{A}}^{(i)}$, $P_j = \{t_1, t_2, \cdots, t_{k-i+1}\}$
4: $\mathcal{S}_{\mathcal{A}} \leftarrow \mathcal{A}(\bigcup_{j=1}^{i} J_{\mathcal{A}}^{(j)})$.
5: **end for**
6: The jobs in $J_{\mathcal{A}}^{(i)}$ released after $J_{\mathcal{A}}^{(i-1)}$. For jobs within $J_{\mathcal{A}}^{(i)}$, the jobs with lower level in $\mathcal{C}$ are released before the jobs with higher level.

Observation 4. *Given an optimal configuration $\mathcal{C}$, there exists constructions of $J_{\mathcal{G}}$ and $J_{\mathcal{A}}$ such that for any job in $\mathcal{C}$, the corresponding jobs in $J_{\mathcal{G}}$ and $J_{\mathcal{A}}$ have the same position in the release orders in $J_{\mathcal{G}}$ and $J_{\mathcal{A}}$.*

Proof. The release orders are the same due to the Line 5 in Algorithm 1 and the Line 6 in Algorithm 2. □

Lemma 5. *Consider the schedules $\mathcal{S}_{\mathcal{A}}$ and $\mathcal{S}_{\mathcal{G}}$, $L_{\mathcal{S}_{\mathcal{G}}}^{(k)} \ge L_{\mathcal{S}_{\mathcal{A}}}^{(k)}$ for all k.*

By Lemma 5 and Theorem 2, we have

Corollary 2. *Given any online algorithm $\mathcal{A}$, $\mathcal{G}(J_{\mathcal{G}}) \le \mathcal{A}(J_{\mathcal{A}})$.*

5 Conclusion

We have shown the optimality of greedy algorithm for online grid scheduling and restricted assignment problem for unit-sized jobs. Nevertheless, we have not been able to derive the precise competitive ratio of the greedy algorithm. It is therefore of immediate interest to find the competitive ratio. As mentioned in the introduction, in the restricted assignment problem for arbitrary sized jobs, the greedy algorithm is almost the best online algorithm. Deriving a similar result for the grid scheduling problem would be of interest. Another direction of research is to consider ℓ_p norm. The assignment problem and restricted assignment problem have been studied in ℓ_p norm [1,5]. As far as we are aware, the general grid problem with arbitrary duration and arbitrary power requirement has not been studied in ℓ_p norm and it would be an interesting direction.

Acknowledgment. The authors would like to thank Marcin Bienkowski for helpful discussion.

References

1. Alon, N., Azar, Y., Woeginger, G.J., Yadid, T.: Approximation schemes for scheduling. In: SODA, pp. 493–500 (1997)
2. Aspnes, J., Azar, Y., Fiat, A., Plotkin, S.A., Waarts, O.: On-line routing of virtual circuits with applications to load balancing and machine scheduling. J. ACM **44**(3), 486–504 (1997)
3. Azar, Y.: On-line load balancing. In: Fiat, A., Woeginger, G.J. (eds.) Online Algorithms. LNCS, vol. 1442, pp. 178–195. Springer, Heidelberg (1998). https://doi.org/10.1007/BFb0029569
4. Azar, Y., Broder, A.Z., Karlin, A.R.: On-line load balancing. Theor. Comput. Sci. **130**(1), 73–84 (1994)
5. Azar, Y., Epstein, L., Richter, Y., Woeginger, G.J.: All-norm approximation algorithms. J. Algorithms **52**(2), 120–133 (2004)
6. Azar, Y., Kalyanasundaram, B., Plotkin, S.A., Pruhs, K., Waarts, O.: On-line load balancing of temporary tasks. J. Algorithms **22**(1), 93–110 (1997)
7. Azar, Y., Naor, J., Rom, R.: The competitiveness of on-line assignments. J. Algorithms **18**(2), 221–237 (1995)
8. Bartal, Y., Fiat, A., Karloff, H.J., Vohra, R.: New algorithms for an ancient scheduling problem. J. Comput. Syst. Sci. **51**(3), 359–366 (1995)
9. Borodin, A., El-Yaniv, R.: Online Computation and Competitive Analysis. Cambridge University Press, Cambridge (1998)
10. Burcea, M., Hon, W., Liu, H., Wong, P.W.H., Yau, D.K.Y.: Scheduling for electricity cost in a smart grid. J. Sched. **19**(6), 687–699 (2016)
11. Caron, S., Kesidis, G.: Incentive-based energy consumption scheduling algorithms for the smart grid. In: SmartGridComm, pp. 391–396. IEEE (2010)
12. Chau, V., Feng, S., Thang, N.K.: Competitive algorithms for demand response management in smart grid. In: Bender, M.A., Farach-Colton, M., Mosteiro, M.A. (eds.) LATIN 2018. LNCS, vol. 10807, pp. 303–316. Springer, Cham (2018). https://doi.org/10.1007/978-3-319-77404-6_23
13. Chen, C., Nagananda, K.G., Xiong, G., Kishore, S., Snyder, L.V.: A communication-based appliance scheduling scheme for consumer-premise energy management systems. IEEE Trans. Smart Grid **4**(1), 56–65 (2013)
14. Ebenlendr, T., Krcál, M., Sgall, J.: Graph balancing: a special case of scheduling unrelated parallel machines. Algorithmica **68**(1), 62–80 (2014)
15. European Commission: Europen smartgrids technology platform (2006). ftp://ftp.cordis.europa.eu/pub/fp7/energy/docs/smartgrids_en.pdf
16. Fang, X., Misra, S., Xue, G., Yang, D.: Smart grid - the new and improved power grid: a survey. IEEE Commun. Surv. Tutorials **14**(4), 944–980 (2012)
17. Farhangi, H.: The path of the smart grid. IEEE Power Energy Mag. **8**(1), 18–28 (2010)
18. Feng, X., Xu, Y., Zheng, F.: Online scheduling for electricity cost in smart grid. In: Lu, Z., Kim, D., Wu, W., Li, W., Du, D.-Z. (eds.) COCOA 2015. LNCS, vol. 9486, pp. 783–793. Springer, Cham (2015). https://doi.org/10.1007/978-3-319-26626-8_58
19. Graham, R.L.: Bounds for certain multiprocessing anomalies. Bell Syst. Tech. J. **45**(9), 1563–1581 (1966)
20. Graham, R.L.: Bounds on multiprocessing timing anomalies. SIAM J. Appl. Math. **17**(2), 416–429 (1969)

21. Hamilton, K., Gulhar, N.: Taking demand response to the next level. IEEE Power Energy Mag. **8**(3), 60–65 (2010)
22. Huo, Y., Leung, J.Y.: Fast approximation algorithms for job scheduling with processing set restrictions. Theor. Comput. Sci. **411**(44–46), 3947–3955 (2010)
23. Ipakchi, A., Albuyeh, F.: Grid of the future. IEEE Power Energy Mag. **7**(2), 52–62 (2009)
24. Kannberg, L., Chassin, D., DeSteese, J., Hauser, S., Kintner-Meyer, M., (U.S.), P.N.N.L., of Energy, U.S.D.: GridWise: The Benefits of a Transformed Energy System. Pacific Northwest National Laboratory (2003)
25. Kolliopoulos, S.G., Moysoglou, Y.: The 2-valued case of makespan minimization with assignment constraints. Inf. Process. Lett. **113**(1–2), 39–43 (2013)
26. Koutsopoulos, I., Tassiulas, L.: Control and optimization meet the smart power grid: scheduling of power demands for optimal energy management. In: E-Energy, pp. 41–50. ACM (2011)
27. Krishnan, R.: Meters of tomorrow [in my view]. IEEE Power Energy Mag. **6**(2), 94–96 (2008)
28. Lam, T.W., Ting, H., To, K., Wong, P.W.H.: On-line load balancing of temporary tasks revisited. Theor. Comput. Sci. **270**(1–2), 325–340 (2002)
29. Lee, K., Leung, J.Y., Pinedo, M.L.: A note on graph balancing problems with restrictions. Inf. Process. Lett. **110**(1), 24–29 (2009)
30. Lenstra, J.K., Shmoys, D.B., Tardos, É.: Approximation algorithms for scheduling unrelated parallel machines. Math. Program. **46**, 259–271 (1990)
31. Liu, F., Liu, H., Wong, P.W.H.: Optimal nonpreemptive scheduling in a smart grid model. CoRR abs/1602.06659 (2016)
32. Liu, F., Liu, H., Wong, P.W.H.: Optimal nonpreemptive scheduling in a smart grid model. In: ISAAC, pp. 53:1–53:13, LIPIcs (2016)
33. Lockheed Martin: SEELoad™ Solution. http://www.lockheedmartin.co.uk/us/products/energy-solutions/seesuite/seeload.html
34. Logenthiran, T., Srinivasan, D., Shun, T.Z.: Demand side management in smart grid using heuristic optimization. IEEE Trans. Smart Grid **3**(3), 1244–1252 (2012)
35. Lui, T., Stirling, W., Marcy, H.: Get smart. IEEE Power Energy Mag. **8**(3), 66–78 (2010)
36. Maharjan, S., Zhu, Q., Zhang, Y., Gjessing, S., Basar, T.: Dependable demand response management in the smart grid: a stackelberg game approach. IEEE Trans. Smart Grid **4**(1), 120–132 (2013)
37. Masters, G.M.: Renewable and Efficient Electric Power Systems. Wiley, Hoboken (2013)
38. Mohsenian-Rad, A.H., Wong, V.W., Jatskevich, J., Schober, R., Leon-Garcia, A.: Autonomous demand-side management based on game-theoretic energy consumption scheduling for the future smart grid. IEEE Trans. Smart Grid **1**(3), 320–331 (2010)
39. Muratore, G., Schwarz, U.M., Woeginger, G.J.: Parallel machine scheduling with nested job assignment restrictions. Oper. Res. Lett. **38**(1), 47–50 (2010)
40. REGEN Energy Inc: ENVIROGRID™ SMART GRID BUNDLE. http://www.regenenergy.com/press/announcing-the-envirogrid-smart-grid-bundle/
41. Salinas, S., Li, M., Li, P.: Multi-objective optimal energy consumption scheduling in smart grids. IEEE Trans. Smart Grid **4**(1), 341–348 (2013)
42. Svensson, O.: Santa claus schedules jobs on unrelated machines. SIAM J. Comput. **41**(5), 1318–1341 (2012)
43. Toronto Hydro Corporation: Peaksaver Program. http://www.peaksaver.com/peaksaver_THESL.html

44. UK Department of Energy & Climate Change: Smart grid: a more energy-efficient electricity supply for the UK (2013). https://www.gov.uk/smart-grid-a-more-energy-efficient-electricity-supply-for-the-uk
45. US Department of Energy: The Smart Grid: An Introduction (2009). http://www.oe.energy.gov/SmartGridIntroduction.htm
46. Verschae, J., Wiese, A.: On the configuration-LP for scheduling on unrelated machines. J. Sched. **17**(4), 371–383 (2014)
47. Wang, C., Sitters, R.: On some special cases of the restricted assignment problem. Inf. Process. Lett. **116**(11), 723–728 (2016)
48. Zpryme Research & Consulting: Power systems of the future: the case for energy storage, distributed generation, and microgrids (2012). http://smartgrid.ieee.org/images/features/smart_grid_survey.pdf

Fair Coresets and Streaming Algorithms for Fair k-means

Melanie Schmidt[1(✉)], Chris Schwiegelshohn[2], and Christian Sohler[1]

[1] Department of Mathematics and Computer Science, University of Cologne, 50923 Köln, Germany
mschmidt@informatik.uni-koeln.de, csohler@uni-koeln.de
[2] Department of Computer, Control and Management Engineering Antonio Ruberti, Sapienza University of Rome, 00185 Roma, RM, Italy
schwiegelshohn@diag.uniroma1.it

Abstract. We study fair clustering problems as proposed by Chierichetti et al. [CKLV17]. Here, points have a sensitive attribute and all clusters in the solution are required to be balanced with respect to it (to counteract any form of data-inherent bias). Previous algorithms for fair clustering do not scale well. We show how to model and compute so-called coresets for fair clustering problems, which can be used to significantly reduce the input data size. We prove that the coresets are composable [IMMM14] and show how to compute them in a streaming setting. This yields a streaming PTAS for fair k-means in the case of two colors (and exact balances). Furthermore, we extend techniques due to Chierichetti et al. [CKLV17] to obtain an approximation algorithm for k-means, which leads to a constant factor algorithm in the streaming model when combined with the coreset.

1 Introduction

Our challenge is to support growth in the beneficial use of big data while ensuring that it does not create unintended discriminatory consequences. (Executive Office of the President, 2016 [MSP16]).

As the use of machine learning methods becomes more and more common in many areas of daily life ranging from automatic display of advertisements on webpages to mortgage approvals, we are faced with the question whether the decisions made by these automatic systems are *fair*, i.e. free of biases by race, gender or other sensitive attributes. While at first glance it seems that replacing human decisions by algorithms will remove any kind of bias as algorithms will only decide based on the underlying data, the problem is that the training data may contain all sorts of biases. As a result, the outcome of an automated decision process may still contain these biases.

This work was done while the first author was at University of Bonn and the third author was at TU Dortmund. The second author acknowledges support by the ERC Advanced Grant 788893 AMDROMA.

E. Bampis and N. Megow (Eds.): WAOA 2019, LNCS 11926, pp. 232–251, 2020.
https://doi.org/10.1007/978-3-030-39479-0_16

Recent findings in algorithmically generated results strengthen this concern. For example, it has been discovered that the COMPAS software that is used to predict the probability of recidivism is much more likely to assign an incorrect high risk score to a black defendant and low risk scores to a white defendant [ALMK16]. This raises the general question how we can guarantee fairness in algorithms.

This questions comes with several challenges. The first challenge is to formally define the concept of fairness. And indeed, it turns out that there are several ways to define *fairness* which result in different optimal solutions [CPF+17], and it has recently been shown that they cannot be achieved simultanuously unless the data has some very special (unlikely) structure [KMR17].

In this paper we build upon the recent work by Chiericetti et al. [CKLV17] and consider fairness of clustering algorithms using the concept of *disparate impact*, which is a notion of (un)fairness introduced to computer science by Feldman et al. [FFM+15]. Disparate impact essentially means that the result of a machine learning task does not correlate strongly with sensitive attributes such as gender, race sexual or religious orientation. More formally and illustrated on the case of a single binary sensitive attribute X and cluster variable C, a clustering algorithm does not show disparate impact if it satisfies the $p\%$ rule (a typical value for p is 0.8) stating that $\frac{\Pr\{C=i|X=0\}}{\Pr\{C=i|X=1\}} \leq p$. If we assume that both attribute values appear with the same frequency, then by Bayes Theorem the above translates to having at most $p\%$ points with a specific attribute value in each cluster.

Chierichetti et al. model fairness based on the disparate impact model in the following way. They assume that every point has one of two colors (red or blue). If a set of points C has r_C red and b_C blue points, then they define its *balance* to be $\min(\frac{r_C}{b_C}, \frac{b_C}{r_C})$. The overall balance of a clustering is then defined as the minimum balance of any cluster in it. Clusterings are then considered fair if their overall balance is close to $1/2$.

An algorithm ensuring fairness has to proceed with care; as mentioned before an algorithm that obliviously optimizes an objective function may retain biases inherent in the training set. Chierichetti et al. avoid this by identifying a set of fair micro-clusters via a suitably chosen perfect matching and running the subsequent optimization on the microclusters. This clever technique has the benefit of always computing a fair clustering, as the union of fair micro clusters is necessarily also fair. However, the min-cost perfect matching is computationally expensive, and it needs random access to the data, which may be undesirable. This raises the following question:

Question 1. Is is possible to perform a fair data analysis efficiently, even when the size of the data set renders random-access unfeasible?

Our Contribution. We address the issue of scaling algorithms by investigating *coresets* for fair clustering problems, specifically for k-means. Given an input set P in d dimensional Euclidean space, the problem consists of finding a set of k centers $c_1, \ldots, c_k$ and a partition of P into k sets $C_1, \ldots, C_k$ such that $\sum_{i=1}^{k} \sum_{p \in C_i} \|p - c_i\|_2^2$ is minimized.

Roughly speaking, a coreset is a summary of a point set which has the property that it approximates the cost function well for any possible candidate solution. The notion was proposed by Har-Peled and Mazumdar [HM04] and has since received a wide range of attention. For clustering coresets, see for example [BFL16,FL11,FMS07,FGS+13,FS05,LS10]. Coresets for geometric clustering are usually composable, meaning that if S_1 is a coreset for P_1 and S_2 is a coreset for P_2, then $S_1 \cup S_2$ is a coreset for $P_1 \cup P_2$ [IMMM14]. Composability is arguably the main appeal of coresets; it enables an easy reduction from coreset constructions to streaming and distributed algorithms which scale to big data. Dealing with fair clustering, composability is not obvious. In this work, we initiate the study of fair coresets and their algorithmic implications:

- The standard coreset definition does not satisfy composability for fair clustering problems. We propose an alternative definition tailored to fair clustering problems and show that this new definition satisfies composability. The definition straightforwardly generalizes to having ℓ color classes and we show how a suitable coreset (of size $O(\ell k \log n \epsilon^{d-1})$ for constant d) may be computed efficiently. This implies a PTAS for fair k-means clustering if k is a constant (we show this for the case of two colors and exact fairness preservation, as will be defined later).
- We extend the coreset construction to the streaming setting. The main challenge to overcome is to reduce the dimension of our fair coresets in a streaming fashion. Our key ingredient to do so is a novel application of the random projections proposed by Cohen et al. [CEM+15], which may be of independent interest.
- Then we describe a constant factor approximation algorithm for fair k-means clustering with two colors (and exact fairness preservation) based on the approach of Chierichetti et al. [CKLV17]. The general technique to obtain a constant factor algorithm is not new, but we do some adaptations to apply it to the k-means case.
- Finally, we extend the practical approximation algorithm k-means++ to the fair k-means setting and empirically evaluate the resulting algorithm and the coreset approach.

Additional Related Work. The research on fairness in machine learning follows two main directions. One is to find proper definitions of fairness. There are many different definitions available including statistical parity [TRT11], disparate impact [FFM+15], disparate mistreatment [ZVGRG17] and many others, e.g. [BHJ+17,HPS16]. For an overview see the recent survey [BHJ+17]. Furthermore, the effects of different definitions of fairness and their relations have been studied in [Cho16,CPF+17,KMR17]. A notion for individual fairness has been developed in [DHP+12]. The other direction is the development of algorithms for fair machine learning tasks. Here the goal is to develop new algorithms that solve learning tasks in such a way that the result satisfies a given fairness condition. Examples include [CKLV17,HPS16,ZVGRG17]. The closest result to our work is the above described paper by Chierichetti et al. [CKLV17].

Polynomial-time approximation schemes for k-means were e.g. developed in [BFL16,FL11,KSS10], assuming that k is a constant. If d is a constant, then [CKM16,FRS16] give a PTAS. If k and d are arbitrary, then the problem is APX-hard [ACKS15,LSW17]. Lloyd's algorithm [Llo57] is an old but very popular local search algorithm for the k-means problem which can converge to arbitrarlity bad solutions. By using k-means++ seeding [AV07] as initialization, one can guarantee that the computed solution is a $O(\log k)$-approximation.

Chierichetti et al. [CKLV17] develop approximation algorithms for fair k-center and k-median with two colors. This approach was further improved by Backurs et al. [BIO+19], who proposed an algorithm to speed up parts of the computation. Rösner and Schmidt [RS18] extend their definition to multiple colors and develop an approximation algorithm for k-center. Bercea et al. [BGK+18] develop an even more generalized notion and provide bicriteria approximations for fair variants of k-center, k-median and also k-means. For k-center, they provide a true 6-approximation. Very recently, Kleindessner et al. [KAM19] proposed a linear-time 2-approximation for fair k-center. This algorithm is not in the streaming setting, but still faster then previously existing approaches for fair clustering.

The fair k-means problem can also be viewed as a k-means clustering problem with size constraints. Ding and Xu [DX15] showed how to compute an exponential sized list of candidate solutions for any of a large class of constrained clustering problems. Their result was improved by Bhattacharya et al. [BJK18].

In addition to the above cited coreset constructions, coresets for k-means in particular have also been studied empirically in various different works, for example [AMR+12,AJM09,FGS+13,FS08,KMN+02]. Dimensionality reductions for k-means are for example proposed in [BZD10,BZMD15,CEM+15,FSS13]. In particular [CEM+15,FSS13] show that any input to the k-means problem can be reduced to $\lceil k/\epsilon \rceil$ dimensions by using singular value decomposition (SVD) while distorting the cost function by no more than an ϵ-factor. Furthermore, [CEM+15] also study random projection based dimensionality reductions. While SVD based reductions result in a smaller size, random projections are more versatile.

1.1 Preliminaries

We use $P \subseteq \mathbb{R}^d$ to denote a set of n points in the d-dimensional vector space $\mathbb{R}^d$. The Euclidean distance between two points $p, q \in \mathbb{R}^d$ is denoted as $\|p - q\|$. The goal of clustering is to find a partition of an input point set P into subsets of 'similar' points called clusters. In k-means clustering we formulate the problem as an optimization problem. The integer k denotes the number of clusters. Each cluster has a center and the cost of a cluster is the sum of squared Euclidean distances to this center. Thus, the problem can be described as finding a set

$C = \{c_1, \ldots, c_k\}$ and corresponding clusters $C_1, \ldots, C_k$ such that $\mathrm{cost}(P, C) = \sum_{i=1}^{k} \sum_{p \in C_i} \|p - c_i\|^2$ is minimized.

It is easy to see that in an optimal (non-fair) clustering each point p is contained in the set C_i such that $\|p - c_i\|^2$ is minimized. The above definition can be easily extended to non-negatively and integer weighted point sets by treating the weight as a multiplicity of a point. We denote the k-means cost of a set S weighted with w and center set C and corresponding clusters $S_1, \ldots, S_k$ by $\mathrm{cost}_w(S, C) = \sum_{i=1}^{k} \sum_{p \in S_i} w(p) \|p - c_i\|^2$. Finally, we recall that the best center for a cluster C_i is its centroid $\mu(C_i) := \frac{1}{|C_i|} \sum_{p \in C_i} p$, which follows from the following proposition.

Proposition 1. *Given a point set $P \subset \mathbb{R}^d$ and a point $c \in \mathbb{R}^d$, the* 1*-means cost of clustering P with c can be decomposed into $\sum_{p \in P} \|p - c\|^2 = \sum_{p \in P} \|p - \mu(P)\|^2 + |P| \cdot \|\mu(P) - c\|^2$.*

Next, we give the coreset definition for k-means as introduced by Har-Peled and Mazumdar.

Definition 1 (Coreset [HM04]). *A set $S \subseteq \mathbb{R}^d$ together with non-negative weights $w : S \rightarrow \mathbb{N}$ is a (k, ϵ)-coreset for a point set $P \subseteq \mathbb{R}^d$ with respect to the k-means clustering problem, if for every set $C \subseteq \mathbb{R}^d$ of k centers we have* $(1 - \epsilon) \cdot \mathrm{cost}(P, C) \leq \mathrm{cost}(S, C) \leq (1 + \epsilon) \cdot \mathrm{cost}(P, C)$.

Fair Clustering. We extend the definition of fairness from [CKLV17] to sensitive attributes with multiple possible values. As in [CKLV17], we model the sensitive attribute by a color. Notice that we can model multiple sensitive attributes by assigning a different color to any combination of possible values of the sensitive attributes. We further assume that the sensitive attributes are not among the point coordinates. Thus, our input set is a set $P \subseteq \mathbb{R}^d$ together with a coloring $c : P \rightarrow \{1, \ldots, \ell\}$.

We define $\xi(j) = |\{p \in P : c(p) = j\}| / |P|$ as the fraction that color j has in the input point set. Then we call a clustering $C_1, \ldots, C_k$ (α, β)-fair, $0 < \alpha \leq 1 \leq \beta$, if for every cluster C_i and every color class $j \in \{1, \ldots, \ell\}$ we have

$$\alpha \cdot \xi(j) \leq \frac{|\{p \in C_i : c(p) = j\}|}{|\{p \in C_i\}|} \leq \beta \cdot \xi(j).$$

For any set $C = \{c_1, \ldots, c_k\}$ of k centers we define $\mathrm{faircost}(P, C)$ to be the minimum of $\sum_{i=1}^{k} \sum_{p \in C_i} \|p - c_i\|^2$ where the minimum is taken over all (α, β)-fair clusterings of P into $C_1, \ldots, C_k$. The optimal (α, β)-fair clustering C' is the one with minimal $\mathrm{faircost}(P, C')$. Alternatively to $\xi(j)$, we could demand that the fraction of all colors is (roughly) $1/\ell$. However, notice that the best achievable fraction is $\xi(j)$. Thus (α, β)-fairness is a strictly more general condition. It is also arguably more meaningful if the data set itself is heavily imbalanced. Consider an instance where the blue points outnumber the red points by a factor of 100. Then the disparity of impact is at least 0.01. A $(1, 1)$-fair clustering then is a clustering where all clusters achieve the best-possible ratio 0.01.

2 Fair Coresets and How to Get Them

First, notice that the *definition* of coresets as given in Definition 1 does not translate well to the realm of fair clustering. Assume we replace cost by faircost in Definition 1. Now consider Fig. 1. We see two point sets P_1 and P_2 with eight points each, which both have an optimum cost of $\Omega(\Delta)$. Replacing the four left and the four right points by one point induces an error of $\mathcal{O}(\epsilon\Delta)$, which is an $\mathcal{O}(\epsilon)$-fraction of the total cost. Thus, the depicted sets S_1 and S_2 are coresets. However, when we combine P_1 and P_2, then the optimum changes. The cost decreases dramatically to $\mathcal{O}(\epsilon)$. For the new optimal solution, $S_1 \cup S_2$ still costs $\Omega(\epsilon\Delta)$, and the inequality in Definition 1 is no longer satisfied.

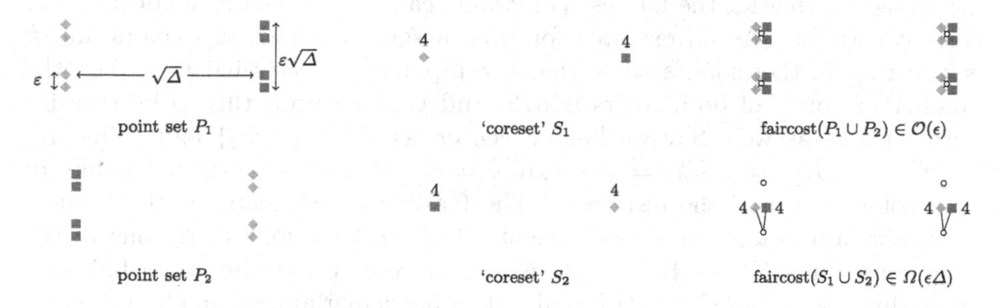

Fig. 1. A simple example of non-composable coresets for the case of $(1, 1)$-fairness.

We thus have to do a detour: We define a stronger, more complicated notion of coresets which regains the property of being composable. Then, we show that a special type of coreset constructions for k-means can be used to compute coresets that satisfy this stronger notion. It is an interesting open question to analyze whether it is possible to design sampling based coreset constructions that satisfy our notion of coresets for fair clustering.

For our detour, we need the following generalization of the standard k-means cost. A *coloring constraint* for a set of k cluster centers $C = \{c_1, \ldots, c_k\}$ and a set of ℓ colors $\{1, \ldots, \ell\}$ is a $k \times \ell$ matrix K. Given a point set P with a coloring $c : P \to \{1, \ldots, \ell\}$ we say that a partition of P into sets $C_1, \ldots, C_k$ satisfies K if $|\{p \in C_i : c(p) = j\}| = K_{ij}$. The cost of the corresponding clustering is $\sum_{i=1}^{k} \sum_{p \in C_i} \|p - c_i\|^2$ as before. Now we define the *color-k-means cost* colcost(P, K, C) to be the minimal cost of any clustering satisfying K. If no clustering satisfies K, colcost$(P, K, C) := \infty$.

Notice that we can prevent the bad example in Fig. 1 by using the color-k-means cost: If colcost(P, K, C) is approximately preserved for the color constraints modeling that each cluster is either completely blue or completely red, then S_1 and S_2 are forbidden as a coresets.

Definition 2. *Let P be a point set with coloring $c : P \to \{1,\dots,\ell\}$. A non-negatively integer weighted set $S \subseteq \mathbb{R}^d$ with a coloring $c' : S \to \{1,\dots,\ell\}$ is a (k,ϵ)-coreset for P for the (α,β)-fair k-means clustering problem, if for every set $C \subseteq \mathbb{R}^d$ of k centers and every coloring constraint K we have*

$$\text{colcost}_w(S,K,C) \in (1 \pm \epsilon) \cdot \text{colcost}(P,K,C),$$

where in the computation of $\text{colcost}(S,K,C)$ *we treat a point with weight w as w unweighted points and therefore a point can be partially assigned to more than one cluster.*

Definition 2 demands that the cost is approximated for *any* possible color constraint. This implies that it is approximated for those constraints we are interested in. Indeed, the fairness constraint can be modeled as a collection of color constraints. As an example for this, assume we have two colors and k is also two; furthermore, assume that the input is perfectly balanced, i.e., the number of points of both colors is $n/2$, and that we want this to be true for both clusters as well. Say we have a center set $C = \{c_1, c_2\}$ and define K^i by $K^i_{11} = i, K^i_{12} = i, K^i_{21} = \frac{n}{2} - i, K^i_{22} = \frac{n}{2} - i$, i.e., K^i assigns i points of each color to c_1 and the rest to c_2. The feasible assignments for the fairness constraint are exactly those assignments that are legal for exactly one of the color constraints K^i, $i \in \{0,\dots,\frac{n}{2}\}$. So since a coreset according to Definition 2 approximates $\text{colcost}(P,C,K^i)$ for all i, it in particular approximately preserves the cost of any fair clustering. This also works in the general case: We can model the (α,β)-fair constraint as a collection of color constraints (and basically any other fairness notion based on the fraction of the colors in the clusters as well).

Proposition 2. *Given a center set C, $|C| = k$, the assignment restriction to be (α,β)-fair can be modeled as a collection of coloring constraints.*

Proof. Recall $\xi(j) = \frac{|\{p \in P \,:\, c(p)=j\}|}{|P|}$. Let $C = \{C_1,\dots,C_k\}$ be a clustering and let K be the coloring constraint matrix induced by C. We observe that the ith row sums up to $|C_i|$ and the jth column sums up to $|\{p \in P \,:\, c(p) = j\}|$. Then C is (α,β)-fair if and only if $\alpha \cdot \xi(j) \le \frac{|\{p \in C_i \,:\, c(p)=j\}|}{|C_i|} = \frac{K_{i,j}}{\sum_{h=1}^{k} K_{i,h}} \le \beta \cdot \xi(j)$ for all $i \in \{1,\dots,k\}$ and $j \in \{1,\dots,\ell\}$.

The main advantage of Definition 2 is that it satisfies composability. The main idea is that for any coloring constraint K, any clustering satisfying K induces specific color constraints K_1 and K_2 for P_1 and P_2; and for these, the coresets S_1 and S_2 also have to satisfy the coreset property. We can thus prove the coreset property for S and K by composing the guarantees for S_1 and S_2 on K_1 and K_2.

Lemma 1 (Composability). *Let $P_1, P_2 \subset \mathbb{R}^d$ be point sets. Let S_1, w_1, c_1 be a (k,ϵ)-coreset for P_1 and let S_2, w_2, c_2 be a (k,ϵ)-coreset for P_2 (both satisfying Definition 2). Let $S = S_1 \cup S_2$ and concatenate w_1, w_2 and c_1, c_2 accordingly to obtain weights w and colors c for S. Then S, w, c is a (k,ϵ)-coreset for $P = P_1 \cup P_2$ satisfying Definition 2.*

Proof. Let $C = \{c_1, \ldots, c_k\} \subset \mathbb{R}^d$ be an arbitrary set of centers, and let $K \in \mathbb{N}^{k \times \ell}$ be an arbitrary coloring constraint for C. We want to show that

$$\text{colcost}_w(S, K, C) \in (1 \pm \epsilon)\text{colcost}(P, K, C).$$

Let $\gamma : P \to C$ be an assignment that minimizes the assignment cost among all assignments that satisfy K, implying that $\text{colcost}(P, K, C) = \sum_{p \in P} ||x - \gamma(x)||^2$. Since γ satisfies K, the number of points of color j assigned to each center $c_i \in C$ is exactly K_{ij}. We split K into two matrices K_1 and K_2 with $K = K_1 + K_2$ by counting the number of points of each color at each center which belong to P_1 and P_2, respectively. In the same fashion, we define two mappings $\gamma_1 : P_1 \to C$ and $\gamma_2 : P_2 \to C$ with $\gamma_1(p) = \gamma(p)$ for all $p \in P_1$ and $\gamma_2(p) = \gamma(p)$ for all $p \in P_2$.

Now we argue that $\text{colcost}(P, C, K) = \text{colcost}(P_1, C, K_1) + \text{colcost}(P_2, C, K_2)$. Firstly, we observe that $\text{colcost}(P, C, K) \leq \text{colcost}(P_1, C, K_1) + \text{colcost}(P_2, C, K_2)$ since γ_1 and γ_2 are legal assignments for the color constraint K_1 and K_2, respectively, and they induce exactly the same point-wise cost as γ. Secondly, we argue that there cannot be cheaper assignments for K_1 and K_2. Assume there where an assignment γ_1' with $\sum_{p \in P_1} ||x - \gamma_1'(x)||^2 < \text{colcost}(P_1, C, K_1)$. Then we could immediately adjust γ to be identical to γ_1' on the points in P_1 instead of γ_1, and this would reduce the cost; a contradiction to the optimality of γ. The same argument holds for γ_2. Thus, $\text{colcost}(P, C, K) = \text{colcost}(P_1, C, K_1) + \text{colcost}(P_2, C, K_2)$ is indeed true.

Now since S_1, w_1, c_1 is a coreset for P_1 and S_2, w_2, c_2 is a coreset for P_2, they have to approximate $\text{colcost}(P_1, C, K_1)$ and $\text{colcost}(P_2, C, K_2)$ well. We get from this that

$$\begin{aligned}
&\text{colcost}_w(S_1, C, K_1) + \text{colcost}_w(S_2, C, K_2) \\
&\in (1 \pm \epsilon) \cdot \text{colcost}(P_1, C, K_1) + (1 \pm \epsilon) \cdot \text{colcost}(P_2, C, K_2) \\
&\in (1 \pm \epsilon) \cdot \text{colcost}(P, C, K).
\end{aligned}$$

Observe that $\text{colcost}_w(S, C, K) \leq \text{colcost}_w(S_1, C, K_1) + \text{colcost}_w(S_2, C, K_2)$ since we can concatenate the optimal assignments for S_1 and S_2 to get an assignment for S. Thus, $\text{colcost}_w(S, C, K) \leq (1 + \epsilon) \cdot \text{colcost}(P, C, K)$. It remains to show that $\text{colcost}_w(S, C, K) \geq (1 - \epsilon) \cdot \text{colcost}(P, C, K)$.

Let $\gamma' : S \to C$ be an assignment that satisfies K and has cost $\text{colcost}_w(S, C, K)$ (for simplicity, we treat S as if it were expanded by adding multiple copies of each weighted point; recall that we allow weights to be split up for S). Let $\gamma_1' : P_1 \to C$ and $\gamma_2' : P_2 \to C$ be the result of translating γ' to P_1 and P_2, and split K into K_1' and K_2' according to γ' as we did above. Then $\text{colcost}_w(S, C, K) = \text{colcost}_w(S_1, C, K_1') + \text{colcost}_w(S_2, C, K_2')$ by the same argumentation as above. Furthermore,

$$\begin{aligned}
\text{colcost}_w(S, C, K) &= \text{colcost}_w(S_1, C, K_1') + \text{colcost}_w(S_2, C, K_2') \\
&\geq (1 - \epsilon)\text{colcost}_w(P_1, C, K_1') + (1 - \epsilon)\text{colcost}_w(P_2, C, K_2') \\
&\geq (1 - \epsilon)\text{colcost}(P, C, K).
\end{aligned}$$

where the first inequality holds by the coreset property and the second is true since we can also use γ' to cluster P, implying the inequality$\text{colcost}_w(P, C, K) \leq \text{colcost}_w(P_1, C, K_1') + \text{colcost}_w(P_2, C, K_2')$. That completes the proof.

We have thus achieved our goal of finding a suitable *definition* of coresets for fair clustering. Now the question is whether we can actually compute sets which satisfy the rather strong Definition 2. Luckily, we can show that a special class of coreset constructions for k-means can be adjusted to work for our purpose. A *coreset construction for k-means* is an algorithm that takes a point set P as input and computes a summary S with integer weights that satisfies Definition 1.

We say that a coreset construction is *movement-based* if

- all weights $w(p), p \in S$ are integers
- there exists a mapping $\pi : P \to S$ with $|\pi^{-1}(p)| = w(p)$ for all $p \in S$ which satisfies that
$$\sum_{x \in P} ||x - \pi(x)||^2 \leq \frac{\varepsilon^2}{16} \cdot \text{OPT}_k,$$
where $\text{OPT}_k = \min_{C \subset \mathbb{R}^d, |C|=k} \text{cost}(P, C)$.

Movement-based coreset constructions compute a coreset by 'moving' points to common places at little cost, and then replacing heaps of points by weighted points. Examples for movement-based coreset constructions can be found in [FGS+13, FS05, HM04]. Now the crucial observation is that we can turn any movement-based coreset construction for k-means (say, a black-box algorithm *ALG*) into an algorithm that computes coresets for fair k-means satisfying Definition 2. The main idea is to run *ALG* to move all points in P to common locations, and then to replace all points *of the same color* at the same location by one coreset point. This may result in up to ℓ points for every location, i.e., the final coreset result may be larger than its colorless counterpart by a factor of at most ℓ. The rest of the proof then shows that Definition 2 is indeed true, following the lines of movement-based coreset construction proofs.

Theorem 1. *Let ALG be a movement-based coreset construction for the k-means problem. Assume that given the input $P \subset \mathbb{R}^d$, $k \in \mathbb{N}$ and $\epsilon \in (0, 1)$, the size of the coreset that ALG computes is bounded by $f(|P|, d, k, \epsilon)$. Then we can construct an algorithm ALG′ which constructs a set S' that satisfies Definition 2, and the size of this set is bounded by $\ell \cdot f(|P|, d, k, \epsilon)$, where ℓ is the number of colors.*

Proof. For any P, *ALG* gives us a set S and a non-negative weight function w such that Definition 1 is true, i.e.,

$$(1 - \epsilon)\text{cost}(P, C) \leq \text{cost}_w(S, C) \leq (1 + \epsilon)\text{cost}(P, C) \tag{1}$$

holds for all set of centers C with $|C| = k$. Since *ALG* is movement-based, the weights are integer; and there exists a mapping $\pi : P \to S$, such that at most

$w(p)$ points from P are mapped to any point $p \in S$, and such that

$$\sum_{x \in P} ||x - \pi(x)||^2 \leq \frac{\varepsilon^2}{16} \cdot \mathrm{OPT}_k \qquad (2)$$

is true. Statement (2) is stronger than (1), and we will only need (2) for our proof. We will, however, need the mapping π to construct ALG'. Usually, the mapping will be at least implicitly computed by ALG. If not or if outputting this information from ALG is cumbersome, we do the following. We assign every point in P to its closest point in S. The resulting mapping has to satisfy 2, since the distance of any point to its closest point in S can only be smaller than in any given assignment. We may now assign more than $w(p)$ points to S. We resolve this by simply changing the weights of the points in S to match our mapping. Since we now have S, w and π satisfying (2), we can proceed as if ALG had given a mapping to us.

Now we do what movement-based coreset constructions do internally as well: We consolidate all points that share the same location. However, since they may not all be of the same color, we possibly put multiple (at most ℓ) copies of any point in S into our coreset S'. More precisely, for every $p \in S$, we count the number $n_{p,i}$ of points of color i. If $n_{p,i}$ is at least one, then we put p into S' with color i and weight $n_{p,i}$. The size of S' is thus at most $\ell \cdot f(|P|, d, k, \epsilon)$.

The proof that S' satisfies Definition 2 is now close to the proof that movement-based coreset constructions work. To execute it, we imagine S' in its expanded form (where every point p is replaced by $n_{p,i}$ points of color i. We call this expanded version P'. Notice that $\mathrm{cost}(P', C) = \mathrm{cost}_w(S', C)$ for all $C \subset \mathbb{R}^d$. We only need P' for the analysis. Notice that π can now be interpreted as a bijective mapping between P and P' and this is how we will use it.

Let C be an arbitrary center set with $|C| = k$ and let K be an arbitrary coloring constraint. We want to show that $(1 - \epsilon) \cdot \mathrm{colcost}(P, K, C) \leq \mathrm{colcost}(P', K, C) \leq (1 + \epsilon) \cdot \mathrm{colcost}(P, K, C)$. If no assignment satisfies K, then $\mathrm{colcost}(P, K, C)$ is infinity, and there is nothing to show. Otherwise, fix an arbitrary optimal assignment $\gamma : P \to C$ of the points in P to C among all assignments that satisfy K. Notice that γ and π are different assignments with different purposes; γ assigning a point in P to its center, and π assigning each point in P to its moved version in P'.

We let $v_c(x) := ||x - \gamma(x)||$ be the distance between $x \in P$ and the center its assigned to. Let v_c be the $|P|$-dimensional vector consisting of all $v_c(x)$ (in arbitrary order). Then we have

$$\mathrm{colcost}(P, C, K) = \sum_{x \in P} ||x - \gamma(x)||^2 = \sum_{x \in P} v_c(x)^2 = ||v_c||^2.$$

Furthermore, we set $v_p(x) = ||\pi(x) - x||$ and let v_p be the $|P|$-dimensional vector of all $v_p(x)$ (ordered in the same way as v_c). We have $\sum_{x \in P} ||\pi(x) - x||^2 \leq \frac{\varepsilon^2}{16} \cdot OPT_k$ by our preconditions.

Now we want to find an upper bound on $\text{colcost}(P', C, K)$. Since we only need an upper bound, we can use γ for assigning the points in P' to C. We already know that γ satisfies K for the points in P; and the points in P' are only moved versions of the points in P. We use this and then apply the triangle inequality:

$$\begin{aligned}\text{colcost}(P', C, K) &\leq \sum_{y \in P'} ||y - \gamma(\pi^{-1}(y))||^2 = \sum_{x \in P} ||\gamma(x) - \pi(x)||^2 \\ &\leq \sum_{x \in P} (||\gamma(x) - x|| + ||x - \pi(x)||)^2 \\ &= \sum_{x \in P} (v_c(x) + v_p(x))^2 = ||v_c + v_p||^2.\end{aligned}$$

Now we can apply the triangle inequality to the vector $v_c + v_p$ to get $||v_c + v_p|| \leq ||v_c|| + ||v_p|| \leq \sqrt{\text{colcost}(P, C, K)} + \sqrt{\frac{\varepsilon^2}{16}} \cdot OPT_k$. So we know that

$$\begin{aligned}\text{colcost}(P', C, K) \leq ||v_c + v_p||^2 &\leq \text{colcost}(P, C, K) + \frac{\varepsilon^2}{16} \cdot OPT_k \\ &\quad + 2\sqrt{\text{colcost}(P, C, K)} \cdot \sqrt{\frac{\varepsilon^2}{16} \cdot OPT_k} \\ &\leq \text{colcost}(P, C, K) + \frac{\varepsilon^2}{16} \cdot OPT_k \\ &\quad + \frac{\varepsilon}{2} \cdot \text{colcost}(P, C, K) \\ &< (1 + \epsilon) \cdot \text{colcost}(P, C, K).\end{aligned}$$

To obtain that also $\text{colcost}(P, C, K) \leq (1 + \epsilon) \cdot \text{colcost}(P', C, K)$, we observe that the above argumentation is symmetric in P and P'. No argument used that P is the original point set and P' is the moved version. So we exchange the roles of P and P' (we repeat the whole argumentation starting at the point where we fix the center set C, so for example, γ is now an optimal assignment of P' to C) to complete the proof.

We can now apply Theorem 1. Movement-based constructions include the original paper due to Har-Peled and Mazumdar [HM04] as well as the practically more efficient algorithm BICO [FGS+13]. For more information on the idea of movement-based coreset constructions, see Sect. 3.1 in the survey [MS18]. For BICO in particular, Lemma 5.4.3 in [Sch14] gives a proof that the construction is movement-based. Using Theorem 1 and Corollary 1 from [FGS+13], we then obtain the following.

Corollary 1. *There is an algorithm in the insertion-only streaming model which computes a (k, ϵ)-coreset for the fair k-means problem according to Definition 2. The size of the coreset and the storage requirement of the algorithm is $m \in O(\ell \cdot k \cdot \log n \cdot \epsilon^{-d+2})$, where ℓ is the number of colors in the input, and where d is assumed to be a constant.*

The running time of the algorithm is $O(N(m)(n + \log(n\Delta)m))$, where Δ is the spread of the input points and $N(m)$ is the time to compute an (approximate) nearest neighbor.

3 Streaming PTAS and Constant-Factor Approximations

In this section, we give a constant-factor approximation algorithm for fair k-means with two colors, assuming that exactly half of the input points are colored with each color, and demanding that this is true for all clusters in the clustering as well. We call this special case *exactly balanced*. We also show how to obtain a PTAS for the case that k is a constant.

We restrict to two colors since for multiple colors, no true approximation algorithms are known even for the related case of k-median, and there is indication that this problem might be very difficult (it is related to solving capacitated k-median/k-means, a notoriously difficult research problem). Notice that the coreset approach works for arbitrary (α, β)-fair k-means.

Fairlet Approach. For two colors, Chierichetti et al. [CKLV17] outline how to transfer approximation algorithms for clustering to the setting of fair clustering, but derive the algorithms only for k-center and k-median. The idea is to first compute a coarse clustering where the microclusters are called *fairlets*, and then to cluster representatives of the fairlets to obtain the final clustering. The following algorithm extends their ideas to compute fairlets for k-means.

Algorithm 1. Fairlet computation

1: Let B be the blue points and R be the red points
2: For any $b \in B$, $r \in R$, set $c(r, b) = ||r - b||^2/2$
3: Consider the complete bipartite graph G on B and R
4: Compute a min cost perfect matching M on G
5: For each edge $(r, b) \in M$, add $\mu(\{r, b\})$ to F
6: Output F

The idea of the algorithm is the following. In any optimal solution, the points can be paired into tuples of a blue and a red point which belong to the same optimal cluster. Clustering the $n/2 \geq k$ tuples with $n/2$ centers cannot be more expensive than the cost of the actual optimal k-means solution. Thus, we would ideally like to know the tuples and partition them into clusters. Since we cannot know the tuples, we instead compute a minimum cost perfect matching between the red and blue points, where the weight of an edge is the 1-means cost of clustering the two points with an optimal center (this is always half their squared distance). The matching gives us tuples; these tuples are the fairlets. The centroid of each fairlet now serves as its representative. The following theorem shows that clustering the representatives yields a good solution for the original problem.

Theorem 2. *There is an algorithm that achieves the following. For any $P \subset \mathbb{R}^d$ which contains $|P|/2$ blue and $|P|/2$ red points, it computes a set of representatives $F \subset P$ of size $|P|/2$, such that an α-approximate solution for the (plain) k-means problem on F yields a $(5.5\alpha + 1)$-approximation for the fair k-means problem on P.*

By combining Theorem 2 with a constant-factor approximation for k-means (the currently best being the one proposed by Ahmadian et al. [ANSW17]), we get the following corollary.

Corollary 2. *There is a $\mathcal{O}(1)$-approximation for exactly balanced fair k-means with two colors.*

Similarly to the fairlet computation, the problem of finding an *fair assignment*, i.e., an cost-wise optimal assignment of points to *given* centers which is fair, can be modeled as a matching problem. This algorithm, as well as algorithms for fairlet computation and fair assignment for *weighted* points are described in the full version of the paper [SSS18].

PTAS. We next give an algorithm to efficiently compute a $(1+\epsilon)$-approximation if k is a constant and not part of the input by extending known ideas to the fair k-means++ case. The main additional step is to use an optimal fair assignment problem algorithm.

We remark that the running time of the below stated algorithm algorithm is worse than that of [BJK18, DX15]. However, it can be easily adapted to work with weighted inputs. While we believe that in principle adapting the algorithms in [BJK18, DX15] to the weighted case is possible, we preferred to stick to the simpler slightly worse result to keep the paper concise.

Theorem 3. *Let $P \subseteq \mathbb{R}^d$ be a weighted point set of n points such that half of the point weight is red and the other half is blue. Then we can compute a $(1+\epsilon)$-approximations to the fair k-means problem in time $n^{O(k/\epsilon)}$.*

We use the well-known fact that every cluster has a subset of $O(1/\epsilon)$ points such that their centroid is a $(1+\epsilon)$-approximation to the centroid of the cluster. We use the following lemma by Inaba et al.

Lemma 2. [IKI94] *Let $P \subseteq \mathbb{R}^d$ be a set of points and let S be a subset of m points drawn independently and uniformly at random from P. Let $c(P) = \frac{1}{|P|}\sum_{p\in P} p$ and $c(S) = \frac{1}{|S|}\sum_{p\in S} p$ be the centroids of P and S. Then with probability at least $1-\delta$ we have*

$$\| \sum_{p\in P} \|p - c(S)\|_2^2 \leq (1 + \frac{1}{\delta m}) \cdot \| \sum_{p \in P} \|p - c(P)\|_2^2 \ .$$

It immediately follows that for $m = \lceil 2/\epsilon \rceil$ there exists a subset S of m points that satisfies the above inequality. The result can immediately be extended to the weighted case. This implies that Algorithm 2 gives a PTAS.

The running time of the algorithm is $n^{O(k/\epsilon)}$ since line two can be implemented in $k^{O(k/\epsilon)}$ time and the partition problem can be solved in $n^{O(1)}$ time. This implies the theorem.

Algorithm 2. PTAS for fair k-means++

Input: (Weighted) point set $P \subseteq \mathbb{R}^d$
1: Consider all subsets $S \subseteq P$ of size $k \cdot \lceil 2/\epsilon \rceil$.
2: Consider all partitions of S into k sets $C_1, \ldots, C_k$ of size $\lceil 2/\epsilon \rceil$.
3: Solve the fair assignment problem for P and $c(C_1), \ldots, c(C_k)$
4: Return the best solution computed above

Streaming PTAS. We would like to extend the PTAS to the streaming setting, using our coreset. Applying Corollary 1 directly incurs an exponential dependency on the dimension d. The standard way to avoid this is to project the data onto the first k/ε principal components, see [CEM+15, FSS13], and then to use a technique called merge-and-reduce. Unfortunately, merge-and-reduce technique requires a rescaling of ε by a factor of $\log n$. In other words, the resulting streaming coreset will have a size $\exp(\left(\frac{\log n}{\varepsilon}\right), k \cdot \frac{\log n}{\varepsilon})$, which is even larger than the input size. To avoid this, we show how to make use of oblivious random projections to reduce the dependency of the dimension for movement-based coreset constructions, and also recover a $(1+\varepsilon)$ approximate solution. This is a novel combination of coreset techniques with a sketching technique due to Cohen et al. [CEM+15] which may be of independent interest.

We review some of the algebraic properties. Given a matrix $A \in \mathbb{R}^{n \times d}$, we define the Frobenius norm as $\|A\|_F = \sqrt{\sum_{i=1}^n \|A_{i*}\|^2}$, where A_{i*} is the ith row of A. For k-means, we will consider the rows of A to be our input points. The spectral norm $\|A\|_2$ is the largest singular value of A.

Let us now consider the n-vector $x = \mathbf{1} \cdot \frac{1}{\sqrt{n}}$. x is a unit vector, i.e. $\|x\|_2 = 1$, and moreover, due to Proposition 1, the rows of $xx^T A$ are $\mu(A)^T$. Hence $\|A - xx^T A\|_F^2$ is the optimal 1-means cost of A. This may be extended to an arbitrary number of centers by considering the n by k clustering matrix X with

$$X_{i,j} = \begin{cases} \sqrt{1/|C_j|} & \text{if } A_{i*} \in \text{ cluster } C_j \\ 0 & \text{otherwise} \end{cases}.$$

XX^T is an orthogonal projection matrix and $\|A - XX^T A\|_F^2$ corresponds to the k-means cost of the clusters $C_1, \ldots, C_k$. If we lift the clustering constraints on X and merely assume X to be orthogonal and rank k, $\|A - XX^T A\|_F^2$ becomes the rank k approximation problem. The connection between rank k approximation and k-means is well established, see for example [BZMD15, DFK+04, FSS13]. Specifically, we aim for the following guarantee.

Definition 3 (Definition 1 of [CEM+15]**).** *$\tilde{A} \in \mathbb{R}^{n \times d'}$ is a rank k-projection-cost preserving sketch of $A \in \mathbb{R}^{n \times d}$ with error $0 < \varepsilon < 1$ if, for all rank k orthogonal projection matrices $XX^T \in \mathbb{R}^{n \times n}$,*

$$\|\tilde{A} - XX^T \tilde{A}\|_F^2 + c \in (1 \pm \varepsilon) \cdot \|A - XX^T A\|_F^2,$$

for some fixed non-negative constant c that may depend on A and $\tilde{A}$, but is independent of XX^T.

Our choice of $\tilde{A}$ is AS, where S is a scaled Rademacher matrix of target dimension $m \in O(k/\varepsilon^2)$, see Theorem 12 of [CEM+15]. In this case $c = 0$.

Algorithm 3. Dimension-Efficient Coreset Streaming

Input: Point set A processed in a stream
1: Initialize $S \in \mathbb{R}^{d\times m}$
2: Maintain a movement-based coreset T of AS
3: Let $\pi^{-1}(T_{i*})$ be the set of rows of AS that are moved to T_{i*}
4: Let $\pi_A^{-1}(T_{i*})$ be the set of corresponding rows of A
5: Maintain $|\pi_A^{-1}(T_{i*})|$ and the linear sum $L(T_{i*})$ of the rows in $\pi_A^{-1}(T_{i*})$
6: Solve (α, β)-fair k-means on T using a γ-approximation algorithm $\rightsquigarrow$ clustering $C_1, \ldots, C_k$
7: For each cluster C_j return the center $\frac{1}{\sum_{T_{i*}\in C_j} |\pi_A^{-1}(T_{i*})|} \cdot \sum_{T_{i*}\in C_j} L(T_{i*})$

We combine oblivious sketches with movement-based coreset constructions in Algorithm 3. The general idea is to run the coreset construction on the rows of AS (which are lower dimensional points). Since the dimensions of AS are n times k/ε^2, this has the effect that we can replace d in the coreset size by $O(k/\varepsilon^2)$. Furthermore, we show that by storing additional information we can still compute an approximate solution for A (the challenge is that although AS will approximately preserve clustering costs, the cluster centers that achieve this cost lie in a different space and cannot be used directly as a solution for A).

Theorem 4. *Let $0 < \varepsilon < 1/2$. Assume there is streaming algorithm ALG that receives the rows of a matrix $A \in \mathbb{R}^{n\times d}$ and maintains an (k, ϵ)-coreset T with the following property: We can replace weighted points in T by a corresponding number of copies to obtain a matrix A' such that $\sum_{i=1}^{n} \|A_{i*} - A_{i'*}\|^2 \leq \frac{\varepsilon^2}{16} \cdot OPT$. Furthermore, assume that ALG uses $f(k, \varepsilon, d, \log n)$ space. If we use ALG in Step 2 of Algorithm 3, then Algorithm 3 will use $f(k, \varepsilon/25, c' \cdot (k/\varepsilon)^2, \log n) \cdot d + O(kd/\epsilon^2)$ space to compute a set of centers C with*

$$\text{faircost}(P, C) \leq \gamma(1 + \varepsilon) \cdot OPT$$

where OPT is the cost of an optimal solution for A and $c' > 0$ is a constant such that the guarantees of Theorem 12 from [CEM+15] *are satisfied.*

Proof. Let X be the optimal clustering matrix on input $A'S$ and Y be the optimal clustering matrix for input A. Let Z be the clustering matrix returned by our (α, β)-fair approximation algorithm on input $A'S$ (or, equivalently, on input T). Let $\varepsilon' = \varepsilon/25$. Since we are using a γ-approximation algorithm, we know that $\|ZZ^T A'S - A'S\|_F^2 \leq \gamma \cdot \|XX^T A'S - A'S\|_F^2$. We also observe that $\|ZZ^T A'S - ZZ^T AS\|_F \leq \|ZZ^T\|_2 \|A'S - AS|_F = \|A'S - AS\|_F$ for an orthogonal projection matrix ZZ^T. Furthermore, we will use that $\|XX^T A'S - A'S\|_F \leq \|XX^T A'S - XX^T AS\|_F + \|XX^T AS - AS\|_F + \|A'S - AS\|_F$ and the fact that the spectral norm and Frobenius norm are conforming, e.g., they satisfy $\|AB\|_F \leq \|A\|_2 \|B\|_F$. We obtain

$$\begin{aligned}
&(1-\varepsilon') \cdot \|ZZ^TA - A\|_F^2 \leq \|ZZ^TAS - AS\|_F^2 \\
&\leq \left(\|ZZ^TA'S - ZZ^TAS\|_F + \|ZZ^TA'S - AS\|_F\right)^2 \\
&\leq (\|ZZ^TA'S - ZZ^TAS\|_F + \|ZZ^TA'S - A'S\|_F \\
&\quad + \|A'S - AS\|_F)^2 \\
&\leq \left(2\|A'S - AS\|_F + \sqrt{\gamma}\|XX^TA'S - A'S\|_F\right)^2 \\
&\leq ((2+\sqrt{\gamma})\|A'S - AS\|_F \\
&\quad + \sqrt{\gamma}(\|XX^TA'S - XX^TAS\|_F + \\
&\quad \|XX^TAS - AS\|_F))^2 \\
&\leq ((2+2\sqrt{\gamma})\|A'S - AS\|_F + \sqrt{\gamma}\|XX^TAS - AS\|_F)^2 \\
&\leq ((\frac{\varepsilon'}{4}(2+2\sqrt{\gamma}) + \sqrt{\gamma})\|XX^TAS - AS\|_F)^2 \\
&\leq ((1+\varepsilon')\sqrt{\gamma})\|XX^TAS - AS\|_F)^2 \\
&\leq (1+\varepsilon')^2\gamma\|YY^TAS - AS\|_F^2 \\
&\leq (1+\varepsilon')^3\gamma\|YY^TA - A\|_F^2
\end{aligned}$$

where the first and the last inequality follows from the guarantee of Definition 3 and Theorem 12 of [CEM+15]. To conclude the proof, observe that $(1+\varepsilon')^3/1-\varepsilon \leq (1+25\varepsilon') = (1+\varepsilon)$ for $0 < \varepsilon < \frac{1}{2}$.

4 Practical Approximation Algorithms

Finally, we give some thought on how to extend the famous k-means++ algorithm [AV07] to the fair k-means setting in a way that results in high quality solutions. The k-means++ algorithm is a combination of an efficient randomized sampling technique that produces a $\mathcal{O}(\log k)$-approximate solution to the k-means problem on the one hand, with a local search heuristic, *Lloyd's algorithm*, which refines the solution to a local optimum, on the other hand.

The straightforward way to adapt k-means++ is to use Theorem 2 (the fairlet approach), run k-means++ on the fairlet representatives and use the resulting centers. There are two variants of this: One can either use the assignment that results directly from clustering the fairlets, or one can only use the computed centers and recompute the assignment by using Algorithm 4. We name the two variants `CKLV-k-means++` and `Reassigned-CKLV-k-means++`.

Algorithm 4. CKLV-k-means++

1: Compute fairlet representatives F with Algorithm 1 (or weighted)
2: Run standard k-means++ on F and assign fairlet points accordingly

Algorithm 5. Reassigned-CKVL

1: Compute a center set C with Algorithm 4
2: Compute an optimal fair assignment of all points in P to C

Compared to k-means++, `Reassigned-CKLV` has the drawback that the computed solution is not really optimal with respect to the original problem: After the reassignment, it may be beneficial to change the centers again. To further refine the solution, we propose to adapt Lloyd's algorithm directly. Lloyd's algorithm is an iterative search heuristic which given initial centers, alternatingly assigns points to their closest center and computes the optimum center for each cluster (the centroid). The two steps are repeated until a stopping criterion is met, for example until the algorithm is converged or has reached a maximum number of iterations. For fair k-means++, we replace the assignment step by a fair assignment step and call the resulting algorithm `fair k-means++`.

Algorithm 6. Fair k-means++

1: Compute initial centers C_0 by Algorithm 4
2: For all $i \geq 0$, unless a stopping criterion is met:
3: Assign every point to a center C_i by computing a fair assignment, partitioning P into $P_i^1, \ldots, P_i^k$
4: Set $C_{i+1} = \{\mu(P_i^j) \mid j \in [k]\}$

4.1 Empirical Evaluation

In the full version [SSS18], we give a short empirical evaluation to demonstrate the practicability of the streaming $\mathcal{O}(\log k)$-approximations and the coreset approach. We compare `CKLV-k-means++` (Algorithm 4), `Reassigned-CKLV` (Algorithm 5) and `fair k-means++` (Algorithm 6). The experiments clearly show that none of the algorithms can scale to big data, with `fair k-means++` being particularly slow due to the repeated fair reassignment. However, using the coreset allows all three algorithms to scale well.

References

[ACKS15] Awasthi, P., Charikar, M., Krishnaswamy, R., Sinop, A.K.: The hardness of approximation of Euclidean k-means. In: 31st International Symposium on Computational Geometry (SoCG), pp. 754–767 (2015)

[AJM09] Ailon, N., Jaiswal, R., Monteleoni, C.: Streaming k-means approximation. In: Proceedings of the 22nd Neural Information Processing Systems (NIPS), pp. 10–18 (2009)

[ALMK16] Angwin, J., Larson, J., Mattu, S., Kirchner, L.: Machine bias: there's software used across the country to predict future criminals and it's biased against blacks. ProPublica, May 2016

[AMR+12] Ackermann, M.R., Märtens, M., Raupach, C., Swierkot, K., Lammersen, C., Sohler, C.: StreamKM++: a clustering algorithm for data streams. ACM J. Exper. Algorithmics **17**, 1–30 (2012). Article no. 2.4

[ANSW17] Ahmadian, S., Norouzi-Fard, A., Svensson, O., Ward, J.: Better guarantees for k-means and Euclidean k-median by primal-dual algorithms. In: Proceedings of the 58th IEEE Annual Symposium on Foundations of Computer Science (FOCS), pp. 61–72 (2017)

[AV07] Arthur, D., Vassilvitskii, S.: k-means++: the advantages of careful seeding. In: Proceedings of the 18th ACM-SIAM Symposium on Discrete Algorithms (SODA), pp. 1027–1035 (2007)

[BFL16] Braverman, V., Feldman, D., Lang, H.: New frameworks for offline and streaming coreset constructions, arXiv preprint arXiv:1612.00889 (2016)

[BGK+18] Bercea, I.O., et al.: On the cost of essentially fair clusterings, CoRR abs/1811.10319 (2018)

[BHJ+17] Berk, R., Heidari, H., Jabbari, S., Kearns, M., Roth, A.: Fairness in criminal justice risk assessments: the state of the art, arXiv:1703.09207 (2017)

[BIO+19] Backurs, A., Indyk, P., Onak, K., Schieber, B., Vakilian, A., Wagner, T.: Scalable fair clustering, CoRR abs/1902.03519 (2019)

[BJK18] Bhattacharya, A., Jaiswal, R., Kumar, A.: Faster algorithms for the constrained k-means problem. Theory Comput. Syst. **62**(1), 93–115 (2018)

[BZD10] Boutsidis, C., Zouzias, A., Drineas, P.: Random projections for k-means clustering.. In: Proceedings of the 24th Neural Information Processing Systems (NIPS), pp. 298–306 (2010)

[BZMD15] Boutsidis, C., Zouzias, A., Mahoney, M.W., Drineas, P.: Randomized dimensionality reduction for k-means clustering. IEEE Trans. Inf. Theory **61**(2), 1045–1062 (2015)

[CEM+15] Cohen, M.B., Elder, S., Musco, C., Musco, C., Persu, M.: Dimensionality reduction for k-means clustering and low rank approximation. In: Proceedings of the Forty-Seventh Annual ACM on Symposium on Theory of Computing (STOC), pp. 163–172 (2015)

[Cho16] Chouldechova, A.: Fair prediction with disparate impact: a study of bias in recidivism prediction instruments, arXiv: 1510.07524v1 (2016)

[CKLV17] Chierichetti, F., Kumar, R., Lattanzi, S., Vassilvitskii, S.: Fair clustering through fairlets. In: Proceedings of the 30th Annual Conference on Neural Information Processing Systems (NIPS), pp. 5036–5044 (2017)

[CKM16] Cohen-Addad, V., Klein, P.N., Mathieu, C.: Local search yields approximation schemes for k-means and k-median in Euclidean and minor-free metrics. In: IEEE 57th Annual Symposium on Foundations of Computer Science (FOCS), pp. 353–364 (2016)

[CPF+17] Corbett-Davies, S., Pierson, E., Feller, A., Goel, S., Huq, A.: Algorithmic decision making and the cost of fairness. In: Proceedings of the 23rd ACM SIGKDD International Conference on Knowledge Discovery and Data Mining, pp. 797–806 (2017)

[DFK+04] Drineas, P., Frieze, A.M., Kannan, R., Vempala, S., Vinay, V.: Clustering large graphs via the singular value decomposition. Mach. Learn. **56**(1–3), 9–33 (2004)

[DHP+12] Dwork, C., Hardt, M., Pitassi, T., Reingold, O., Zemel, R.: Fairness through awareness. In: ITCS 2012, pp. 214–226 (2012)

[DX15] Ding, H., Xu, J.: A unified framework for clustering constrained data without locality property. In: Proceedings of the Twenty-Sixth Annual ACM-SIAM Symposium on Discrete Algorithms, SODA 2015, San Diego, CA, USA, 4–6 January 2015, pp. 1471–1490 (2015)

[FFM+15] Feldman, M., Friedler, S.A., Moeller, J., Scheidegger, C., Venkatasubramanian, S.: Certifying and removing disparate impact. In: Proceedings of the 21th ACM SIGKDD International Conference on Knowledge Discovery and Data Mining, pp. 259–268 (2015)

[FGS+13] Fichtenberger, H., Gillé, M., Schmidt, M., Schwiegelshohn, C., Sohler, C.: BICO: BIRCH meets coresets for k-means clustering. In: Bodlaender, H.L., Italiano, G.F. (eds.) ESA 2013. LNCS, vol. 8125, pp. 481–492. Springer, Heidelberg (2013). https://doi.org/10.1007/978-3-642-40450-4_41

[FL11] Feldman, D., Langberg, M.: A unified framework for approximating and clustering data. In: Proceedings of the 43rd Annual ACM Symposium on Theory of Computing (STOC), pp. 569–578 (2011)

[FMS07] Feldman, D., Monemizadeh, M., Sohler, C.: A PTAS for k-means clustering based on weak coresets. In: Proceedings of the 23rd Symposium on Computational Geometry, pp. 11–18 (2007)

[FRS16] Friggstad, Z., Rezapour, M., Salavatipour, M.R.: Local search yields a PTAS for k-means in doubling metrics. In: IEEE 57th Annual Symposium on Foundations of Computer Science (FOCS), pp. 365–374 (2016)

[FS05] Frahling, G., Sohler, C.: Coresets in dynamic geometric data streams. In: Proceedings of the 37th Annual ACM Symposium on Theory of Computing (STOC), pp. 209–217 (2005)

[FS08] Frahling, G., Sohler, C.: A fast k-means implementation using coresets. Int. J. Comput. Geom. Appl. **18**, 605–625 (2008)

[FSS13] Feldman, D., Schmidt, M., Sohler, C.: Turning big data into tiny data: constant-size coresets for k-means, PCA and projective clustering. In: Proceedings of the Twenty-Fourth Annual ACM-SIAM Symposium on Discrete Algorithms (SODA), pp. 1434–1453 (2013)

[HM04] Har-Peled, S., Mazumdar, S.: On coresets for k-means and k-median clustering. In: Proceedings of the 36th Annual ACM Symposium on Theory of Computing (STOC), pp. 291–300 (2004)

[HPS16] Hardt, M., Price, E., Srebro, N.: Equality of opportunity in supervised learning. In: Proceedings of the 30th International Conference on Neural Information Processing Systems (NIPS) (2016)

[IKI94] Inaba, M., Katoh, N., Imai, H.: Applications of weighted Voronoi diagrams and randomization to variance-based k-clustering (extended abstract). In: Proceedings of the Tenth Annual Symposium on Computational Geometry (SoCG), pp. 332–339 (1994)

[IMMM14] Indyk, P., Mahabadi, S., Mahdian, M., Mirrokni, V.S.: Composable coresets for diversity and coverage maximization. In: Proceedings of the 33rd ACM SIGMOD-SIGACT-SIGART Symposium on Principles of Database Systems (PODS), pp. 100–108 (2014)

[KAM19] Kleindessner, M., Awasthi, P., Morgenstern, J.: Fair k-center clustering for data summarization, CoRR abs/1901.08628 (2019)

[KMN+02] Kanungo, T., Mount, D.M., Netanyahu, N.S., Piatko, C.D., Silverman, R., Wu, A.Y.: An efficient k-means clustering algorithm: analysis and implementation. IEEE Trans. Pattern Anal. Mach. Intell. **24**(7), 881–892 (2002)

[KMR17] Kleinberg, J.M., Mullainathan, S., Raghavan, M.: Inherent trade-offs in the fair determination of risk scores. In: 8th Innovations in Theoretical Computer Science Conference (ITCS), pp. 43:1–43:23 (2017)

[KSS10] Kumar, A., Sabharwal, Y., Sen, S.: Linear-time approximation schemes for clustering problems in any dimensions. J. ACM **57**(2), 51–532 (2010)

[Llo57] Lloyd, S.P.: Least squares quantization in PCM. Bell Laboratories Technical Memorandum (1957). Later published as [Llo82]

[Llo82] Lloyd, S.P.: Least squares quantization in PCM. IEEE Trans. Inf. Theory **28**(2), 129–137 (1982)

[LS10] Langberg, M., Schulman, L.J.: Universal epsilon-approximators for integrals. In: Proceedings of the 21st ACM-SIAM Symposium on Discrete Algorithms (SODA), pp. 598–607 (2010)

[LSW17] Lee, E., Schmidt, M., Wright, J.: Improved and simplified inapproximability for k-means. Inf. Process. Lett. **120**, 40–43 (2017)

[MS18] Munteanu, A., Schwiegelshohn, C.: Coresets-methods and history: a theoreticians design pattern for approximation and streaming algorithms. KI **32**(1), 37–53 (2018)

[MSP16] Munoz, C., Smith, M., Patil, D.: Big data: a report on algorithmic systems, opportunity, and civil rights. Executive Office of the President. The White House (2016)

[RS18] Rösner, C., Schmidt, M.: Privacy preserving clustering with constraints. In: Proceedings of the 45th International Colloquium on Automata, Languages, and Programming (ICALP) (2018)

[Sch14] Schmidt, M.: Coresets and streaming algorithms for the k-means problem and related clustering objectives. Ph. D. thesis, Universität Dortmund (2014)

[SSS18] Schmidt, M., Schwiegelshohn, C., Sohler, C.: Fair coresets and streaming algorithms for fair k-means clustering, CoRR abs/1812.10854 (2018)

[TRT11] Thanh, B.L., Ruggieri, S., Turini, F.: K-NN as an implementation of situation testing for discrimination discovery and prevention. In: Proceedings of the 17th ACM SIGKDD International Conference on Knowledge Discovery and Data Mining, pp. 502–510 (2011)

[ZVGRG17] Zafar, M.B., Valera, I., Rodriguez, M.G., Gummadi, K.P.: Fairness beyond disparate treatment & disparate impact: learning classification without disparate mistreatment. In: Proceedings of the 26th International Conference on World Wide Web (WWW), pp. 1171–1180 (2017)

Correction to: Approximation and Online Algorithms

Evripidis Bampis and Nicole Megow

Correction to: E. Bampis and N. Megow (Eds.): *Approximation and Online Algorithms*, LNCS 11926, https://doi.org/10.1007/978-3-030-39479-0

In the original version of the book, the affiliation of Nicole Megow was wrong. The affiliation has been corrected to:

Department for Mathematics and Computer Science, University of Bremen, Bremen, Germany.

The updated version of the book can be found at https://doi.org/10.1007/978-3-030-39479-0

E. Bampis and N. Megow (Eds.): WAOA 2019, LNCS 11926, p. C1, 2020.
https://doi.org/10.1007/978-3-030-39479-0_17

Author Index